Signaling and Communication in Plants

Series Editor
František Baluška
Department of Plant Cell Biology, IZMB, University of Bonn, Bonn, Germany

More information about this series at
http://www.springer.com/series/8094

Kapuganti Jagadis Gupta • Abir U. Igamberdiev
Editors

Reactive Oxygen and Nitrogen Species Signaling and Communication in Plants

Volume 23

Springer

Editors
Kapuganti Jagadis Gupta
Department of Plant Sciences
University of Oxford
Oxford, United Kingdom

Abir U. Igamberdiev
Memorial University of Newfoundland
St John's, Newfoundland and Labrador
Canada

ISSN 1867-9048 ISSN 1867-9056 (electronic)
ISBN 978-3-319-10078-4 ISBN 978-3-319-10079-1 (eBook)
DOI 10.1007/978-3-319-10079-1
Springer Cham Heidelberg New York Dordrecht London

Library of Congress Control Number: 2014958304

Printed on acid-free paper

Springer is part of Springer Science+Business Media (www.springer.com)

Preface

Reactive oxygen species (ROS) and nitric oxide (NO), which is the main representative of reactive nitrogen species (RNS), are important free radical molecules that are formed as by-products of metabolism and participate in signalling events. ROS and RNS alone or together play role in a wide array of plant processes such as plant–microbe interactions, responses to abiotic stress, stomatal regulation, and a range of developmental processes. Due to their short half-life, high diffusion capacity, and ability to react rapidly with different components in the cell, they participate in various processes connected with signalling and communication in plants.

The spatial and temporal regulation of ROS and RNS production and scavenging is an important aspect of their signalling function. ROS and RNS are produced in plant cell in various compartments that include chloroplasts, mitochondria, peroxisomes, the endoplasmic reticulum (ER), and plasma membranes. ROS and RNS fulfil numerous essential functions right from germination of seeds to plant senescence. This book describes different signalling and communication processes governed by these molecules. In the first chapters, starting from the general overview of the editors K.J. Gupta and A.U. Igamberdiev, the production of NO and ROS is discussed in relation to operation of various enzyme systems and their compartmentalization. E. Urarte et al. describe the principal role of xanthine oxidoreductase in production and turnover of ROS and NO in plants. F. Minibayeva and R.P. Beckett particularly focus on the role of plant peroxidases in metabolism of reactive nitrogen species and other nitrogenous compounds. Hypoxia in plants is often associated with flooding and represents an important topic relevant to plant productivity. In this context, K.H. Hebelstrup and I.M. Møller review the data on ROS and RNS signalling between mitochondria and the rest of the cell under hypoxic conditions. Plants also contain various scavenging systems for ROS and RNS. The extent of ROS and RNS formation and their level depend on their production and scavenging systems. V.N. Popov provides the overview of ROS production in mitochondria and of their role in induction of the systems that participate in ROS avoidance and scavenging. If ROS are produced in high concentrations they can be toxic for cellular growth and metabolism, and in the next chapter V. Mittova et al. focus on the operation of various antioxidant systems in wild and cultivated varieties of tomato, showing the importance of ROS scavenging systems in stress tolerance.

Nitrogen is an essential component of proteins. Plants assimilate nitrogen from soil in the form of nitrate or ammonium. In some soils ammonium concentrations are very high. Ammonium toxicity is associated with redox imbalance and increased ROS levels. In this context, A. Podgórska and B. Szal describe the connection between NH_4^+ nutrition, ROS-producing reactions, and antioxidant systems. Plant organisms, when they are subjected to allelopathic compounds, respond by induction of oxidative stress, manifested as overproduction of ROS and alteration in cellular antioxidant systems. This is discussed in the chapter of A. Gniazdowska et al.

Seed germination is a developmental stagein which plant life originates from the quiescent embryo. There are various events associated with germination ranging from the initial uptake of water by dry seed to the emergence of radicle through the seed coat. Seed germination is accompanied by intensive production of ROS (superoxide anion, hydrogen peroxide, etc.) and RNS (NO and its derivatives). In this context, N.V. Bykova et al. provide a detailed overview of ROS and RNS in bioenergetics, metabolism, and signalling processes associated with seed germination. M. Elhiti and C. Stasolla summarize the information related to the role of ROS homeostasis and signalling in the induction and development of in vitro produced embryos. U. Krasuska et al. Describe the pathways of NO biosynthesis in germinating seeds, potential modes of NO action, and its cross-talk with plant growth regulators that determine seed dormancy and germination.

Nitric oxide participates in cell signalling via post-translational modification of various proteins. One of the key processes is S-nitrosylation. NO signalling in the nucleus is important due to activation of various genes involved in plant response to biotic and abiotic stress and in plant development. In this regard, A. Sehrawat and R. Deswal provide the information on S-nitrosylation of nuclear proteins and on its role in regulation of gene expression. The chapter of Corpas et al. describes in detail the processes of nitration and S-nitrosylation in plants. *S*-nitrosoglutathione reductase is considered as a key enzyme of the regulation of intracellular levels of *S*-nitrosoglutathione and indirectly also of protein *S*-nitrosothiols. Petřivalský et al. describe the role of this important enzyme in their chapter. ROS and RNS interact and cross-talk with calcium signalling. This information, which is crucial for understanding ROS, RNS communication network, is covered by S. Sharma.

Overall the book aims to cover various important aspects of reactive oxygen and reactive nitrogen species signalling and communication in plants. Respected scientists from several countries have contributed for this book, and the editors are extremely grateful to all contributors. We express our heartfelt gratitude to the technical editors and book publishing staff of Springer for their continuous support and timely advice during the course of the preparation of this volume.

Wellington Square, Oxford Kapuganti J. Gupta
St. John's, NL, Canada Abir U. Igamberdiev

Contents

Compartmentalization of Reactive Oxygen Species and Nitric Oxide Production in Plant Cells: An Overview

Kapuganti J. Gupta and Abir U. Igamberdiev

1 Introduction

In recent years, the evidence has been increasing that reactive oxygen species (ROS) and reactive nitrogen species (RNS) play a vital role in plants by controlling major physiological processes such as growth, development, resistance to biotic and abiotic environmental stimuli, and progression of programmed cell death. ROS and RNS are the by-products of plant metabolism: the various sites for ROS and RNS production include chloroplasts, mitochondria, peroxisomes, the endoplasmic reticulum (ER), and plasma membranes. The extent of ROS and RNS formation and their level depend on their production and scavenging systems. The enzymes that generate ROS and the main RNS species nitric oxide (NO) have various localizations and can be found in different cell compartments. The spatial and temporal location of ROS and RNS production is important in signalling. For instance, the mitochondrial ROS play role in retrograde signalling and communication between mitochondria and nucleus, while the mitochondrial NO may regulate ATP production. The plasma membrane-derived ROS participate in signalling during biotic and abiotic stresses, whereas the plasma membrane-originated NO plays role in nitrate sensing. In this chapter we briefly discuss metabolic and signalling roles of ROS and NO in relation to their compartmentalization.

K.J. Gupta (✉)
Department of Plant Sciences, University of Oxford, South Parks Road, Oxford OX1 3RB, UK
e-mail: jagadis.kapuganti@plants.ox.ac.uk

A.U. Igamberdiev
Department of Biology, Memorial University of Newfoundland, St. John's, NL, Canada A1B 3X9
e-mail: igamberdiev@mun.ca

K.J. Gupta, A.U. Igamberdiev (eds.), *Reactive Oxygen and Nitrogen Species Signaling and Communication in Plants*, Signaling and Communication in Plants 23,
DOI 10.1007/978-3-319-10079-1_1

2 Compartmentalization of ROS Production

There are several forms of ROS including free radicals such as $O_2^{\bullet -}$ (superoxide radical), $OH^{\bullet}$ (hydroxyl radical), and non-radical (molecular) forms: H_2O_2 (hydrogen peroxide) and $^1O_2$ (singlet oxygen). ROS are unavoidable by-products of aerobic metabolism being produced in various cellular compartments including chloroplasts, mitochondria, peroxisomes, plasma membrane, and apoplast. We will overview below the sites of ROS production with the emphasis on superoxide and hydrogen peroxide which are most abundant and have important metabolic and signalling roles (Fig. 1).

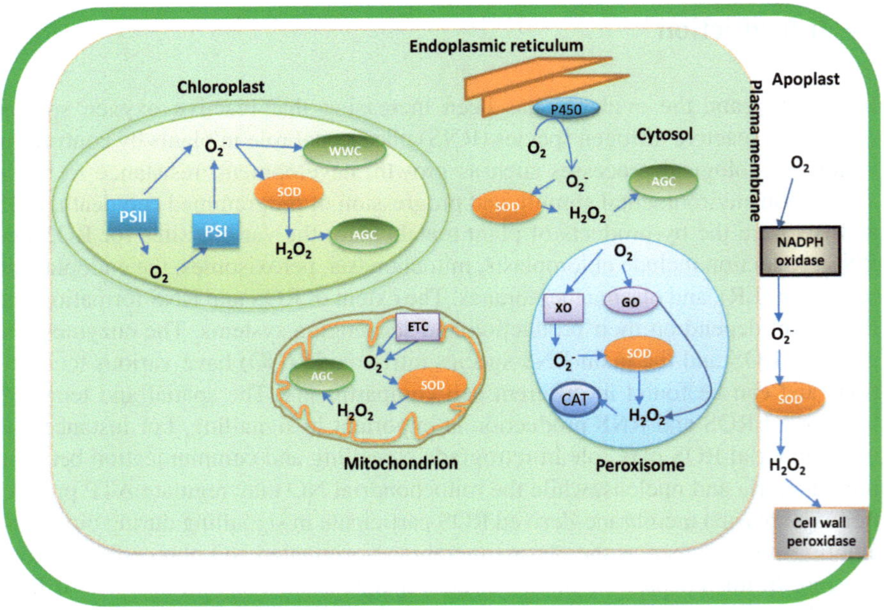

Fig. 1 ROS-generating pathways in various compartments of plant cell. In chloroplasts superoxide production takes place at PSI and PSII; it is converted by SOD to hydrogen peroxide which is scavenged in the ascorbate–glutathione cycle (AGC), also the water–water cycle (WWC) is involved. In plasma membrane NADPH oxidase generates superoxide which is converted by SOD to hydrogen peroxide and the latter is used by cell wall peroxidase. In mitochondria the complexes I and III are sites for ROS production, SOD and AGC are scavengers. In peroxisomes, glycolate oxidase (GO), acyl-CoA oxidase and xanthine oxidase (XO) are major sites of ROS production, catalase (CAT) and SOD are scavengers. In endoplasmic reticulum cytochome P-450 generates superoxide which is scavenged via cytosolic SOD and AGC

2.1 ROS in Chloroplasts

In the light, chloroplasts are the major sites for ROS production due to intensive electron transport during photosynthesis. Both photosystems (PSI and PSII) in chloroplast thylakoids are the major sites for the production of singlet oxygen and superoxide. Electron transport chain (ETC) actively participates in ROS production upon overloading of electron flow in these organelles, which is facilitated by the formation of oxygen in PSII and its abundance. The main reaction associated with ROS production in chloroplasts is Mehler reaction. In this reaction the electron flow is diverted from ferredoxin to O_2, reducing it to superoxide anion. The plastoquinone pool associated with PSII also provides a leakage to O_2 producing superoxide. The singlet oxygen is another ROS which is a by-product of photosynthesis, formed mainly in PSII even under low light. The steady-state level of H_2O_2 in chloroplasts was calculated as ~0.5 μM increasing in stress conditions to 5–15 μM, while the rate of superoxide production is estimated to be ~240 μM s^{-1}, increasing up to 720 μM s^{-1} under stress (Polle 2001). The powerful scavenging mechanisms can keep ROS level under control, but under various stresses, the limitation of carbon fixation in the Calvin cycle leads to a decrease in utilization of NADPH and results in deviation of electrons from ETC to O_2 to form $O_2^{\bullet -}$. The protection from ROS in chloroplasts is achieved by involvement of superoxide dismutases, highly intensive ascorbate–glutathione cycle, chlororespiration, and other mechanisms.

The production of superoxide in PSI becomes an efficient alternative mechanism of photons and electrons sink when it is integrated to be a part of so called water–water cycle, in which SOD and reactions of the ascorbate–glutathione cycle participate (Asada 1999). In the water–water cycle the photoreduction of O_2 to water in PSI takes place by the electrons generated in PSII from water. Its efficiency is based on the intensity of SOD and ascorbate peroxidase activities which are several orders of magnitude higher than the rate of superoxide production. Also the reduction of oxidized forms of ascorbate either by the reduced ferredoxin or by dehydroasorbate and monodehydroascorbate reductases is very fast and efficient. As a result the water–water cycle effectively scavenges photoproduced O_2^- and H_2O_2 and suppresses the production of $OH^\bullet$ radicals, thus preventing their interaction with target molecules and hence photoinhibition. Thus, the water–water cycle not only scavenges ROS but also efficiently dissipates excess photon energy and electrons.

2.2 Mitochondria as a Source for ROS

The inner membrane of mitochondria contains protein complexes where electron transfer leads to generation of proton gradient across membrane and this process is coupled to the production of ATP. The reduction level of ubiquinone is directly linked to the leakage of electrons to oxygen which results in formation of superoxide anion. Downstream to the ubiquinone pool two pathways of electron transport operate. One is the cytochrome pathway via the complexes III and IV and the other is the alternative cyanide-resistant pathway via the alternative oxidase (AOX) which is activated upon the increase of reduction level of ubiquinone and thus prevents overproduction of superoxide (Maxwell et al. 1999). Electron transfer via the cytochrome pathway leads to ATP production but no production of ATP takes place when electrons transfer via AOX pathway due to lack proton pumping sites in the pathway. Under normal respiratory conditions the production of ROS takes place due to the leakage of electrons to oxygen. Under stress conditions this process intensifies and the excess of ROS production takes place. Plants possess various antioxidant systems keeping ROS at low levels but if ROS production exceeds the capacity of antioxidant systems, then ROS become deleterious, causing damage to proteins, lipids and nucleic acids.

In mitochondria the complexes I and III are major sites for ROS production (Møller 2001). The primary electron donor to oxygen for ROS formation is semiubiquinone, however, under stress conditions electrons transfer takes place directly from the complexes I and III to oxygen (Raha and Robinson 2000; Sweetlove and Foyer 2004). The production of ROS in mitochondria occurs when the rate of electron transfer exceeds the capacities of AOX and COX pathways due to generation of excess of electrons (Møller 2001; Rhoads et al. 2006). For instance, the addition of excess of a substrate to mitochondria leads to increased production of ROS (Maxwell et al. 1999). Under stress conditions the ubiquinone pool is over-reduced, which leads to ROS production. Gupta et al. (2014a) has shown that infection of Arabidopsis roots with *Fusarium oxysporum* results in the increased ROS levels due to over-reduction of ubiquinone pool through inhibition of the complex IV by generated NO. Carbon monoxide or cyanide inhibition of complex IV can also increase ROS production in mitochondria (Piantadosi 2008), also the suppression of AOX results in the increased ROS generation (Parsons et al. 1999).

The increased production of ROS leads to lipid peroxidation which can cause cellular damage by reacting with proteins, other lipids, and nucleic acids. The polyunsaturated fatty acids of membrane lipids are prone to ROS production of aldehydes, alkenals, and hydroxyalkenals, malondialdehyde (MDA) and 4-hydroxy-2-nonenal (HNE). For instance, 4-hydroxy-2-nonenal (HNE), a toxic product of lipid peroxidation, inhibits oxygen consumption of mitochondria and leads to reduced ATP production (Taylor et al. 2002)

ROS also damage proteins by various mechanisms, e.g. they inhibit both pyruvate dehydrogenase (PDC), 2-oxoglutarate dehydrogenase (OGDC), and glycine decarboxylase (GDC) complexes through modification of lipoic acid moieties. The mechanisms of protein damage by ROS involve (1) Oxidation of various protein moieties such as Cys, Met, Arg, Lys, Pro, and Thr residues (Dean et al. 1997); (2) Cleavage and degradation of protein backbones (Dean et al. 1997); (3) Tyrosine nitration due to joint reaction of ROS and NO (Sehrawat et al. 2013); (4) Direct oxidation of metallic proteins such as aconitase (Verniquet et al. 1991; Gupta et al. 2012). Proteomic approaches have revealed that there are numerous proteins in mitochondria prone to ROS modification (Sweetlove et al. 2002; Kristensen et al. 2004; Taylor et al. 2005). For instance, H_2O_2 or menadione treatments lead to damage to pyruvate decarboxylase complex, ATP synthase, and various enzymes of the TCA cycle.

Mitochondrial DNA is also damaged by ROS, which has severe consequences in mitochondrial DNA replication and repair. DNA damage also affects retrograde signalling (Rhoads et al. 2006).

2.3 ROS Generation in Peroxisomes

Peroxisomes generate hydrogen peroxide and superoxide anion as a consequence of their metabolic activity involving such processes as the photorespiratory cycle, fatty acid β-oxidation, the glyoxylate cycle, and metabolism of ureides (Corpas et al. 2001; Igamberdiev and Lea 2002). Peroxisomes have extremely high metabolic plasticity in their enzyme composition in development and stress conditions. In leaf peroxisomes, xanthine oxidase (XOD) generates superoxide radicals via oxidation of xanthine or hypoxanthine to uric acid. Superoxide dismutase dismutates superoxide anion to H_2O_2 and catalase degrades H_2O_2. Photorespiratory formation of glycolate by flavin-containing glycolate oxidase in peroxisomes accounts for the majority of H_2O_2 production in leaves (Noctor et al. 2002; Mittler et al. 2004). Catalase is present only in peroxisomes being the most abundant enzyme in these organelles and constituting their core crystal structure, and its function is to scavenge hydrogen peroxide formed mainly in the reactions of flavin-dependent oxidases. The affinity of catalase to H_2O_2 is significantly lower than that of other H_2O_2-scavenging systems, e.g. of ascorbate peroxidase, therefore it can only reduce H_2O_2 level to low millimolar concentrations (Igamberdiev and Lea 2002). Under stress conditions the suppression of catalase in peroxisomes leads to increased production of ROS (del Río et al. 1996).

The peroxisomal membrane is involved in NAD(P)H-dependent O_2^- production in the process of electron transfer involving flavoproteins, NAD(P)H and cytochrome *b* to oxygen (Sandalio et al. 1988; López-Huertas et al. 1999). Monodehydroascorbate reductase (MDHAR) can be an important participant in the balance of ROS in the proximity of peroxisomal membranes by oxidizing NADH and facilitating ascorbate peroxidase reaction (Corpas et al. 2001).

2.4 Plasma Membrane-Mediated ROS Production

Plasma membrane is a major site for ROS production due to presence of NADPH oxidase (Apel and Hirt 2004). This protein is encoded by the respiratory burst oxidase homolog (RBOH) gene family (Torres et al. 2005), it is integral to plasma membrane and composed of six transmembrane domains with two heme groups, C-terminal FAD and NADPH hydrophilic domains and two N-terminal calcium-binding (EF-hand) domains (Marino et al. 2012). The protein is involved in root hair and pollen development (Potocký et al. 2007; Foreman et al. 2003) and in seed development (Müller et al. 2009), for instance the mutation of Arabidopsis *AtRbohB* leads to altered seed germination (Müller et al. 2009) due to reduced levels of ROS required for cell wall loosening during the germination process. NADPH oxidase is also important in development of programmed cell death (PCD) (Torres et al. 2002). The mechanism of ROS production by NADPH oxidase involves the electron transfer from cytoplasmic NADPH to O_2 to form $O_2^{\bullet -}$, and the latter can be dismutated to H_2O_2.

Mutation in RBOH leads to reduced levels of ROS, failure to induce PCD and to develop the systemic acquired resistance (SAR) in plants (Alvarez et al. 1998). Application of the NADPH oxidase inhibitor diphenyleneiodonium (DPI), leads to reduced production of H_2O_2 (Laloi et al. 2004). NADPH oxidase is involved in signal perception during stress which leads to elevation of calcium and activation of MAP kinases. In tomato RBOH is involved in wounding response (Sagi et al. 2004). The abscisic acid-induced stomatal closure is also associated with RBOH (Kwak et al. 2003). RBOH is also involved in the establishment of symbiotic relations, e.g. of *Medicago truncatula–Sinorhizobium meliloti* symbiotic interaction and nodule functioning (Marino et al. 2011). Downregulation of *MtRbohA* expression leads to a decrease in nodule nitrogen fixation activity. The expression of *MtRbohA* expression is strongly increased under hypoxic conditions (Marino et al. 2011).

Due to operation of RBOH proteins, under many stress conditions the increased ROS accumulation is observed in the apoplast (Hernández et al. 2001). In Arabidopsis guard cells, the apoplast AtRbohD is one of the compartments for ROS production. AtRbohF is another ROS generating enzyme expressed in mesophyll cells of leaves (Kwak et al. 2003). Other candidate enzymes responsible for ROS production in apoplast are polyamine oxidases (Mittler 2002). Apoplastic ROS play role in elongation of leaves which leads to reduced growth under the osmotic stress (Rodríguez et al. 2004).

3 Nitric Oxide Compartmentalization in Plants

Reactive nitrogen species (RNS) are derived from nitric oxide (NO) and include NO itself, peroxynitrite (the product of NO reaction with superoxide) and its derivatives, other reactive nitrogen oxides (NO_2 and N_2O_3). NO-derivatives of small molecules such as nitrosoglutathione are also included in this group. Since the primary process of RNS generation is NO production, we will discuss below localization of production and scavenging of this compound in the plant cell. NO is a free radical that plays role under various biotic and abiotic stresses, growth, and development. There are at least seven pathways responsible for NO production in plants (Gupta et al. 2011), which are classified into the oxidative and reductive pathways. The reductive pathways include participation of the cytosolic nitrate reductase (NR), mitochondrial nitrite NO:reductase, plasma membrane nitrite: NO reductase and xanthine oxidoreductase (XOR),whereas polyamine, hydroxylamine,

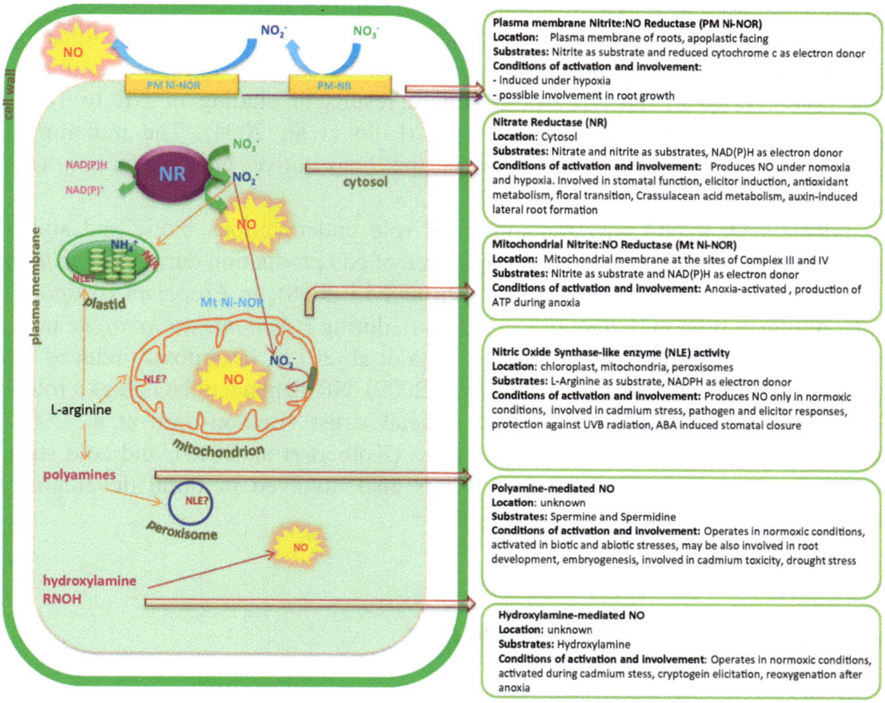

Fig. 2 NO-generating pathways in various compartments of plant cell

and L-arginine-dependent pathways are oxidative in nature. The pathways of NO production and scavenging are located in various compartments and therefore they might fulfil the needs required specifically in each compartment (Fig. 2).

3.1 *Cytosolic NO Production*

Nitrate reductase (NR) is the major enzyme that is able to produce NO in the cytosol. Using NADH (or NADPH with lower efficiency) as the electron donor it catalyses reduction of nitrate to nitrite, but it can also reduce nitrite to NO (Dean and Harper 1988; Yamasaki and Sakihama 2000). In Arabidopsis NR is encoded by the two genes *Nia1* and *Nia2*. The mutants of NR, *nia1* and *nia2* are impaired in NO production (Gupta et al. 2012). Under various stress conditions accumulation of nitrite takes place which leads to NO formation because nitrite is a limiting factor for NO production (Planchet et al. 2005). Concentration of nitrite should reach high levels to make NR a source of NO generation. The *K*m value for nitrite in NO production by NR is of the order of 100 μM (Yamasaki and Sakihama 2000), which is slightly lower than the *K*m value for nitrite in NO synthesis by mitochondria (175 μM) (Gupta et al. 2005). If the cytosolic pH drops, this leads to inhibition of nitrite reductase, followed by accumulation of nitrite and NO synthesis. Another factor that influences NO production from NR is its posttranslational modification. Phosphorylation and dephosphorylation regulate NR activity. Phosphorylation of a conserved serine in the NR by NR-kinase results in binding of NR to 14-3-3 proteins, which leads to its inactivation (Lillo et al. 2004). The mutation in phosphorylation site of NR results in the constitutive NO production (Lea et al. 2004).

NR-dependent NO plays an important role under various biotic and abiotic stresses. For instance, NR is the major source of NO production during *Verticillium dahlia* pathogenicity in Arabidopsis (Shi and Li 2008), in *Fusarium oxysporum* infection in Arabidopsis (Gupta et al. 2014a), during *Pseudomonas syringae* infection in tobacco (Modolo et al. 2005; Gupta et al. 2013), in chitosan-induced NO production in guard cells (Srivastava et al. 2009). NR-dependent NO plays a role in several abiotic stresses such as heavy metal stress (Besson-Bard et al. 2009), hypoxia (Gupta et al. 2012), osmotic stress (Kolbert et al. 2010), and cold stress (Zhao et al. 2009). NR-dependent NO is also involved in floral development (Seligman et al. 2008).

3.2 *Mitochondrial NO Production*

Recent studies have shown that mitochondria are one of the major NO sources and that cytochrome *c* oxidase (COX) and complex III are the plausible sites for NO production in the hypoxic plant cell (Planchet et al. 2005; Stoimenova et al. 2007). The reaction of NO formation at complex IV involves nitrite reduction at the binuclear centre $Fe_{a3}Cu_B$, while the reaction at complex III can be related to leakage of electrons to nitrite from the complex similarly as the leakage to O_2 leads to formation of superoxide (Igamberdiev et al. 2010). The *K*m (nitrite) for this mitochondrial nitrite: NO reductase reaction in roots is 175 μM for root mitochondria, the 50 % inhibition of NO production is observed at 0.05 % O_2 (Gupta et al. 2005). The rates of hypoxic NO production by plant mitochondria are in the range of 1–20 nmol mg^{-1} protein per hour for barley, pea, Arabidopsis (Gupta et al. 2011), however they may be much higher because the most part of NO is immediately scavenged. The scavenging experiments conducted by Gupta et al. (2005) revealed that even under hypoxia mitochondria scavenge 70 % of added NO suggesting that mitochondria are not only producers but also major sinks for NO. NO can also diffuse from mitochondria into cytosol where it is oxygenated to nitrate by the class 1 non-symbiotic hemoglobin (Igamberdiev et al. 2006), which is saturated even at nanomolar oxygen levels. The overall sequence of reactions which includes production of NO from nitrite by mitochondrial complexes III and IV, NO scavenging to nitrate by non-symbiotic hemoglobin, and further reduction of nitrate to nitrite by NR is known as the hemoglobin/nitric oxide (Hb/NO) cycle, playing an important role in bioenergetics of the hypoxic plant cell (Igamberdiev and Hill 2004; Gupta and Igamberdiev 2011).

In animal systems it is shown that COX is not only a source but also an important target for NO (Cleeter et al. 1994). Its inhibition by NO occurs via the competitive binding of NO to the Fe^{2+}-heme group at O_2-binding site of the binuclear centre $Fe_{a3}Cu_B$. This leads to generation of ferrous-heme-nitrosyl complex (Cleeter et al. 1994). The binding has physiological relevance because it keeps oxygen concentration above certain levels and improves oxidative phosphorylation capability (Clerc et al. 2007). COX can scavenge NO back to nitrite during the transition from hypoxia to normoxia (Brunori et al. 2006), and it can also scavenge peroxynitrite formed in the reaction between NO and superoxide (Pearce et al. 2002).

Recently, using the non-symbiotic hemoglobin-overexpressing (nHb+) plants, it was shown that the inhibition of respiration by NO in plants is important for oxygen and ROS homeostasis. Inhibition of COX leads to increase of the internal oxygen concentrations for keeping ROS levels at low level and control of carbohydrate consumption (Gupta et al. 2014b). The reduced levels of NO in nHb+ plants lead to the increased rates of ROS production. A decreased electron flow in the mitochondrial ETC in complex I-deficient plants leads to lower NO production, which affects stomatal conductance and delays growth and morphogenesis (Shah et al. 2013).

In animal systems it was shown that the reduction of nitrite to NO by COX leads to proton translocation (Castello et al. 2006), the same was indirectly established for plants (Stoimenova et al. 2007). It was shown that oxidation of NADH and NADPH under hypoxic conditions leads to low but continuous levels of ATP production, which is very important for hypoxic survival. This reaction was sensitive to myxothiazol and KCN treatment, i.e. the complexes III and IV were involved. It was found that the anoxia-tolerant rice mitochondria produced more NO and ATP under hypoxia than anoxia-intolerant barley mitochondria (Stoimenova et al. 2007). The rates of ATP production were determined as 7–9 nmol min^{-1} mg^{-1} (mitochondrial protein) for barley and 15–17 nmol min^{-1} mg^{-1} (protein) for rice. These rates constitute 3–5 % of the mitochondrial ATP production in ambient oxygen concentrations. The nitrite-dependent ATP generation was insensitive to rotenone suggesting that complex I is not involved in NAD(P)H oxidation in these conditions, and that likely the alternative NADH and NADPH dehydrogenases facing the external site of inner mitochondrial membrane participate in this reaction.

3.3 Peroxisomes as a Source for NO

Plant peroxisomes can be a major site of NO production in the oxidative pathways. It is established that pea leaf peroxisomes generate NO in the nitric oxide synthase-like (NOS like) activity (Barroso et al. 1999). The presence of this reaction was determined based on conversion of L-arginine to citrulline and NO, the level of activity was 170 pmol of L-[^{3}H]citrulline min^{-1} mg^{-1} (peroxisomal protein). This activity was stimulated by calcium, NADPH was required for the reaction and this activity was inhibited by the arginine analogs (Barroso et al. 1999).

Peroxisomes can be involved in NO formation also in the reductive pathways. The peroxisomal enzyme xanthine oxidoreductase (XOR) is able to reduce nitrite to NO (Godber et al. 2000; Corpas et al. 2008). Under hypoxic conditions, the purified XOD was shown to reduce nitrite to NO and this reaction requires NADH or xanthine as electron donors (Corpas et al. 2008). In plants XOR plays a role in NO production in various conditions, e.g. upon phosphate deficiency in cluster roots of white lupin (*Lupinus albus*) (Wang et al. 2010).

3.4 Plasma Membrane NO Production from Nitrite

Plasma membranes of roots contain a protein that possesses nitrite: NO reductase (Ni-NOR) activity. This enzyme has pH optimum at 6.1 (Stöhr et al. 2001). The Ni-NOR has a capacity to reduce the nitrite pool generated in apoplast by the plasma membrane-bound nitrate reductase. Since Ni-NOR is present in plasma membrane, together with plasma membrane NR it has a capacity to sense nitrate in soil, and NO is a signal in this sensing process (Meyer and Stöhr 2002). This enzyme is shown to produce NO in roots in response to infection by mycorrhizal fungi (Moche et al. 2010).

Conclusion

ROS and RNS are the key signalling molecules in plant cells, but when they are produced in excess they cause cellular damage. Under hypoxia, the turnover of nitric oxide complements and partially replaces oxygenic respiration, thus NO serves not only as a signal but also as an important metabolite. The cellular localization of ROS and RNS production and scavenging is based on several organelle-specific systems. Their proper operation is important for keeping the balance of oxidative metabolism in the cell. Any shift of this balance, e.g. under abiotic and biotic stresses, results in switching on mechanisms in which ROS and RNS play a signalling role, and initiating the integrated response with inclusion of hormones, regulatory proteins and responsive genetic elements, which provides adaptation of plants to changing environment.

References

Alvarez ME, Pennell RI, Meijer PJ, Ishikawa A, Dixon RA, Lamb C (1998) Reactive oxygen intermediates mediate a systemic signal network in the establishment of plant immunity. Cell 92:773–784

Apel K, Hirt H (2004) Reactive oxygen species: metabolism, oxidative stress, and signal transduction. Annu Rev Plant Biol 55:373–399

Asada K (1999) The water-water cycle in chloroplasts: scavenging of active oxygens and dissipation of excess photons. Annu Rev Plant Physiol Plant Mol Biol 50:601–639

Barroso JB, Corpas FJ, Carreras A, Sandalio LM, Valderrama R, Palma JM, Lupiáñez JA, del Río LA (1999) Localization of nitric-oxide synthase in plant peroxisomes. J Biol Chem 274:36729–36733

Besson-Bard A, Gravot A, Richaud P, Auroy P, Duc C, Gaymard F, Taconnat L, Renou JP, Pugin A, Wendehenne D (2009) Nitric oxide contributes to cadmium toxicity in Arabidopsis by promoting cadmium accumulation in roots and by up-regulating genes related to iron uptake. Plant Physiol 149:1302–1315

Brunori M, Forte E, Arese M, Mastronicola D, Giuffrè A, Sarti P (2006) Nitric oxide and the respiratory enzyme. Biochim Biophys Acta 1757:1144–1154

Castello PR, David PS, McClure T, Crook Z, Poyton RO (2006) Mitochondrial cytochrome oxidase produces nitric oxide under hypoxic conditions: implications for oxygen sensing and hypoxic signaling in eukaryotes. Cell Metab 3:277–287

Cleeter MW, Cooper JM, Darley-Usmar VM, Moncada S, Schapira AH (1994) Reversible inhibition of cytochrome c oxidase, the terminal enzyme of the mitochondrial respiratory chain, by nitric oxide. Implications for neurodegenerative diseases. FEBS Lett 345:50–54

Clerc P, Rigoulet M, Leverve X, Fontaine E (2007) Nitric oxide increases oxidative phosphorylation efficiency. J Bioenerg Biomembr 39:158–166

Corpas FJ, Barroso JB, del Río LA (2001) Peroxisomes as a source of reactive oxygen species and nitric oxide signal molecules in plant cells. Trends Plant Sci 6:145–150

Corpas FJ, Palma JM, Sandalio LM, Valderrama R, Barroso JB, Del Río LA (2008) Peroxisomal xanthine oxidoreductase: characterization of the enzyme from pea (*Pisumsativum* L.) leaves. J Plant Physiol 165:1319–1330

Dean JV, Harper JE (1988) The conversion of nitrite to nitrogen oxide(s) by the constitutive NAD (P)H-nitrate reductase enzyme from soybean. Plant Physiol 88:389–395

Dean RT, Fu S, Stocker R, Davies MJ (1997) Biochemistry and pathology of radical-mediated protein oxidation. Biochem J 324:1–18

del Río LA, Palma JM, Sandalio LM, Corpas FJ, Pastori GM, Bueno P, López-Huertas E (1996) Peroxisomes as a source of superoxide and hydrogen peroxide in stressed plants. Biochem Soc Trans 24:434–438

Foreman J, Demidchik V, Bothwell JH, Mylona P, Miedema H, Torres MA, Linstead P, Costa S, Brownlee C, Jones JD, Davies JM, Dolan L (2003) Reactive oxygen species produced by NADPH oxidase regulate plant cell growth. Nature 422:442–446

Godber BL, Doel JJ, Sapkota GP, Blake DR, Stevens CR, Eisenthal R, Harrison R (2000) Reduction of nitrite to nitric oxide catalysed by xanthine oxidoreductase. J Biol Chem 275:7757–7763

Gupta KJ, Igamberdiev AU (2011) The anoxic plant mitochondrion as a nitrite: NO reductase. Mitochondrion 11:537–543

Gupta KJ, Stoimenova M, Kaiser WM (2005) In higher plants, only root mitochondria, but not leaf mitochondria reduce nitrite to NO, in vitro and in situ. J Exp Bot 56:2601–2609

Gupta KJ, Fernie AR, Kaiser WM, van Dongen JT (2011) On the origins of nitric oxide. Trends Plant Sci 16:160–168

Gupta KJ, Shah JK, Brotman Y, Jahnke K, Willmitzer L, Kaiser WM, Bauwe H, Igamberdiev AU (2012) Inhibition of aconitase by nitric oxide leads to induction of the alternative oxidase and to a shift of metabolism towards biosynthesis of amino acids. J Exp Bot 63:1773–1784

Gupta KJ, Brotman Y, Segu S, Zeier T, Zeier J, Persijn ST, Cristescu SM, Harren FJ, Bauwe H, Fernie AR, Kaiser WM, Mur LA (2013) The form of nitrogen nutrition affects resistance against *Pseudomonas syringae* pv. *phaseolicola* in tobacco. J Exp Bot 64:553–568

Gupta KJ, Mur LAJ, Brotman Y (2014a) *Trichoderma asperelloides* suppresses nitric oxide generation elicited by *Fusarium oxysporum* in *Arabidopsis* roots. Mol Plant Microbe Interact 27:307–314

Gupta KJ, Hebelstrup KH, Kruger NJ, Ratcliffe GR (2014b) Nitric oxide is required for homeostasis of oxygen and reactive oxygen species in barley roots under aerobic conditions. Mol Plant 7:747–750

Hernández JA, Ferrer MA, Jiménez A, Barceló AR, Sevilla F (2001) Antioxidant systems and $O_2^{\bullet -}/H_2O_2$ production in the apoplast of pea leaves. Its relation with salt-induced necrotic lesions in minor veins. Plant Physiol 127:817–831
Igamberdiev AU, Hill RD (2004) Nitrate, NO and haemoglobin in plant adaptation to hypoxia: an alternative to classic fermentation pathways. J Exp Bot 55:2473–2482
Igamberdiev AU, Lea PJ (2002) The role of peroxisomes in the integration of metabolism and evolutionary diversity of photosynthetic organisms. Phytochemistry 60:651–674
Igamberdiev AU, Bykova NV, Hill RD (2006) Nitric oxide scavenging by barley hemoglobin is facilitated by a monodehydroascorbate reductase-mediated ascorbate reduction of methemoglobin. Planta 223:1033–1040
Igamberdiev AU, Bykova NV, Shah JK, Hill RD (2010) Anoxic nitric oxide cycling in plants: participating reactions and possible mechanisms. Physiol Plant 138:393–404
Kolbert Z, Ortega L, Erdei L (2010) Involvement of nitrate reductase (NR) in osmotic stress-induced NO generation of *Arabidopsis thaliana* L. roots. J Plant Physiol 167:77–80
Kristensen BK, Askerlund P, Bykova NV, Egsgaard H, Møller IM (2004) Identification of oxidised proteins in the matrix of rice leaf mitochondria by immunoprecipitation and two-dimensional liquid chromatography-tandem mass spectrometry. Phytochemistry 65:1839–1851
Kwak JM, Mori IC, Pei ZM, Leonhardt N, Torres MA, Dangl JL, Bloom RE, Bodde S, Jones JD, Schroeder JI (2003) NADPH oxidase *AtrbohD* and *AtrbohF* genes function in ROS-dependent ABA signaling in Arabidopsis. EMBO J 22:2623–2633
Laloi C, Apel K, Danon A (2004) Reactive oxygen signalling: the latest news. Curr Opin Plant Biol 7:323–328
Lea US, Ten Hoopen F, Provan F, Kaiser WM, Meyer C, Lillo C (2004) Mutation of the regulatory phosphorylation site of tobacco nitrate reductase results in high nitrite excretion and NO emission from leaf and root tissue. Planta 219:59–65
Lillo C, Meyer C, Lea US, Provan F, Oltedal S (2004) Mechanism and importance of post-translational regulation of nitrate reductase. J Exp Bot 55:1275–1282
López-Huertas E, Corpas FJ, Sandalio LM, Del Río LA (1999) Characterization of membrane polypeptides from pea leaf peroxisomes involved in superoxide radical generation. Biochem J 337:531–536
Marino D, Andrio E, Danchin EG, Oger E, Gucciardo S, Lambert A, Puppo A, Pauly N (2011) A Medicago truncatula NADPH oxidase is involved in symbiotic nodule functioning. New Phytol 189:580–592
Marino D, Dunand C, Puppo A, Pauly N (2012) A burst of plant NADPH oxidases. Trends Plant Sci 17:9–15
Maxwell DP, Wang Y, McIntosh L (1999) The alternative oxidase lowers mitochondrial reactive oxygen production in plant cells. Proc Natl Acad Sci USA 96:8271–8276
Meyer C, Stöhr C (2002) Soluble and plasma membrane bound enzymes involved in nitrate and nitrite metabolism. In: Foyer CH, Noctor G (eds) Photosynthetic nitrogen assimilation and associated carbon metabolism. Kluwer, Dordrecht, pp 49–62
Mittler R (2002) Oxidative stress, antioxidants and stress tolerance. Trends Plant Sci 7:405–410
Mittler R, Vanderauwera S, Gollery M, Van Breusegem F (2004) Reactive oxygen gene network of plants. Trends Plant Sci 9:490–498
Moche M, Stremlau S, Hecht L, Göbel C, Feussner I, Stöhr C (2010) Effect of nitrate supply and mycorrhizal inoculation on characteristics of tobacco root plasma membrane vesicles. Planta 231:425–436
Modolo LV, Augusto O, Almeida IM, Magalhaes JR, Salgado I (2005) Nitrite as the major source of nitric oxide production by *Arabidopsis thaliana* in response to *Pseudomonas syringae*. FEBS Lett 579:3814–3820
Møller IM (2001) Plant mitochondria and oxidative stress: electron transport, NADPH turnover, and metabolism of reactive oxygen species. Annu Rev Plant Physiol Plant Mol Biol 52:561–591

Müller K, Carstens AC, Linkies A, Torres MA, Leubner-Metzger G (2009) The NADPH-oxidase AtrbohB plays a role in Arabidopsis seed after-ripening. New Phytol 184:885–897

Noctor G, Veljovic-Jovanovic S, Driscoll S, Novitskaya L, Foyer CH (2002) Drought and oxidative load in the leaves of C3 plants: a predominant role for photorespiration? Ann Bot 89:841–850

Parsons HL, Yip JYH, Vanlerberghe GC (1999) Increased respiratory restriction during phosphate-limited growth in transgenic tobacco cells lacking alternative oxidase. Plant Physiol 121:1309–1320

Pearce LL, Kanai AJ, Birder LA, Pitt BR, Peterson J (2002) The catabolic fate of nitric oxide: the nitric oxide oxidase and peroxynitrite reductase activities of cytochrome oxidase. J Biol Chem 277:13556–13562

Piantadosi CA (2008) Carbon monoxide, reactive oxygen signaling, and oxidative stress. Free Radic Biol Med 45:562–569

Planchet E, Gupta KJ, Sonoda M, Kaiser WM (2005) Nitric oxide emission from tobacco leaves and cell suspensions: rate limiting factors and evidence for the involvement of mitochondrial electron transport. Plant J 41:732–743

Polle A (2001) Dissecting the superoxide dismutase–ascorbate peroxidase–glutathione pathway in chloroplasts by metabolic modeling. Computer simulations as a step towards flux analysis. Plant Physiol 126:445–462

Potocký M, Jones MA, Bezvoda R, Smirnoff N, Zárský V (2007) Reactive oxygen species produced by NADPH oxidase are involved in pollen tube growth. New Phytol 174:742–751

Raha S, Robinson BH (2000) Mitochondria, oxygen free radicals, disease and ageing. Trends Biochem Sci 25:502–508

Rhoads DM, Umbach AL, Subbaiah CC, Siedow JN (2006) Mitochondrial reactive oxygen species. Contribution to oxidative stress and interorganellar signaling. Plant Physiol 141:357–366

Rodríguez AA, Córdoba AR, Ortega L, Taleisnik E (2004) Decreased reactive oxygen species concentration in the elongation zone contributes to the reduction in maize leaf growth under salinity. J Exp Bot 55:1383–1390

Sagi M, Davydov O, Orazova S, Yesbergenova Z, Ophir R, Stratmann JW, Fluhr R (2004) Plant respiratory burst oxidase homologs impinge on wound responsiveness and development in *Lycopersicon esculentum*. Plant Cell 16:616–628

Sandalio LM, Fernández VM, Rupérez FL, del Río LA (1988) Superoxide free radicals are produced in glyoxysomes. Plant Physiol 87:1–4

Sehrawat A, Gupta R, Deswal R (2013) Nitric oxide-cold stress signalling cross-talk, evolution of a novel regulatory mechanism. Proteomics 13:1816–1835

Seligman K, Saviani EE, Oliveira HC, Pinto-Maglio CAF, Salgado I (2008) Floral transition and nitric oxide emission during flower development in Arabidopsis thaliana is affected in nitrate reductase-deficient plants. Plant Cell Physiol 49:1112–1121

Shah JK, Cochrane DW, De Paepe R, Igamberdiev AU (2013) Respiratory complex I deficiency results in low nitric oxide levels, induction of hemoglobin and upregulation of fermentation pathways. Plant Physiol Biochem 63:185–190

Shi FM, Li YZ (2008) *Verticillium dahliae* toxins-induced nitric oxide production in Arabidopsis is major dependent on nitrate reductase. BMB Rep 41:79–85

Srivastava N, Gonugunta VK, Puli MR, Raghavendra AS (2009) Nitric oxide production occurs downstream of reactive oxygen species in guard cells during stomatal closure induced by chitosan in abaxial epidermis of *Pisum sativum*. Planta 229:757–765

Stöhr C, Strube F, Marx G, Ullrich WR, Rockel P (2001) A plasma membrane-bound enzyme of tobacco roots catalyses the formation of nitric oxide from nitrite. Planta 212:835–841

Stoimenova M, Igamberdiev AU, Gupta KJ, Hill RD (2007) Nitrite-driven anaerobic ATP synthesis in barley and rice root mitochondria. Planta 226:465–474

Sweetlove LJ, Foyer CH (2004) Roles for reactive oxygen species and antioxidants in plant mitochondria. In: Day DA, Millar AH, Whelan J (eds) Plant mitochondria: from genome to function, vol 1, Advances in Photosynthesis and Respiration. Kluwer, Dordrecht, pp 307–320

Sweetlove LJ, Heazlewood JL, Herald V, Holtzapffel R, Day DA, Leaver CJ, Millar AH (2002) The impact of oxidative stress on Arabidopsis mitochondria. Plant J 32:891–904

Taylor NL, Day DA, Millar AH (2002) Environmental stress causes oxidative damage to plant mitochondria leading to inhibition of glycine decarboxylase. J Biol Chem 277:42663–42668

Taylor NL, Heazlewood JL, Day DA, Millar AH (2005) Differential impact of environmental stresses on the pea mitochondrial proteome. Mol Cell Proteomics 4:1122–1133

Torres MA, Dangl JL, Jones JDG (2002) Arabidopsis gp91phox homologues *AtrbohD* and *AtrbohF* are required for accumulation of reactive oxygen intermediates in the plant defense response. Proc Natl Acad Sci USA 99:517–522

Torres MA, Jones JD, Dangl JL (2005) Pathogen-induced, NADPH oxidase-derived reactive oxygen intermediates suppress spread of cell death in *Arabidopsis thaliana.* Nat Genet 37:1130–1134

Verniquet F, Gaillard J, Neuburger M, Douce R (1991) Rapid inactivation of plant aconitase by hydrogen peroxide. Biochem J 276:643–648

Wang BL, Tang XY, Cheng LY, Zhang AZ, Zhang WH, Zhang FS, Liu JQ, Cao Y, Allan DL, Vance CP, Shen JB (2010) Nitric oxide is involved in phosphorus deficiency-induced cluster-root development and citrate exudation in white lupin. New Phytol 187:1112–1123

Yamasaki H, Sakihama Y (2000) Simultaneous production of nitric oxide and peroxynitrite by plant nitrate reductase: in vitro evidence for the NR-dependent formation of active nitrogen species. FEBS Lett 468:89–92

Zhao MG, Chen L, Zhang LL, Zhang WH (2009) Nitric reductase-dependent nitric oxide production is involved in cold acclimation and freezing tolerance in Arabidopsis. Plant Physiol 151:755–767

Established and Proposed Roles of Xanthine Oxidoreductase in Oxidative and Reductive Pathways in Plants

Estibaliz Urarte, Raquel Esteban, Jose Fernando Moran, and Florian Bittner

Abbreviations

ABA	Abscisic acid
AO	Aldehyde oxidase
GOGAT cycle	Glutamine oxoglutarate aminotransferase cycle
H_2O_2	Hydrogen peroxide
NO	Nitric oxide
NR	Nitrate reductase
$O_2^{\bullet-}$	Superoxide radical
$ONOO^-$	Peroxynitrite
SOD	Superoxide dismutase
RNS	Reactive nitrogen species
ROS	Reactive oxygen species
SHAM	Salicylhydroxamic acid
XDH	Xanthine dehydrogenase
XO	Xanthine oxidase
XOR	Xanthine oxidoreductase

E. Urarte • R. Esteban • J.F. Moran
Institute of Agrobiotechnology, IdAB-CSIC-UPNA-Government of Navarre, 31192 Pamplona, Spain

F. Bittner (✉)
Department of Plant Biology, Braunschweig University of Technology, 38106 Braunschweig, Germany
e-mail: f.bittner@tu-braunschweig.de

K.J. Gupta, A.U. Igamberdiev (eds.), *Reactive Oxygen and Nitrogen Species Signaling and Communication in Plants*, Signaling and Communication in Plants 23,
DOI 10.1007/978-3-319-10079-1_2

1 Introduction

Despite the current description of more than 50 molybdoenzymes in living organisms (Mo-enzymes; Bittner and Mendel 2010), only five of them are found in plants: nitrate reductase (NR), aldehyde oxidase (AO), sulfite oxidase (SO), mitochondrial amidoxime reducing component (mARC), and xanthine oxidoreductase (XOR). This review will focus on XOR, a FAD-, molybdenum-, iron-, and sulfur-containing hydroxylase fundamental in nucleic acid metabolism.

XOR (EC 1.17.1.4., formerly EC 1.1.1.204) is a facultatively oxygen consuming enzyme, which catalyzes the conversion of the purines hypoxanthine and xanthine into uric acid, with the concomitant formation of either NADH or superoxide radical (Bittner and Mendel 2010). The monomer can be subdivided into three distinct domains: a N-terminal domain binding two iron–sulfur clusters of the [2Fe–2S]-type, a domain harboring a flavin adenine dinucleotide (FAD)-binding site, and a C-terminal domain required for molybdenum cofactor (Moco) binding and dimerization. Currently, the crystal structure of plant XOR is not available. However, due to its similarity with bovine, human, and rat XOR, the structure of plant XOR may be assumed based on these models. XOR is a ubiquitous enzyme among the kingdom *Plantae* and for the last three decades it has been deeply studied by different approaches. At early stages, studies with inhibitors like allopurinol provided a useful tool, followed in the last years by *Arabidopsis* knockout plants (Watanabe et al. 2010).

The study on XOR gave origin to the idea of the free radical chemistry in living organisms, mostly based on work by Fridovich and Handler on O_2 consumption by XOR (1958). 10 years later McCord and Fridovich (1968) envisaged that a free radical product derived from oxygen may be produced in biological systems during the reaction of XOR with O_2. The importance of this enzyme lies thus on the fact that it participates in the regulation of reactive oxygen species (ROS) production (Montalbini 1992a). This ROS production in plants by XOR might have outstanding physiological consequences in plant signaling, in the processes of host–pathogen relationships (Montalbini 1992a, b), during abiotic stresses (Zdunek-Zastocka and Lips 2003), or for natural plant senescence (Hesberg et al. 2004). By a common regulatory sulfuration step, the ratio of inactive and active XOR and AO enzymes can be changed rapidly in order to increase the amount of certain important plants hormones such as abscisic acid (ABA) as was shown in drought and salt stressed *Arabidopsis thaliana* (Xiong et al. 2001). Under physiological aspects, this sulfuration step provides an efficient way of regulating the amount of active XOR and AO within the cell and thus, to adopt it to the physiological demands of the plant (Mendel 2011).

This chapter reviews recent insights into the study of the enzyme XOR in plants, with a specific focus on the established and possible physiological substrates, inhibitors and feasible function in plants.

2 History and Evolution of XOR

Due to its high abundance in milk and the simplicity of purification procedures, mammalian XOR is among the most-intensively studied enzymes and was originally referred to as the "Schardinger enzyme" that catalyzes the reduction of methylene blue when formaldehyde served as substrate (Schardinger 1902). Nearly two decades later, extracts of yeast and animal tissues were demonstrated to harbor certain substances that are likewise involved in the reduction of methylene blue in milk (Hopkins 1921), and only 1 year later these substances have been identified as hypoxanthine and xanthine, which lead to the renaming of the enyzme into xanthine oxidase (XO) (Morgan et al. 1922; Booth 1935 and references therein). Later, the term XO became representative for a specific condition of the enzyme (along with xanthine dehydrogenase, XDH), and the term xanthine oxidoreductase (XOR) was introduced as a more general and ubiquitous name for this enzyme that is found in all kingdoms of life. In particular mammalian XOR attracted, and still attracts, many researchers due to its implications in ischemia/reperfusion injury and reactive oxygen species (ROS) generation, but also because of its possible involvement in nitric oxide (NO) and peroxynitrite formation, and because of XOR-related diseases such as xanthinuria and hyperuricemia/gout (reviewed by Harrison 2002; Agarwal et al. 2011). Moreover, mammalian XOR was proven to have another distinct function that is associated with the formation of milk-fat droplets but totally unrelated to enzymatic activity of XOR (Vorbach et al. 2002). However, XOR is generally recognized as a key enzyme in purine degradation, where it catalyzes the oxidation of hypoxanthine to xanthine and of xanthine to uric acid. In primates and many other animals, the latter reaction represents the last step in this pathway, thus demanding excretion of poorly soluble uric acid with the urine to prevent severe consequences due to crystal formation in blood, serum, tissues, and muscles. In plants however, uric acid is likewise the product of XOR catalysis but far away from being the final product of purine degradation (Fig. 1).

Instead, uric acid is further decomposed to allantoin and allantoate, two important nitrogen storage and transport compounds at least in tropical legumes (Smith and Atkins 2002). Subsequently, allantoate is converted to glyoxylate via three enzymatic steps, which include the formation of ureidoglycine and ureidoglycolate and the release of carbon dioxide and ammonia (Werner and Witte 2011; Werner et al. 2013, and references therein). It is assumed that the latter compounds have the potential of being reused in photosynthesis or reassimilated by the photorespiratory glycolate and GOGAT cycles. Remarkably, the XOR substrate xanthine is the first common intermediate in the degradation of all purine nucleotides, thus representing an important bottleneck in this pathway and making XOR a key enzyme in purine catabolism.

In prokaryotes, XOR has been found exclusively in the XDH form and enables bacteria such as *Comamonas acidovorans* to grow on minimal medium supplemented with hypoxanthine or xanthine as the sole source of carbon and nitrogen (Xiang and Edmondson 1996; Ivanov et al. 2004). In contrast, other

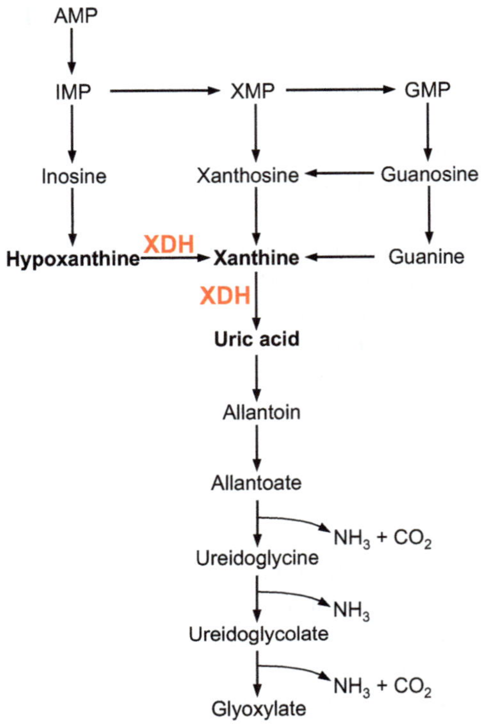

Fig. 1 General scheme of purine degradation in higher plants. The purine degradation pathway is depicted starting with adenosine monophosphate (AMP) at the crosspoint between purine de novo synthesis and catabolism. Only the relevant intermediates are shown according to Zrenner et al. (2006), Werner and Witte (2011) and Werner et al. (2013), presumed back-reactions are not considered. *IMP* inosine monophosphate, *GMP* guanosine monophosphate, and *XMP* xanthosine monophosphate

bacteria like *Escherichia coli* turned out to be unable to live on purines as sole carbon or nitrogen source (Xi et al. 2000), indicating that XOR in these bacteria is involved in purine salvage rather than being part of a complete purine catabolic pathway. The *C. acidovorans* enzyme consists of two 58 kDa α subunits, which each bind two nonidentical [2Fe–2S] iron–sulfur clusters, and two 87 kDa β subunits each binding one FAD and one Moco in its dinucleotide form. The fully constituted and active enzyme thus is an $(\alpha\beta)_2$ heterodimer with a molecular mass of about 290 kDa indistinguishable from the mass of the eukaryotic XOR enzymes. While several other bacteria such as *Rhodobacter capsulatus* (Leimkühler et al. 1998) basically harbor the same type of XOR with identical subunit composition, XOR enzymes from other prokaryotic species such as *Pseudomonas putida*, *Veillonella atypica*, and *Eubacterium barkeri* have been demonstrated to consist of only one or even three subunits, which associate to functionally active enzymes with molecular masses of up to 550 kDa (Woolfolk 1985; Hettrich and Lingens 1991; Gremer and Meyer 1996; Schräder et al. 1999). Differences between these enzymes do also concern variations of the bound redox groups, the composition of redox groups and their oligomerization states, but also the ligands bound to the molybdenum centers as well as the preferred co-substrate, which in some cases is NAD^+ and in others is $NADP^+$. Interestingly, in some bacteria the oxidation of hypoxanthine to xanthine and of xanthine to uric acid involves two enzymes, with purine hydroxylase catalyzing the first reaction and XOR catalyzing the latter (Self 2002). Yet, among the diverse bacterial XOR forms, the one found in *R. capsulatus* appeared to be more similar to its eukaryotic counterparts than to other prokaryotic XOR enzymes with respect to molecular mass, domain composition, bound redox groups, catalytical properties, and 3D structure (Leimkühler et al. 1998; Truglio et al. 2002), suggesting that eukaryotic XOR has evolved from an ancient bacterial XOR form as found in *R. capsulatus*.

Besides prokaryotes and eukaryotes, XOR homologues can be found also in archaea, with their physiological roles being largely unraveled in most cases. It is noteworthy that XOR is not present in all species but absent in those, which are specialized in a way that the function of XOR is not required (some bacteria and most archaea) or which have adopted to parasitic living styles that either enable the respective organism to acquire all necessary purine intermediates from their host or that have established alternative pathways. In eukaryotes, the loss of XOR genes seems to be accompanied always by the complete loss of molybdenum metabolism as is found in some yeasts species like *Saccharomyces cerevisiae* and *Schizosaccharomyces pombe* but not in other yeasts such as *Pichia* (*Komagataella*) *pastoris* (Zhang et al. 2011). Remarkably, the loss of XOR and molybdenum metabolism in eukaryotes appears to be limited to unicellar organisms, whereas multicellular eukaryotes generally appear to have an essential demand for molybdenum metabolism including XOR, which might be ascribed to the specialization of cells and tissues and the need for well-orchestrated pathways.

3 Distribution of XOR Among the Kingdom *Plantae* and Subcellular Localization

XOR occurs ubiquitously within the kingdom *Plantae* and it has been isolated from all organs of plants: root nodules (Triplett et al. 1982; Boland et al. 1983; Nguyen et al. 1986), roots (Barabás et al. 2000), leaves (Nguyen and Feierabend 1978; Montalbini 2000; Vitória and Mazzafera 1999), seedlings and whole plants (Suzuki and Takahashi 1975; Tajima and Yamamoto 1975; Kumar and Taneja 1977; Sauer et al. 2002), and fruits (Vitória and Mazzafera 1999; Taylor and Cowan 2001). Indeed, XOR activity has been reported in a great variety of plant species, including soybean, pea, wheat, maize, tea, lentil, tomato, coffee, alfalfa, avocado, tobacco, and *Arabidopsis* (i.e., Montalbini 1998; Sauer et al. 2002; Nakagawa et al. 2007; Corpas et al. 2009; Shakirov et al. 2012). Despite this widespread distribution and the reasonable number of studies of XOR in plants, the exact subcellular localization has been discussed rather controversially. Initially, XOR was found associated to microbodies (Nguyen 1986), glyoxysomes, and peroxisomes (Sandalio et al. 1988; Del Río et al. 1989), but later immunolabeling techniques revealed that XOR was present in the cytosol of infected and uninfected cells of legume root nodules, and that this feature was related to the process of ureide biosynthesis (Datta et al. 1991). Parallel and later studies supported the hypothesis of XOR being involved in stress response, and other locations within the plant cell were explored. In this respect, XOR was once more localized in peroxisomes, which are organelles with a potent oxidative metabolism that are additionally well known as ROS and reactive nitrogen species (RNS) generators (Corpas et al. 2008, 2009). In pea leaf peroxisomes, XOR was found to act with both oxidizing substrates, NAD^+ and O_2, where the activity with the latter was surprisingly twofold more pronounced, as indicated by enzymatic assays (Corpas et al. 2008). However, this recent article localizes XOR also in the cytosol, suggesting that XOR has several localizations within the plant cell.

4 Structure of XOR and Its Redox Centers

Unfortunately, no crystal structure of a plant XOR is available until now and a detailed description with regard to structure–function relationships has thus been precluded. Nevertheless, based on the high sequence similarity, the identical domain composition, and nearly identical substrate preferences, the well-known structures of bovine (Enroth et al. 2000), human (Yamaguchi et al. 2007) and rat XOR (Nishino et al. 2005) might well serve as model structures for plant and other eukaryotic XOR enzymes.

All eukaryotic XOR enzymes are homodimeric proteins of 290–300 kDa with each monomer of approximately 145 kDa being constituted of three distinct domains (Fig. 2a). The N-terminal domain of 20 kDa is responsible for binding of two nonidentical iron–sulfur clusters of the [2Fe–2S] type and is followed by a larger middle domain of 40 kDa that binds FAD. The C-terminal domain of 85 kDa exhibits two functions, one of which is the binding of one Moco while the other is to mediate the dimerization of two identical, monomeric XOR polypeptides (Fig. 2b). Iron-sulfur clusters, FAD and Moco are all prosthetic groups that are permanently bound by the enzyme and participate in catalysis and electron transfer. As is known for other molybdenum-dependent enzymes, Moco has a very particular role in forming an essential part of the active site of XOR where substrates are bound and converted and where the respective product is released. Notably, in the crystal structures of mammalian XOR proteins both molybdenum centers are about 50 Å apart, indicating that the two XOR monomers act independently from each other.

Once a reducing substrate such as xanthine or hypoxanthine is bound in the active site of XOR, the oxidative hydroxylation of these substrates takes place at the molybdenum center, whereby the molybdenum becomes protonated and reduced

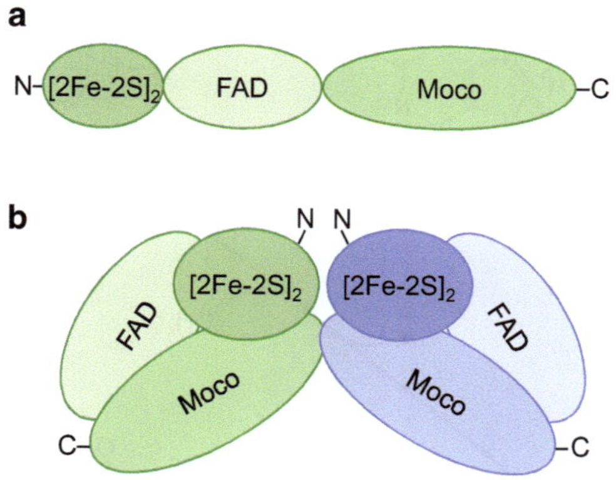

Fig. 2 Schematic presentation of the structure of xanthine oxidoreductase. (**a**) Linear presentation of the domain arrangement in one XOR polypeptide. Each domain is separated from the others by flexible hinge regions. (**b**) Simplified scheme of the three-dimensional structure of eukaryotic XOR according to Enroth et al. (2000). The two monomers of the XOR homodimer are shown in *green* and *blue*, respectively, with the dimerization interface being located within the large Moco domain. In contrast to what the linear structure suggests, electron transfer in the fully folded enzyme is directed from Moco via the two [2Fe–2S] clusters to the FAD, which is enabled by folding of the N-terminal Fe–S cluster domain between the Moco- and FAD-binding domains

from MoVI to MoIV in the reductive half-reaction of the catalytic sequence (Fig. 3). Electrons thus liberated from the substrate are passed on via the two iron–sulfur clusters to the FAD site, where they are finally removed either by NAD^+ or O_2 in the oxidative half-reaction of the catalytic sequence. In this reaction, a particular glutamate residue strictly conserved among all XOR and AO proteins (e.g., Glu1261 in bovine XOR, Glu1297 in *Arabidopsis* AtXDH1) is assumed to function as an active site base required for the activation of substrates (Hille et al. 2011). While the ultimate source of the oxygen atom incorporated by XOR into the hydroxyl group of the respective product has been identified as water (Murray et al. 1966), it has been demonstrated that a hydroxyl ligand of the molybdenum center (Mo–OH) represents the catalytically labile oxygen within the protein (Hille and Sprecher 1987), which is supported by the crystal structure of bovine XOR in which the Mo–OH group points toward the substrate-binding site (Enroth et al. 2000). Accordingly, substrates are coordinated to the catalytically introduced hydroxyl group of the molybdenum atom in a Mo–O–substrate fashion, thereby converting the respective substrate into hydroxylated product (e.g., hypoxanthine into xanthine or xanthine into uric acid, respectively). The product is finally released from the enzyme upon displacement by solvent-derived hydroxide, with electron transfer from the molybdenum center to the iron-sulfur clusters and FAD.

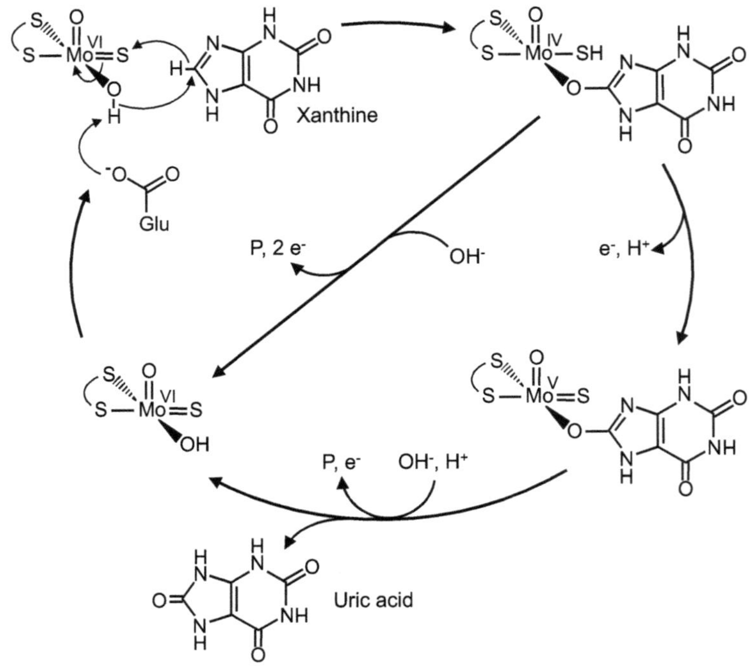

Fig. 3 General reaction mechanism of xanthine oxidoreductase with xanthine as substrate. Hydroxylation of xanthine is likely to be initiated by (1) proton abstraction of the hydroxyl group of the molybdenum center (Mo–OH) supported by a strictly conserved and catalytically active glutamate residue in the active site of XOR enzymes. (2) By base-assisted nucleophilic attack of the formed $Mo–O^{-}$ on the C8 position of xanthine and by concomitant hydride transfer to the Mo=S group a $Mo^{IV}O(SH)$–O–substrate intermediate is generated, which (3) breaks down by electron transfer to the other redox centers (Fe–S and FAD) and deprotonation, with the occurence of a Mo^{V} species. Alternatively, the intermediate can be broken down by hydroxide displacement of the product from the molybdenum center and subsequent transfer of electrons to the other redox centers [according to Hille et al. (2011)]

5 Established and Possible Physiological Substrates of Plant XOR

For all XOR proteins, irrespective of their origin, hypoxanthine and xanthine are the preferred substrates (Fig. 4a), which points out the involvement of XOR enzymes in purine catabolism (see above). Besides these purine substrates however, XOR enzymes catalyze the conversion of many other substrates, with some of them having medicinal or pharmaceutical relevance (Kitamura et al. 2006), some others being discussed to be of physiological importance, and those, whose importance for the respective organism is questionable. In case of plant XOR, one class of alternative substrates is represented by aldehydes, whose conversion is likely to be facilitated in a fashion basically identical to the oxidation of hypoxanthine and

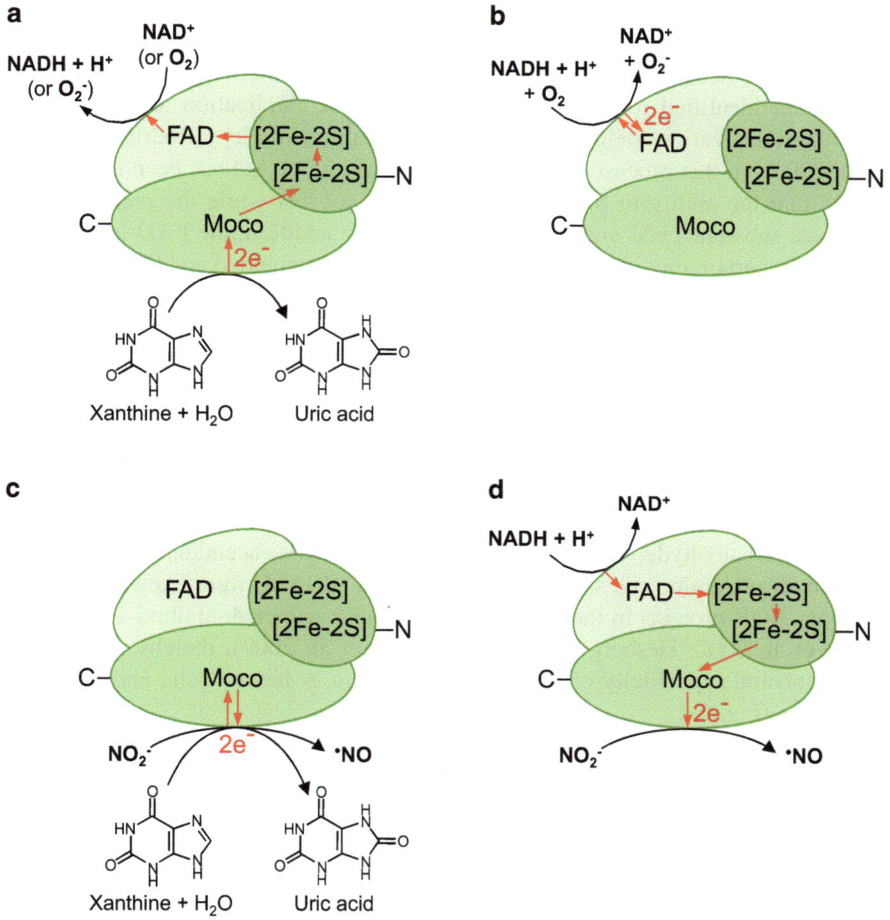

Fig. 4 The diverse subactivities of plant xanthine oxidoreductase. (**a**) Major activity of XOR with xanthine as reducing substrate, NAD^+ as the preferred oxidizing substrate and O_2 as alternative oxidizing substrate, which are reduced to NADH or superoxide anions (O_2^-), respectively. This activity involves all four redox groups. In place of xanthine, various other compounds such as hypoxanthine and aldehydes can likewise serve as reducing substrate. (**b**) NADH oxidase activity of XOR, which takes place solely at the FAD domain and involves oxidation of NADH with concomitant reduction of O_2 to O_2^-. (**c**) Nitrite reductase activity of XOR as exhibited by the substrate(xanthine)-reduced enzyme. Electrons derived from xanthine are not passed to other redox centers but to nitrite (NO_2^-), which is thereby reduced to nitric oxide ($^{\bullet}NO$). This activity is strictly dependent on low-oxygen conditions. (**d**) Nitrite reductase activity of XOR after reduction of the enzyme by NADH. This activity involves all four redox-active groups and resembles the inverse reaction of (**a**) with regard to electron flow. It is assumed that this activity also occurs in the presence of O_2. For clarity, only one monomer of the otherwise homodimeric enzyme is presented

xanthine, respectively (Coelho et al. 2012). It is thus not surprising that aldehyde oxidase (AO) proteins, which share a high degree of sequence similarity with XOR and which present identical domain structures and prosthetic group composition, have been identified to derive from XOR by gene duplication and subsequent neo-functionalization before the origin of multicellularity (Rodríguez-Trelles et al. 2003). In this process, the substrate-binding site of AO has been modified in a way that the ability to bind purine substrates got lost while the preference for aldehyde substrates was significantly enhanced. In addition, the FAD-binding site underwent certain modifications in which NAD^+ was precluded and O_2 was favored as final electron acceptor. Even though AO is now specialized for oxidation of aldehydes, it appears reasonable to assume that XOR contributes to a certain extent to the oxidation of certain aldehydes in vivo. In fact, among the various aldehyde substrates tested for plant XOR the ABA precursor abscisic aldehyde was shown to be oxidized with greatest efficiency of about 30 % relative to the native substrates xanthine/hypoxanthine (Hesberg et al. 2004), thus allowing to speculate that XOR has a minor function in ABA biosynthesis in vivo, e.g., under conditions of decreased AO activity. Albeit with lower efficiency, other compounds such as indole-3-carbaldehyde, 1-naphthaldehyde, heptaldehyde, acetaldehyde, benzaldehyde, hydroxylamine and salicylhydroxamic acid (SHAM) were likewise oxidized by plant XOR proteins to the respective carboxylic acids (Montalbini 1998, 2000; Sauer et al. 2002; Hesberg et al. 2004; Rümer et al. 2009), thereby proving the broad substrate specificity of plant XOR for cyclic, *N*-heterocyclic and noncyclic aldehydes (Fig. 5).

A typical feature of molybdenum-containing enzymes seems to be the capacity to catalyze the one-electron reduction of nitrite to nitric oxide (NO) under low-oxygen conditions as has been shown for plant nitrate reductase (summarized by Meyer et al. 2005; Gupta et al. 2011a), the human mitochondrial reducing components (Sparacino-Watkins et al. 2014), rat AO (Li et al. 2008, 2009), and also mammalian XOR (summarized by Cantu-Medellin and Kelley 2013). Importantly, in contrast to the oxidation of xanthine/hypoxanthine and aldehydes electrons are consumed rather than produced upon nitrite reduction, meaning that the molybdenum center of the enzyme needs to be reduced before nitrite can be converted into NO (Fig. 4c, d). For rat XOR, an interesting observation concerns the capability of the enzyme to reduce nitrite to NO under both anaerobic and aerobic conditions, with the NO formation rate under aerobic conditions and with NADH as reducing substrate reaching about 70 % of the anaerobic rate (Li et al. 2008). Though not in the context of NO production, also plant XOR has been demonstrated to react with NADH (Yesbergenova et al. 2005; Zarepour et al. 2010) and it can thus be speculated that similar to mammalian XOR, plant XOR is able to catalyze the NADH-dependent reduction of nitrite in the presence of oxygen (Fig. 4d). Moreover, as NO is involved in the regulation of many important processes in plants including seed germination, root growth, respiration, stomatal closure, as well as in a variety of response reactions to abiotic and biotic stresses (Gupta et al. 2011a and references therein), XOR-derived NO might well be of physiological relevance, in particular due to the abundance of XOR in nearly all plant tissues and at all developmental stages.

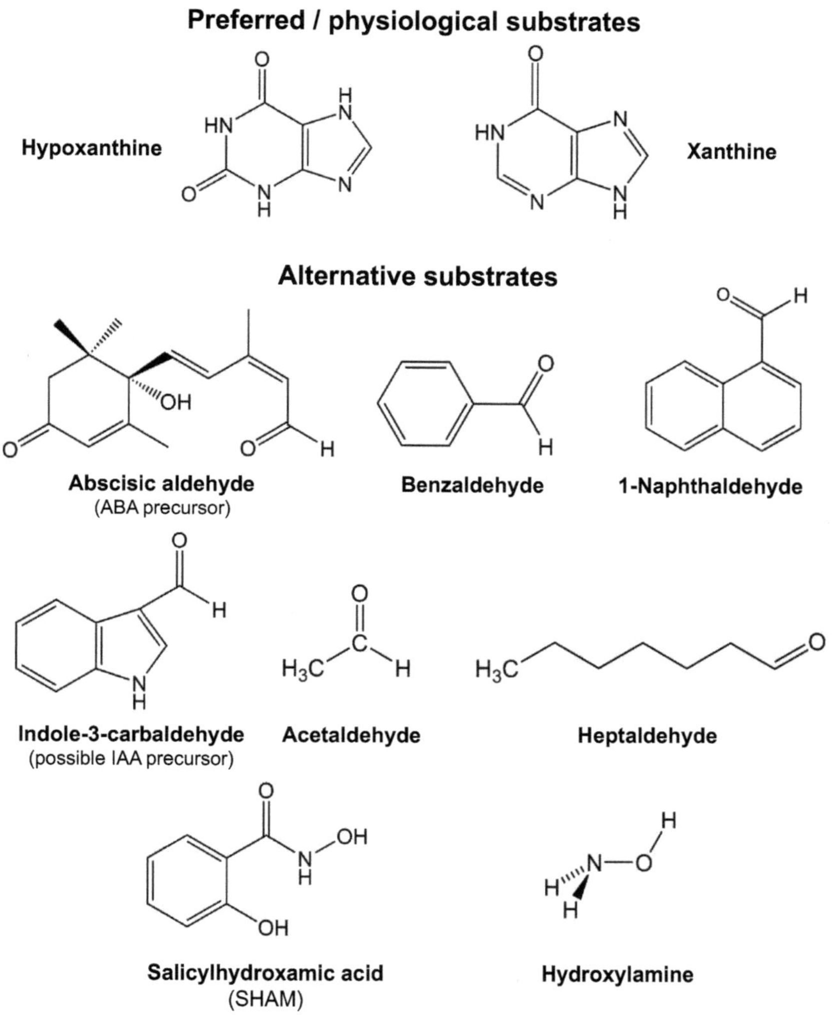

Fig. 5 Established and possible physiological substrates of plant xanthine oxidoreductase. Xanthine and hypoxanthine are the preferred substrates for XOR, which points out the involvement of XOR enzymes in purine catabolism. Albeit with lower efficiency, other compounds such as benzaldehyde, indole-3-carbaldehyde, abscisic aldehyde, 1-naphthaldehyde, heptaldehyde, acetaldehyde, hydroxylamine, and SHAM were likewise oxidized by plant XOR proteins. For more detailed information see Sect. 5

Another possible physiological substrate of XOR is represented by NADH, which has been demonstrated to be oxidized not only by mammalian (Harrison 2002; Maia et al. 2007) but also by *Arabidopsis* and tomato XOR (Yesbergenova et al. 2005; Zarepour et al. 2010). The reaction takes place exclusively at the FAD domain of the enzyme and is believed not to involve the other redox centers Moco and iron–sulfur clusters (Fig. 4b). Once NADH is oxidized and its electrons are transferred to FAD, the reduced flavin tends to transfer its electrons to O_2 with concomitant formation of superoxide anions ($O_2^{\bullet-}$). Interestingly, the rate of NADH oxidation by the XOR enzyme AtXDH1 from *Arabidopsis* is extremely high as compared to the NADH oxidation rates of mammalian XOR enzymes, suggesting that XOR-dependent NADH oxidation might be an important source of $O_2^{\bullet-}$ in plant cells (Zarepour et al. 2010). In combination with the use of NAD^+ as preferred electron acceptor, the intrinsic NADH oxidase activity of plant XOR might also be involved in the regulation of the cellular NAD^+/NADH ratio. In this respect, it was demonstrated that NAD^+ and NADH indeed compete for the same binding site at the FAD domain of plant XOR, with NAD^+ being prefered over NADH when simultaneously offered to AtXDH1. This indicates that, as long as the respective components are available, the physiological activity with xanthine/hypoxanthine as substrates and NAD^+ as electron acceptor dominates over the alternative NADH oxidase activity of XOR.

6 Inhibitors of XOR

Within the last years research on the possible physiological roles of XOR in plants has concentrated on the analysis of T-DNA insertion lines and RNAi knockdown plants from *Arabidopsis* (Nakagawa et al. 2007; Brychkova et al. 2008; Watanabe et al. 2010). However, before such mutants became available other approaches aimed to modulate XOR activity by use of appropriate and more or less specific inhibitors mostly in other plant species. As such inhibitors provide powerful tools, in particular, when lacking a XOR mutant in a nonmodel species, some of these inhibitors are briefly presented and discussed in the following section with respect to their suitability:

The presently best established and most-specific inhibitor of XOR enzymes from all organisms including plants is allopurinol. This inhibitor has been established in the therapy of patients suffering from hyperuricemia/gout, which is characterized by abnormally high levels of uric acid in the blood, serum, muscles, and urine, often leading to urate crystals in these tissues. In those cases, where hyperuricemia and the elevation of uric acid levels is due to an extremely high XOR activity allopurinol is used as a substrate analogue that competes with hypoxanthine/xanthine for the substrate-binding site in XOR. Once allopurinol is bound in the active site it is oxidized to oxypurinol (also referred to as alloxanthine), which remains covalently bound to the reduced molybdenum atom in the active site of XOR (Okamoto and Nishino 2008). In this way, oxypurinol serves as a suicide inhibitor of XOR, acting in a mechanism-based fashion. Treatment of plants, plant extracts and purified plant XOR enzymes with allopurinol has been proven to be very efficient and has thus been reported frequently and for diverse plant species including *Arabidopsis* (Hesberg et al. 2004; Yesbergenova et al. 2005; Nakagawa et al. 2007; Brychkova et al. 2008; Zarepour et al. 2010), avocado (Taylor and Cowan 2004), cowpea (Woo et al. 1980; Triplett 1986; Atkins et al. 1988), fava bean (Montalbini 2000), french bean (Quiles et al. 2009), lima bean (Triplett 1986), soybean (Fujihara and Yamaguchi 1978; Triplett et al. 1980; Boland and Schubert 1982), lentil (Montalbini 2000), pea (Sauer et al. 2002), and wheat (Montalbini 1998). Since during allopurinol treatment side effects have never been reported, allopurinol can indeed be considered the most-suitable XOR inhibitor in plants. The relatively new inhibitors Febuxostat and Y-700 likewise act near to the active site of XOR, but in contrast to allopurinol/oxypurinol these compounds serve as structure-based inhibitors by interacting with several amino acids in the active site tunnel (Okamoto and Nishino 2008), thereby precluding hypoxanthine and xanthine from the substrate-binding site. Although the use of these inhibitors in plants has not yet been reported, Febuxostat and Y-700 are likely to act as efficient and specific in inhibition of plant XOR as allopurinol.

Unlike allopurinol, cyanide treatment of XOR in plant extracts is much less specific, even though XOR activity is very efficiently blocked. The reason for this is the interaction of cyanide with a variety of micro- and macromolecules, including proteins. However, purified XOR (and also XOR in plant extracts) can be very effectively and reversibly inactivated by cyanide, which under oxidative conditions removes the terminal, or so-called cyanolyzable sulfur ligand of the molybdenum atom in the active site of XOR (Wahl and Rajagopalan 1982; Schwartz et al. 1997; Wollers et al. 2008). By anaerobic treatment with sulfide and reducing agents such as dithionite the sulfur can be reintroduced, whereby the enzyme's activity is largely restored. In vivo, the absence of the terminal/cyanolyzable sulfur in XOR is caused by a deficiency in the Moco sulfurase gene (*ABA3* in *Arabidopsis*, *Flacca* in tomato, *ABA1* in tobacco), which has been demonstrated to be essentially required for generating and providing the sulfur ligand of the Moco in both XOR and AO (Schwartz et al. 1997; Bittner et al. 2001; Sagi et al. 2002; Wollers et al. 2008). Like XOR, AO therefore is likewise susceptible to cyanide treatment and subsequent re-sulfuration.

Another inhibitor that affects the activity of XOR is the molybdate (MoO_4^{2-}) analogue tungstate (WO_4^{2-}) (Bentley et al. 1981; Arst et al. 1982; Kawada et al. 1982; Schieber and Edmondson 1993; Nielsen et al. 1996). While molybdate is the natural form of molybdenum required for synthesis of Moco in eukaryotic and most prokaryotic molybdenum-dependent enzymes, tungstate is used instead of molybdate by a number of enzymes in archaea and certain bacteria (Grimaldi et al. 2013). Due to their similar size, charge and physicochemical properties, tungstate can substitute molybdate, which usually results in inactivation of the respective Mo-enzyme, with the strength of inactivation depending on various parameters such as concentration, affected enzyme, and plant species. As tungstate affects not only XOR but all molybdenum-dependent enzymes and also many physiological processes such as seedling growth, biomass, cell cycle, components of the cytoskeleton, and gene regulation (Adamakis et al. 2012), tungstate generally appears to be a relative inappropriate inhibitor for studying XOR-related functions in plants and its use in plants is thus documented by only very few reports.

In contrast to all the afore-mentioned inhibitors, diphenyleneiodonium (DPI) does not attack the active site or the molybdenum center but rather the flavin moiety in the FAD domain, where it forms a covalent complex with the FAD itself (O'Donnell et al. 1994). However, DPI is far from being a specific inhibitor of XOR enzymes but rather interacts with basically any type of enzyme that has a reduced FAD or FMN molecule bound. DPI is thus a very potent inhibitor of cellular $O_2^{\bullet -}$ generating systems, which often carry flavins as redox groups and catalyze one-electron transfer reactions, as is the case for several enzymes of the respiratory chain and the photosystems, NAD(P)H oxidases and XOR. Nevertheless, DPI is a very powerful tool to study the contribution of the flavin cofactor to a specific sub-activity of a flavin-dependent enzyme such as XOR (Yesbergenova et al. 2005; Zarepour et al. 2010), but its use is largely limited to purified enzymes.

7 Interconversion of XOR into Xanthine Dehydrogenase and Xanthine Oxidase Forms

An often-discussed issue concerns the discrimination of eukaryotic XOR enzymes into xanthine dehydrogenase (XDH) forms and xanthine oxidase (XO) forms. Basically, all XOR enzymes are initially produced in the XDH form and the term XDH is thus likewise adequate for XOR enzymes. Yet, mammalian XOR can be converted from its initial XDH form into the XO form either reversibly by disulfide bond formation between the sulfhydryl groups of two conserved cysteine residues (Cys535 and Cys992 in rat XOR; Nishino and Nishino 1997; Nishino et al. 2005) or irreversibly by limited proteolysis, the latter resulting in cleavage between the three functional domains of XOR with maintenance of catalytic activity (Amaya et al. 1990) (Fig. 6). Under physiological conditions, the XDH form appears to dominate with 80 % over the XO form with 20 % in tissues (Della Corte et al. 1969; Stirpe et al. 2002), while the XO form clearly predominates in milk, most likely due to a specific function of XO in the formation of milk-fat droplets (Jeong et al. 2013). Both, reversible and irreversible interconverion result in conformational changes

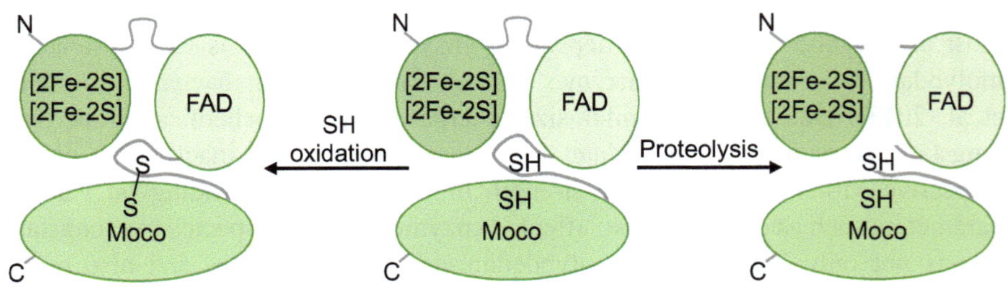

Fig. 6 Schematic presentation of interconversion of mammalian xanthine oxidoreductase into dehydrogenase and oxidase forms. Mammalian XOR is converted from the initially existing dehydrogenase form (*middle*) into the oxidase form either irreversibly by limited proteolysis within the hinge regions between the domains (*right*) or reversibly by oxidation of conserved cysteine thiols (*left*). For clarity, only one monomer of the otherwise homodimeric enzyme is presented

near to the NAD^+-binding site at the FAD domain, thereby preventing access of NAD^+ to the FAD cofactor (Nishino et al. 2005). The interconversion between XDH and XO therefore affects the preference for the oxidizing substrates NAD^+ and O_2: whereas XDH possesses high reactivity toward NAD^+ and low reactivity toward O_2 as electron acceptor, XO uses O_2 very efficiently and has negligible reactivity toward NAD^+ (Hille and Nishino 1995). In contrast to mammalian XOR proteins, other animal XOR enzymes such as those from chicken and insects do not possess the conserved cysteine couple required for reversible conversion into XO and are exclusively found in the XDH form (Nishino and Nishino 1989; Komoto et al. 1999). It should thus be considered that interconversion between XDH and XO might be a regulatory element in mammals required to separate and control the specific functions of XOR in tissues and milk. In fact, also all plant XOR proteins analyzed so far from nodules of bean (Boland 1981), pea seedlings (Sauer et al. 2002), wheat leaves (Montalbini 1998), *Arabidopsis* (Hesberg et al. 2004), and leaves of various legumes (Montalbini 2000) gave no indications for a possible interconversion into XO and it therefore appears likely that plant XOR enzymes are present exclusively in the XDH form. This is supported by the fact that the two conserved cysteine residues required for reversible conversion of XDH into XO by disulfide bond formation are absent in all accessible sequences of plant XOR proteins. Furthermore, irreversible proteolytic cleavage of plant XOR into several distinct domains that maintain catalytic activity as known from the mammalian enzymes has not been reported until now. Although not absolutely conclusive, these observations indicate that plant XOR proteins exist exclusively in the XDH form and are not, neither reversibly nor irreversibly, converted into XO. It should be noted in this respect that some research aimed to discriminate between XDH and XO forms in plant cell extracts. In all cases, this discrimination was based on the misleading assumption that XDH exclusively transfers its electrons to NAD^+, whereas XO exclusively transfers its electrons to $O_2^{\bullet}$. However, due to the fact that XDH is capable of transferring its hypoxanthine/xanthine-derived electrons alternatively also to O_2 such discrimination between XDH and XO forms is presently impossible.

8 Plant XOR: An Important Source of ROS and NO in Plants

In 1969 McCord and Fridovich discovered that a basal production of free radicals like $O_2^{\bullet-}$ may take place in biological systems as shown for XOR catalysis (McCord and Fridovich 1969; Fridovich 1998). Since then, ROS and RNS became acknowledged as a group of highly reactive molecules playing key roles in cellular signaling. In the past decades, their involvement in numerous plant physiological processes has been ascertained. For instance, $O_2^{\bullet-}$ appears to be principally involved in programmed cell death (PCD) associated to hypersensitive response

(HR) (Doke and Ohashi 1988; Jabs et al. 1996). Hydrogen peroxide (H_2O_2) is also an active participant of the HR (Desikan et al. 1998; Yoda et al. 2003), further mediating stomatal closure (Pei et al. 2000), mechanical wounding response (Orozco-Cárdenas et al. 2001), and plant–pathogen interaction (Mittler et al. 1999). In the last 20 years NO has come to light as an essential second messenger in a wide range of phenomena such as seed germination, root growth, leaf expansion, senescence, and stress response (Delledonne et al. 2001; Besson-Bard et al. 2008; Wang et al. 2010; Gupta et al. 2011a, b). Other nitrogen oxidized species, represented by nitrite (NO_2^-), nitrate (NO_3^-), or peroxynitrite ($ONOO^-$), have also been described to act as signals in various responses within plant cells but further research is required in order to clarify these issues (Scheible et al. 1997; Zhang and Forde 2000; Serrano et al. 2012; Konishi and Yanagisawa 2013). It has been ascertained that the production of these molecules may be derived from NO via different reactions, so it is considered that their participation in diverse plant signaling pathways is fundamental (Dordas et al. 2003; Begara-Morales et al. 2013).

In mammals, XOR is able to reduce O_2 with the subsequent generation of $O_2^{\bullet -}$, and it may also reduce NO_2^- to NO. In principal, both species can rapidly react, which results in $ONOO^-$ formation (Harrison 2002). In plants, in addition to purine catabolism, XOR is often related to plant–pathogen interaction phenomena, PCD and natural senescence (Bittner and Mendel 2010), with these reactions representing a spotlight of $O_2^{\bullet -}$ production within the cells. There are early descriptions of $O_2^{\bullet -}$ production by XOR in peroxisomes of watermelon and pea (Sandalio et al. 1988; Del Río et al. 1989). Later on, *Arabidopsis* recombinant AtXDH1 was described as a $O_2^{\bullet -}$ producer in vitro, transferring around 20 % of the electrons from xanthine to O_2 (Hesberg et al. 2004). Enzymatic AtXDH1 activity rises at senescence *in planta* giving one more hint of $O_2^{\bullet -}$ involvement in this physiological process. Tomato XOR was also found to produce $O_2^{\bullet -}$ and it was observed that both tomato and *Arabidopsis* XOR from leaves were capable of producing $O_2^{\bullet -}$ with either xanthine/hypoxanthine or NADH as reducing substrate (Yesbergenova et al. 2005). Subsequently, $O_2^{\bullet -}$ production by AtXDH1 was characterized and it was established that NADH-dependent $O_2^{\bullet -}$ generation was pH dependent, suggesting a certain regulation of the process (Zarepour et al. 2010). AtXDH1 exhibited the highest rates of NADH oxidase activity among all enzymes studied so far and, besides, this reaction was not dependent of the sulfuration state of the Moco center, which enables the plant to control $O_2^{\bullet -}$ contents and maintain a constitutive cellular pool of this molecule (Zarepour et al. 2010).

Superoxide dismutases (SODs) are powerful antioxidant metalloenzymes whose primary activity is the dismutation of $O_2^{\bullet -}$ resulting in H_2O_2 and molecular oxygen (Fridovich 1995). As it was mentioned before, an excess of H_2O_2 within the cell is often related with HR and PCD, as well as with nodule formation and nodule/leaf senescence (Levine et al. 1994; Escuredo et al. 1996; Evans et al. 1999; Grant and Loake 2000; Santos et al. 2001; Alesandrini et al. 2003). In this context, and taking into account that plant XOR is involved in many of these processes, its implication in H_2O_2 production could be explained, although none of the plant XOR enzymes

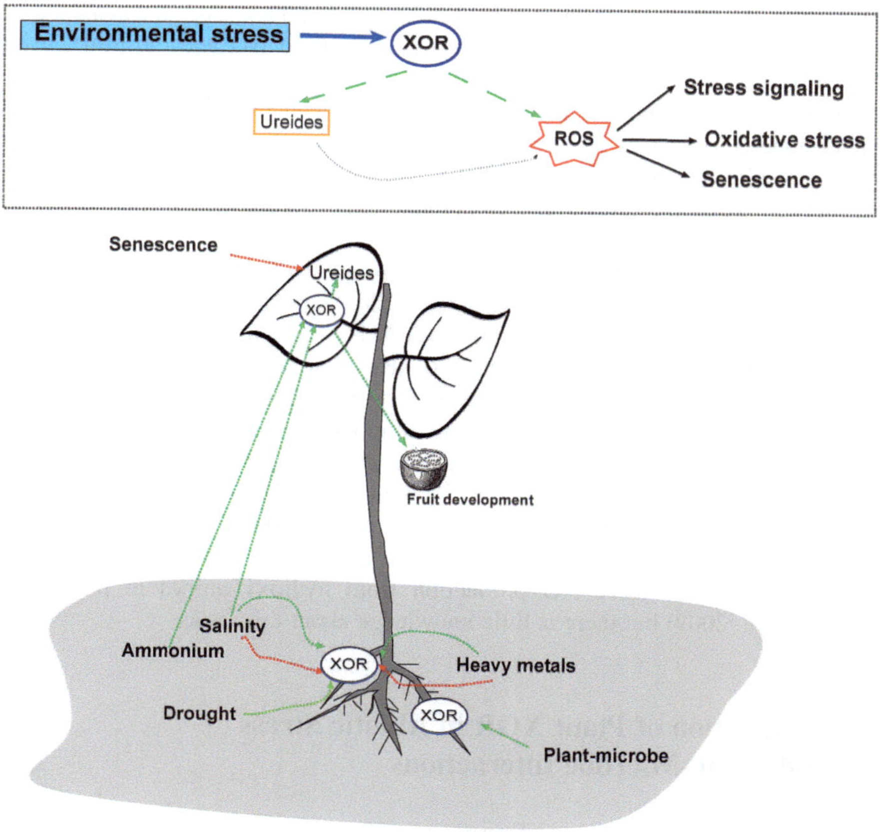

Fig. 7 Schematic view of the implication of plant xanthine oxidoreductase in abiotic stress and plant–microbe interactions. Plant XOR is implicated in the ROS generation in plant tissues under abiotic or biotic stresses. These ROS are involved in oxidative stress, stress signaling, and senescence. At the same time, XOR produces ureides that alleviate ROS. XOR levels/activities increase under some of stresses, e.g., ammonium, drought, plant–microbe interactions (indicated by the *green dashed arrows*), whereas in other cases, a reduction of XOR levels/activities has been demonstrated (indicated by the *red dashed arrows*). Besides, XOR is also involved in fruit development

analyzed to date where shown to directly produce this ROS (Fig. 7). Rubio et al. (2004) proved that some of the H_2O_2 in legume root nodules is produced through two consecutive reactions involving an NAD(P)H oxidase-like $O_2^{\bullet-}$-generating enzyme and a CuZnSOD, and complementary scientific data gave light to the possibility that XOR may also be involved in cytosolic H_2O_2 generation (Datta et al. 1991; Zarepour et al. 2010).

In the late 80s NO was revealed as the endothelium-derived relaxing factor (EDRF), responsible for the relaxation of the smooth muscle (Ignarro et al. 1987). In consequence, there was a clear paradigm shift regarding this molecule, drastically changing the vision of NO by the scientific community. The most important

NO source in metazoans is the enzyme nitric oxide synthase (NOS), a hemeprotein that transfers five electrons and oxidizes L-arginine to NO and citrulline, using NADPH and O_2 as cosubstrates (Halliwell and Gutteridge 2007). A decade ago, XOR was also discovered as a NO-generating enzyme in animal systems, especially under anaerobiosis (i.e., ischemia) based on its capacity to reduce NO_2^- in the presence of either xanthine or NADH (Godber et al. 2000; Harrison 2002). Concerning oxidative pathways of NO production in plants, the enzymatic reaction analogous to those in animal cells involving NOS has been greatly discussed, including the retraction of several papers which had previously affirmed the existence of a mammalian-like NOS (Travis 2004; Crawford et al. 2006; Zemojtel et al. 2006). Reductive NO formation from NO_2^- has long been studied in plant systems and cytosolic nitrate reductase (cNR) is currently considered the major source for NO in plants (Planchet and Kaiser 2006). Knowing that plants are able to grow and complete their life cycle using ammonium as the sole nitrogen source (Ariz et al. 2011), the existence of oxidative reactions for NO production in plants is essential. XOR was initially proposed as a reductive NO source, being the catalyst of NO_2^- reduction to NO in analogy to animals, but definitive evidence is still lacking (Del Río et al. 2004). Concerning this issue, there is some evidence of XOR involvement in oxidative NO production from hydroxylamines in plant cells (Rümer et al. 2009) but there is little knowledge about this issue.

9 Implication of Plant XOR in Abiotic Stress and Plant–Microbe Interactions

9.1 Abiotic Stresses and Senescence

Plant growth is ultimately modulated by the environment with the drawback of suffering stress and affection of the metabolic functions under extreme conditions, such as extreme temperatures, drought, or salinity. Under these circumstances, XOR together with other molybdoenzymes (AO, NR) may constitute one of the earliest metabolic adjustments towards abiotic stress in plants. However, the involvement of XOR in the response mechanisms of plants to stress and plant-microbe signaling remains unknown; hence, these are reviewed and discussed herein (Fig. 7).

The involvement of XOR in salt tolerance is controversial. Shakirov et al. (2012) described that the different activities of XOR isoforms increased with salinity. XOR activity also increased in leaves of wheat genotypes with differing salt tolerance. However, root XOR activity decreased with increasing concentrations of NaCl in the salt-tolerant genotype and was enhanced in the salt-sensitive genotype (Xu et al. 2012). On the other hand, the expression of the XOR gene *AtXDH1* in *Arabidopsis thaliana* was induced by salt stress; however, XOR activity was not affected (Hesberg et al. 2004). The different implications of XOR activity in different species and different organs under salinity stress suggest that XOR is subject of various regulatory mechanisms (Xu et al. 2012). When salinity is combined with ammonium (NH_4^+) nutrition, XOR activity was enhanced in maize nodal roots (Barabás et al. 2000) and ryegrass, where it correlated with a higher content of ureides in plant tissues (Sagi et al. 1998). In pea roots however, salinity did not significantly affect XOR activity, and there was even a sharp depression of XOR activity in leaves (Zdunek-Zastocka and Lips 2003). The changes of root and leaf XOR activity in pea plants seem to correlate with ureide biosynthesis that is induced by NH_4^+, the product of N fixation, and rather not by salinity.

Under drought stress, Yesbergenova et al. (2005) demonstrated that XOR was induced by water deprivation but also by the main hormone that regulates the response to drought, ABA. It is suggested that XOR, besides the main producer AO, takes part in the synthesis of this phytohormone, as it was demonstrated by Sagi et al. (1999) in Moco-deficient tomato mutants lacking AO and XOR activities, and hence exhibiting a reduced capacity to produce ABA. Besides, it has been reported that XOR responds to drought with transcriptional upregulation and with a concomitant increase of the enzymatic activity (Hesberg et al. 2004). This is in agreement with Watanabe et al. (2010), who demonstrated that XOR-suppressed lines of *Arabidopsis* having a deprived purine metabolism were more susceptible to drought-induced oxidative damage. These results indicate that XOR has critical roles in maintaining plant growth and development under both normal and stressful growth conditions.

In leaf peroxisomes cadmium treatment reduced XOR activity (Romero-Puertas et al. 1999; Corpas et al. 2008), whereas cadmium stress enhanced XOR activities in roots of *Phragmites australis* (Jiang and Wang 2007). These results suggest the possible implication of XOR in plant response mechanisms to abiotic stress by heavy metals, and it is discussed that in peroxisomes the ratio of NAD^+ and O_2 as oxidizing substrates for XOR is regulated in a way which is dependent on the physiological stage and stress conditions (Corpas et al. 2008).

Another example for the possible implication of XOR in the response to abiotic stresses is the enhanced chlorophyll degradation in leaves in the absence of XOR, which might result from ROS production during dark stress (Guo and Crawford 2005; Brychkova et al. 2008). Apart from environmental stresses, it has also been proposed that XOR contributes to plant hormone homeostasis and to the control of fruit size in avocado (Cowan et al. 2001).

Senescence is a natural phenomenon occurring in plant tissues, which is known to induce oxidative processes, including lipid peroxidation and protein carbonylation among others. Although the precise role of XOR in senescence remains unclear, the fact that this enzyme is a natural producer of $O_2^{\bullet-}$ (Hesberg et al. 2004; Yesbergenova et al. 2005) gives an idea on the kind of involvement XOR may have in cell ageing phenomena. Furthermore, there is evidence of an augmentation of enzymatic levels both in mammalian heart (Willems et al. 2003) and plant XOR in ageing and senescence procedures. In pea leaves, natural senescence showed an increase in peroxisomal XOR activity, along with the enzymatic activities of uricase (UO) and superoxide dismutase (SOD) (Pastori and Del Río

1997). These facts indicate that purine catabolism may be taking place in this organelle during leaf senescence. In *Arabidopsis*, mRNA levels of *AXDH1* notably increased in aging (6-week-old) and senescent (8-week-old) leaves, but enzymatic activity only rose significantly in senescent tissues (Hesberg et al. 2004). Further studies revealed that *Arabidopsis* XOR gene silencing lines showed an accelerated senescence phenotype caused by a decrease of XOR levels, while wild-type plants presented an age-dependent increase of XOR levels (Nakagawa et al. 2007). Additionally, research with *AtXDH1* mutants confirmed that XOR regulates ureide levels during nutrient mobilization, the latter being a process highly associated with senescence (Brychkova et al. 2008). Finally, senescence of *Phalaenopsis* flower was likewise correlated with an increase of XOR activity (Tewari et al. 2009). In view of the controversial results, Zarepour et al. (2010) proposed that in *Arabidopsis thaliana* XOR is involved in a general and constitutive function during purine degradation but probably not at the level of stress adaptation.

9.2 XOR and Plant–Microbe Interactions

Plants identify and defend against many invading pathogens by inducing the HR, which leads to localized PCD at the site of infection, confining the disease and preventing it from spreading through the whole plant. The HR is characterized by triggering oxidative burst including production of ROS like $O_2^{\bullet-}$ and H_2O_2. As mentioned previously, plant XOR is often related to pathogenic defence and HR, as these processes require ROS formation to be triggered (Montalbini 1995; Montalbini and Della Torre 1996). In this respect, it was proven that allopurinol treatment decreased both ureide levels and rust infection in wheat leaves (Montalbini 1992a), the HR induced by the incompatible interaction between bean and *Uromyces phaseoli* (Montalbini 1992b), and additionally the HR triggered by tobacco necrosis virus in tobacco leaves (Montalbini and Della Torre 1996). Moreover, it was observed that XOR activity increased with the infection of wheat leaves by *Puccinia recondita* (Montalbini 1995). A recent work pointed out that XOR activity and H_2O_2 contents rapidly increased in leaves of a rust-sensitive wheat, whereas in the resistant variety such changes were not observable (Sarsenbaev et al. 2013). All these facts suggest that XOR is a key enzyme in the plant–pathogen interaction process.

It is well known that XOR is one of the key enzymes during purine metabolism in the nitrogen-fixing nodules of tropical legumes, where it has a key role in ureide biosynthesis and major nitrogen storage and transport (Smith and Atkins 2002). In fact, inhibition of nodule functioning has been observed when plants were supplied with allopurinol (Atkins et al. 1988). As mentioned earlier, allopurinol is a competitive inhibitor of XOR which impairs hypoxanthine/xanthine oxidation to uric acid, thus preventing ureide synthesis in the nodule and resulting in nitrogen starvation, chlorosis, and reduced growth (Atkins et al. 1988). No connection between XOR and root nodulation has been established so far, although it is widely known that in legumes ROS levels increase during the rhizobial infection (Santos et al. 2001; Rubio et al. 2004). In this regard, the elimination of $O_2^{\bullet-}$ by SOD during the legume–Rhizobium symbiosis is known to be essential for the establishment of symbiosis and nodule development (Santos et al. 2000). SOD from either plant legume or bacterials cells localized in the nucleus of the plant nodule cells evidenced that the elimination of $O_2^{\bullet-}$ exerted a positive signaling during the symbiosis initial stages (Rubio et al. 2009; Asensio et al. 2011). All these facts suggest that XOR isoforms may have a role in this process.

9.3 Does XOR Have a Relevant Function in Plant Stress and Plant–Microbe Interactions?

Although it has been broadly suggested that XOR may have a function in plant stress, there is still no overwhelming evidence to assure that this enzyme is responsible for the resistance to environmental stresses. It is proposed that the role of XOR during stress might be related to the need for a more efficient use of available carbon skeletons, which are required to synthesize organic nitrogen compounds with a low C/N ratio for their transport through the xylem to the shoot (Sagi et al. 1998). However, as described in this review, the implication of XOR under stress conditions seems to be quite complex. In all these stresses, it remains unclear whether the enzymatic activity is physiologically relevant for plant protection from stressful conditions. The question whether XOR is involved in plant injury (due to ROS production and consequent damage of plant tissues) or whether it is protective (due to the production of ureides to alleviate ROS) is still not clarified. Further studies are needed in order to fill the knowledge gap in the involvement of XOR in plant adaptation to environmental stresses.

Concluding Remarks

In the last 50 years, the study of XOR has led to a detailed view of this enzyme and to a better understanding of its biochemical features and physiological significance in plant metabolism. Important aspects have been elucidated or are in the process, such as its key role in ureide biosynthesis, its subcellular localization, and its capacity of ROS production under certain circumstances. However, questions dealing with the possible involvement of XOR in the response to stress conditions and with XOR as a tangible NO generator within plant cells remain unclear. The advanced study of these reactions appears fundamental and future research should focus on the exact reactions that ascribe XOR a function in stress response and production of oxidized nitrogen species, including NO, under NH_4^+ nutrition.

Acknowledgments and Funding The authors acknowledge the support of the research grants AGL2010-16167 to J.F.M. from the Spanish Ministry of Science and Innovation and Bi 1075/5-1 to F.B. by the Deutsche Forschungsgemeinschaft. R.E. received a JAE-Doctor grant from the Spanish Research Council (CSIC). We are greatly thankful to all the researchers who worked and have been involved in the XOR enzyme study, for advancing knowledge in this field.

References

Adamakis IDS, Panteris E, Eleftheriou EP (2012) Tungsten toxicity in plants. Plants 1:82–99

Agarwal A, Banerjee A, Banerjee UC (2011) Xanthine oxidoreductase: a journey from purine metabolism to cardiovascular excitation-contraction coupling. Crit Rev Biotechnol 31:264–280

Alesandrini F, Mathis R, Van de Sype G (2003) Possible roles for a cysteine protease and hydrogen peroxide in soybean nodule development and senescence. New Phytol 158:131–138

Amaya Y, Yamazaki K, Sato M et al (1990) Proteolytic conversion of xanthine dehydrogenase from the NAD-dependent type to the O_2-dependent type. Amino acid sequence of rat liver xanthine dehydrogenase and identification of the cleavage sites of the enzyme protein during irreversible conversion by trypsin. J Biol Chem 265:14170–14175

Ariz I, Artola E, Cabrera Asensio A et al (2011) High irradiance increases NH_4^+ tolerance in *Pisum sativum*: higher carbon and energy availability improve ion balance but not N assimilation. J Plant Physiol 168:1009–1015

Arst HN Jr, Tollervey DW, Sealy-Lewis HM (1982) A possible regulatory gene for the molybdenum-containing cofactor in Aspergillus nidulans. J Gen Microbiol 128:1083–1093

Asensio AC, Marino D, James EK, Ariz I, Arrese-Igor C, Aparicio-Tejo PM, Arredondo-Peter R, Moran JF (2011) Expression and localization of a Rhizobium-derived cambialistic superoxide dismutase in pea (Pisum sativum) nodules subjected to oxidative stress. Mol Plant Microbe Interact 24:1247–1257

Atkins CA, Sanford PJ, Storer PJ, Pate JS (1988) Inhibition of nodule functioning in cowpea by a xanthine oxidoreductase inhibitor, allopurinol. Plant Physiol 88:1229–1234

Barabás NK, Omarov RT, Erdei L, Lips SH (2000) Distribution of the Mo-enzymes aldehyde oxidase, xanthine dehydrogenase and nitrate reductase in maize (*Zea mays* L.) nodal roots as affected by nitrogen and salinity. Plant Sci 15:49–58

Begara-Morales JC, Chaki M, Sánchez-Calvo B et al (2013) Protein tyrosine nitration in pea roots during development and senescence. J Exp Bot 64:1121–1134

Bentley MM, Williamson JH, Oliver MJ (1981) The effects of molybdate, tungstate and lxd on aldehyde oxidase and xanthine dehydrogenase in *Drosophila melanogaster*. Can J Genet Cytol 23:597–609

Besson-Bard A, Pugin A, Wendehenne D (2008) New insights into nitric oxide signalling in plants. Annu Rev Plant Biol 59:21–39

Bittner F, Mendel RR (2010) Cell biology of molybdenum. Plant Cell Monogr 17:119–143

Bittner F, Oreb M, Mendel RR (2001) ABA3 is a molybdenum cofactor sulfurase required for activation of aldehyde oxidase and xanthine dehydrogenase in *Arabidopsis thaliana*. J Biol Chem 276:40381–40384

Boland MJ (1981) NAD^+: xanthine dehydrogenase from nodules of navy beans: partial purification and properties. Biochem Int 2:567–574

Boland MJ, Schubert KR (1982) Purine biosynthesis and catabolism in soybean root nodules: incorporation of 14C from 14CO2 into xanthine. Arch Biochem Biophys 213:486–491

Boland MJ, Blevins DG, Randall DD (1983) Soybean nodule xanthine dehydrogenase: a kinetic study. Arch Biochem Biophys 222:435–441

Booth VH (1935) The identity of xanthine oxidase and the Schardinger enzyme. Biochem J 29:1732–1748

Brychkova G, Alikulov Z, Fluhr R, Sagi M (2008) A critical role for ureides in dark and senescence-induced purine remobilization is unmasked in the AtXDH1 *Arabidopsis* mutant. Plant J 54:496–509

Cantu-Medellin N, Kelley EE (2013) Xanthine oxidoreductase-catalyzed reduction of nitrite to nitric oxide: insights regarding where, when and how. Nitric Oxide 34:19–26

Coelho C, Mahro M, Trincão J et al (2012) The first mammalian aldehyde oxidase crystal structure: insights into substrate specificity. J Biol Chem 287:40690–40702

Corpas FJ, Palma JM, Sandalio LM et al (2008) Peroxisomal xanthine oxidoreductase: characterization of the enzyme from pea (*Pisum sativum* L.) leaves. J Plant Physiol 165:1319–1330

Corpas FJ, Hayashi M, Mano S et al (2009) Peroxisomes are required for in vivo nitric oxide accumulation in the cytosol following salinity stress of arabidopsis plants. Plant Physiol 151:2083–2094

Cowan AK, Cripps RF, Richings EW, Taylor NJ (2001) Fruit size: towards an understanding of the metabolic control of fruit growth using avocado as a model system. Physiol Plant 111:127–136

Crawford NM, Galli M, Tischner R et al (2006) Response to Zemojtel et al: plant nitric oxide synthase: back to square one. Trends Plant Sci 11:526

Datta DB, Triplett EW, Newcomb EH (1991) Localization of xanthine dehydrogenase in cowpea root nodules: implications for the interaction between cellular compartments during ureide biogenesis. Proc Natl Acad Sci USA 88:4700–4702

Del Río LA, Fernández VM, Rupérez FL et al (1989) NADH induces the generation of superoxide radicals in leaf peroxisomes. Plant Physiol 89:728–731

Del Río LA, Corpas FJ, Barroso JB (2004) Nitric oxide and nitric oxide synthase activity in plants. Phytochemistry 65:783–792

Della Corte E, Gozzetti G, Novello F, Stirpe F (1969) Properties of the xanthine oxidase from human liver. Biochim Biophys Acta 191:164–166

Delledonne M, Zeier J, Marocco A, Lamb C (2001) Signal interactions between nitric oxide and reactive oxygen intermediates in the plant hypersensitive disease resistance response. Proc Natl Acad Sci USA 98:13454–13459

Desikan R, Burnett EC, Hancock JT et al (1998) Harpin and hydrogen peroxide induce the expression of a homologue of gp91-phox in *Arabidopsis thaliana* suspension cultures. J Exp Bot 49:1767–1771

Doke N, Ohashi Y (1988) Involvement of an O_2^--generating system in the induction of necrotic lesions on tobacco leaves infected with tobacco mosaic virus. Physiol Mol Plant Pathol 32:163–175

Dordas C, Rivoal J, Hill RD (2003) Plant haemoglobins, nitric oxide and hypoxic stress. Ann Bot 91:173–178

Enroth C, Eger BT, Okamoto K et al (2000) Crystal structures of bovine milk xanthine dehydrogenase and xanthine oxidase: structure-based mechanism of conversion. Proc Natl Acad Sci USA 97:10723–10728

Escuredo PR, Minchin FR et al (1996) Involvement of activated oxygen in nitrate-induced senescence of pea root nodules. Plant Physiol 11:1187–1195

Evans PJ, Gallesi D, Mathieu C et al (1999) Oxidative stress occurs during soybean nodule senescence. Planta 208:73–79

Fridovich I (1995) Superoxide radical and superoxide dismutases. Annu Rev Biochem 64:97–112

Fridovich I (1998) The trail to superoxide dismutase. Protein Sci 7:2688–2690

Fridovich I, Handler P (1958) Xanthine oxidase. III. Sulfite oxidation as an ultra sensitive assay. J Biol Chem 233:1578–1580

Fujihara S, Yamaguchi M (1978) Effects of allopurinol [4-hydroxypyrazolo(3,4-d)pyrimidine] on the metabolism of allantoin in soybean plants. Plant Physiol 62:134–138

Godber BLJ, Doel JJ, Sapkota GP et al (2000) Reduction of nitrite to nitric oxide catalysed by xanthine oxidoreductase. J Biol Chem 275:7757–7763

Grant JJ, Loake GJ (2000) Role of reactive oxygen intermediates and cognate redox signaling in disease resistance. Plant Physiol 124:21–29

Gremer L, Meyer O (1996) Characterization of xanthine dehydrogenase from the anaerobic bacterium *Veillonella atypica* and identification of a molybdopterin-cytosine-dinucleotide containing molybdenum cofactor. Eur J Biochem 238:862–866

Grimaldi S, Schoepp-Cothenet B, Ceccaldi P et al (2013) The prokaryotic Mo/W-bisPGD enzymes family: a catalytic workhorse in bioenergetic. Biochim Biophys Acta 1827:1048–1085

Guo FQ, Crawford NM (2005) Arabidopsis nitric oxide synthase1 is targeted to mitochondria and protects against oxidative damage and dark-induced senescence. Plant Cell 17:3436–3450

Gupta KJ, Alisdair RF, Kaiser WM, van Dongen J (2011a) On the origins of nitric oxide. Trends Plant Sci 16:160–168

Gupta KJ, Hincha DK, Mur LAJ (2011b) NO way to treat a cold. New Phytol 189:360–363

Halliwell B, Gutteridge JMC (eds) (2007) Free radicals in biology and medicine. Clarendon, Oxford

Harrison R (2002) Structure and function of xanthine oxidoreductase: where are we now? Free Radic Biol Med 33:774–797

Hesberg C, Hänsch R, Mendel RR, Bittner F (2004) Tandem orientation of duplicated xanthine dehydrogenase genes from *Arabidopsis thaliana*. J Biol Chem 279:13547–13554

Hettrich D, Lingens F (1991) Microbial metabolism of quinoline and related compounds. VIII. Xanthine dehydrogenase from a quinoline utilizing *Pseudomonas putida* strain. Biol Chem Hoppe Seyler 372:203–211

Hille R, Nishino T (1995) Flavoprotein structure and mechanism. 4. Xanthine oxidase and xanthine dehydrogenase. FASEB J 9:995–1003

Hille R, Sprecher H (1987) On the mechanism of action of xanthine oxidase. Evidence in support of an oxo transfer mechanism in the molybdenum-containing hydroxylases. J Biol Chem 262:10914–10917

Hille R, Nishino T, Bittner F (2011) Molybdenum enzymes in higher organisms. Coord Chem Rev 255:1179–1205

Hopkins FG (1921) On an autoxidisable constituent of the cell. Biochem J 15:286–305

Ignarro LJ, Buga GM, Wood KS et al (1987) Endothelium-derived relaxing factor produced and released from artery and vein is nitric oxide. Proc Natl Acad Sci USA 84:9265–9269

Ivanov NV, Trani M, Edmondson DE (2004) High-level expression and characterization of a highly functional Comamonas acidovorans xanthine dehydrogenase in Pseudomonas aeruginosa. Protein Expr Purif 37:72–82

Jabs T, Dietrich RA, Dangl JL (1996) Initiation of runaway cell death in an *Arabidopsis* mutant by extracellular superoxide. Science 273:1853–1856

Jeong J, Lisinski I, Kadegowda AK et al (2013) A test of current models for the mechanism of milk-lipid droplet secretion. Traffic 14:974–986

Jiang X, Wang C (2007) Cadmium distribution and its effects on molybdate-containing hydroxylases in *Phragmites australis*. Aquat Bot 86:353–360

Kawada J, Shirakawa Y, Yoshimura Y, Nishida M (1982) Thyroid xanthine oxidase and its role in thyroid iodine metabolism in the rat: difference between effects of allopurinol and tungstate. J Endocrinol 95:117–241

Kitamura S, Sugihara K, Ohta S (2006) Drug-metabolizing ability of molybdenum hydroxylases. Drug Metab Pharmacokinet 21:83–98

Komoto N, Yukuhiro K, Tamura T (1999) Structure and expression of tandemly duplicated xanthine dehydrogenase genes of the silkworm (*Bombyx mori*). Insect Mol Biol 8:73–83

Konishi M, Yanagisawa S (2013) Arabidopsis NIN-like transcription factors have a central role in nitrate signalling. Nat Commun 4:1617

Kumar R, Taneja VA (1977) Xanthine oxidase in lentil (*Lens esculenta*) seedlings. Biochim Biophys Acta 485:489–491

Leimkühler S, Kern M, Solomon PZ et al (1998) Xanthine dehydrogenase from the phototrophic purple bacterium *Rhodobacter capsulatus* is more similar to its eukaryotic counterparts than to prokaryotic molybdenum enzymes. Mol Microbiol 27:853–869
Levine A, Tenhaken R, Dixon R, Lamb C (1994) H_2O_2 from the oxidative burst orchestrates the plant hypersensitive disease resistance response. Cell 79:583–593
Li H, Cui H, Kundu TK et al (2008) Nitric oxide production from nitrite occurs primarily in tissues not in the blood: critical role of xanthine oxidase and aldehyde oxidase. J Biol Chem 283:17855–17863
Li H, Kundu TK, Zweier JL (2009) Characterization of the magnitude and mechanism of aldehyde oxidase-mediated nitric oxide production from nitrite. J Biol Chem 284:33850–33858
Maia L, Duarte RO, Ponces-Freire A et al (2007) NADH oxidase activity of rat and human liver xanthine oxidoreductase: potential role in superoxide production. J Biol Inorg Chem 12:777–787
McCord JM, Fridovich I (1968) The reduction of cytochrome *c* by milk xanthine oxidase. J Biol Chem 243:5753–5760
McCord JM, Fridovich I (1969) An enzymic function for erythrocuprein (Hemocuprein). J Biol Chem 244:6049–6055
Mendel RR (2011) Cell biology of molybdenum in plants. Plant Cell Rep 30:1787–1797
Meyer C, Lea US, Provan F et al (2005) Is nitrate reductase a major player in the plant NO (nitric oxide) game? Photosynth Res 83:181–189
Mittler R, Herr EH, Orvar BL et al (1999) Transgenic tobacco plants with reduced capability to detoxify reactive oxygen intermediates are hyperresponsive to pathogen infection. Proc Natl Acad Sci USA 96:14165–14170
Montalbini P (1992a) Inhibition of hypersensitive response by allopurinol applied to the host in the incompatible relationship between *Phaseolus vulgaris* and *Uromyces phaseoli*. J Phytopathol 134:218–228
Montalbini P (1992b) Ureides and enzymes of ureide synthesis in wheat seeds and leaves and effect of allopurinol on *Puccinia recondita* f. sp. *tritici* infection. Plant Sci 87:225–231
Montalbini P (1995) Effect of rust infection on purine catabolism enzyme levels in wheat leaves. Physiol Mol Plant Pathol 46:275–292
Montalbini P (1998) Purification and some properties of xanthine dehydrogenase from wheat leaves. Plant Sci 134:89–102
Montalbini P (2000) Xanthine dehydrogenase from leaves of leguminous plants: purification, characterization and properties of the enzyme. J Plant Physiol 156:3–16
Montalbini P, Della Torre G (1996) Evidence of a two-fold mechanism responsible for the inhibition by allopurinol of the hypersensitive response induced in tobacco by tobacco necrosis virus. Physiol Mol Plant Pathol 48:273–287
Morgan EJ, Stewart CP, Hopkins FG (1922) On the anaerobic and aerobic oxidation of xanthine and hypoxanthine by tissues and by milk. Proc R Soc Lond B 94:109–131
Murray KN, Watson JG, Chaykin S (1966) Catalysis of the direct transfer of oxygen from nicotinamide N-oxide to xanthine by xanthine oxidase. J Biol Chem 241:4798–4801
Nakagawa A, Sakamoto S, Takahashi M et al (2007) The RNAi-mediated silencing of xanthine dehydrogenase impairs growth and fertility and accelerates leaf senescence in transgenic *Arabidopsis* plants. Plant Cell Physiol 48:1484–1495
Nguyen J (1986) Plant xanthine dehydrogenase: its distribution, properties and function. Physiol Végétale 24:263–281
Nguyen J, Feierabend J (1978) Some properties and sub cellular localization of xanthine dehydrogenase in pea leaves. Plant Sci Lett 13:125–132
Nguyen J, Machal L, Vidal J et al (1986) Immunochemical studies on xanthine dehydrogenase of soybean root nodules: ontogenic changes in the level of enzyme and immunocytochemical localization. Planta 167:190–195
Nielsen VG, Tan S, Weinbroum A et al (1996) Lung injury after hepatoenteric ischemia-reperfusion: role of xanthine oxidase. Am J Respir Crit Care Med 154:1364–1369

Nishino T, Nishino T (1989) The nicotinamide adenine dinucleotide-binding site of chicken liver xanthine dehydrogenase. Evidence for alteration of the redox potential of the flavin by NAD binding or modification of the NAD-binding site and isolation of a modified peptide. J Biol Chem 264:5468–5473

Nishino T, Nishino T (1997) The conversion from the dehydrogenase type to the oxidase type of rat liver xanthine dehydrogenase by modification of cysteine residues with fluorodinitrobenzene. J Biol Chem 272:29859–29864

Nishino T, Okamoto K, Kawaguchi Y et al (2005) Mechanism of the conversion of xanthine dehydrogenase to xanthine oxidase: identification of the two cysteine disulfide bonds and crystal structure of a non-convertible rat liver xanthine dehydrogenase mutant. J Biol Chem 280:24888–24894

O'Donnell VB, Smith GCM, Jones OTG (1994) Involvement of phenyl radicals in Iodonium compound inhibition of flavoenzymes. Mol Pharm 46:778–785

Okamoto K, Nishino T (2008) Crystal structures of mammalian xanthine oxidoreductase bound with various inhibitors: allopurinol, febuxostat, and FYX-051. J Nippon Med Sch 75:2–3

Orozco-Cárdenas ML, Narváez-Vásquez J, Ryan CA (2001) Hydrogen peroxide acts as a second messenger for the induction of defense genes in tomato plants in response to wounding, systemin, and methyl jasmonate. Plant Cell 13:179–191

Pastori GM, Del Río LA (1997) Natural senescence of pea leaves: an activated oxygen-mediated function of peroxisomes. Plant Physiol 113:411–418

Pei ZM, Mutrata Y, Benning G et al (2000) Calcium channels activated by hydrogen peroxide mediate abscisic acid signalling in guard cell. Nature 406:731–734

Planchet E, Kaiser WM (2006) Nitric oxide production in plants: facts and fictions. Plant Signal Behav 1:46–51

Quiles FA, Raso MJ, Pineda M, Piedras P (2009) Ureide metabolism during seedling development in French bean (*Phaseolus vulgaris*). Physiol Plant 135:19–28

Rodríguez-Trelles F, Tarrío R, Ayala FJ (2003) Convergent neofunctionalization by positive Darwinian selection after ancient recurrent duplications of the xanthine dehydrogenase gene. Proc Natl Acad Sci USA 100:13413–13417

Romero-Puertas MC, McCarthy I, Sandalio LM et al (1999) Cadmium toxicity and oxidative metabolism of pea leaf peroxisomes. Free Radic Res 31:S25–S31

Rubio MC, James EK, Clemente MR et al (2004) Localization of superoxide dismutases and hydrogen peroxide in legume root nodules. Mol Plant Microbe Interact 17:1294–1305

Rubio MC, Becana M, Kanematsu S et al (2009) Immunolocalization of antioxidant enzymes in high-pressure frozen root and stem nodules of *Sesbania rostrata*. New Phytol 183:395–407

Rümer S, Gupta KJ, Kaiser WM (2009) Plant cells oxidize hydroxylamines to NO. J Exp Bot 60:2065–2072

Sagi M, Omarov RT, Lips SH (1998) The Mo-hydroxylases xanthine dehydrogenase and aldehyde oxidase in ryegrass as affected by nitrogen and salinity. Plant Sci 135:125–135

Sagi M, Fluhr R, Lips SH (1999) Aldehyde oxidase and xanthine dehydrogenase in a *flacca* tomato mutant with deficient abscisic acid and wilty phenotype. Plant Physiol 120:571–578

Sagi M, Scazzocchio C, Fluhr R (2002) The absence of molybdenum cofactor sulfuration is the primary cause of the *flacca* phenotype in tomato plants. Plant J 31:305–317

Sandalio LM, Fernández VM, Rupérez FL, Del Río LA (1988) Superoxide free radicals are produced in glyoxysomes. Plant Physiol 87:1–4

Santos R, Herouart D, Puppo A, Touati D (2000) Critical protective role of bacterial superoxide dismutase in *Rhizobium*-legume symbiosis. Mol Microbiol 38:750–759

Santos R, Hérouart D, Sigaud S et al (2001) Oxidative burst in alfalfa-*Sinorhizobium meliloti* symbiotic interaction. Mol Plant Microbe Interact 14:86–89

Sarsenbaev KN, Abiyev SA, Kozhamzharova LS et al (2013) Effect of leaf rust on the molybdenum-containing enzymes activity in spring wheat varieties differing in resistance to infection. World Appl Sci J 25:1308–1313

Sauer P, Frébortová J, Sebela M et al (2002) Xanthine dehydrogenase of pea seedlings: a member of the plant molybdenum oxidoreductase family. Plant Physiol Biochem 40:393–400
Schardinger F (1902) Über das Verhalten der Kuhmilch gegen Methylenblau und seine Verwendung zur Unterscheidung von ungekochter und gekochter Milch. Zeitschrift für Untersuchung der Nahrungs- und Genussmittel 5:1113–1121
Scheible W-R, González-Fontes A, Lauerer M et al (1997) Nitrate acts as a signal to induce organic acid metabolism and repress starch metabolism in tobacco. Plant Cell 9:783–798
Schieber A, Edmondson DE (1993) Studies on the induction and phosphorylation of xanthine dehydrogenase in cultured chick embryo hepatocytes. Eur J Biochem 215:307–314
Schräder T, Rienhöfer A, Andreesen JR (1999) Selenium-containing xanthine dehydrogenase from *Eubacterium barkeri*. Eur J Biochem 264:862–871
Schwartz SH, Léon-Kloosterziel KM, Koornneef M, Zeevaart JA (1997) Biochemical characterization of the aba2 and aba3 mutants in *Arabidopsis thaliana*. Plant Physiol 114:161–166
Self WT (2002) Regulation of purine hydroxylase and xanthine dehydrogenase from Clostridium purinolyticum in response to purines, selenium, and molybdenum. J Bacteriol 184:2039–2044
Serrano I, Romero-Puertas MC, Rodríguez Serrano M et al (2012) Role of peroxynitrite in programmed cell death induced in self-incompatible pollen. Plant Signal Behav 7:779–781
Shakirov ZS, Khakimov SA, Shomurodov KF (2012) Effect of salinity and drought on symbiotical and biochemical properties of *Onobrychis* and alfalfa. Agric Sci 3:444–454
Smith PMC, Atkins CA (2002) Purine biosynthesis. Big in cell division, even bigger in nitrogen assimilation. Plant Physiol 128:793–802
Sparacino-Watkins CE, Tejero J, Sun B et al (2014) Nitrite reductase and NO synthase activity of the mitochondrial molybdopterin enzymes mARC1 and mARC2. J Biol Chem. doi:10.1074/jbc.M114.555177
Stirpe F, Ravaioli M, Battelli MG et al (2002) Xanthine oxidoreductase activity in human liver disease. Am J Gastroenterol 97:2079–2085
Suzuki T, Takahashi E (1975) Metabolism of xanthine and hypoxanthine in the tea plant (*Thea sinensis* L.). Biochem J 146:79–85
Tajima S, Yamamoto Y (1975) Enzymes of purine catabolism in soybean plants. Plant Cell Physiol 16:271–282
Taylor N, Cowan K (2001) Plant hormone homeostasis and the control of avocado fruit size. Plant Growth Regul 35:247–255
Taylor NJ, Cowan AK (2004) Xanthine dehydrogenase and aldehyde oxidase impact plant hormone homeostasis and affect fruit size in 'Hass' avocado. J Plant Res 117:121–130
Tewari RK, Kumar P, Kim S et al (2009) Nitric oxide retards xanthine oxidase-mediated superoxide anion generation in Phalaenopsis flower: an implication of NO in the senescence and oxidative stress regulation. Plant Cell Rep 28:267–279
Travis J (2004) NO-making enzyme no more: cell, PNAS papers retracted. Science 306:960
Triplett EW (1986) Two indirect methods for detecting ureide synthesis by nodulated legumes. Plant Physiol 81:566–571
Triplett EW, Blevins DG, Randall DD (1980) Allantoic acid synthesis in soybean root nodule cytosol via xanthine dehydrogenase. Plant Physiol 65:1203–1206
Triplett EW, Blevins DG, Randall DD (1982) Purification and properties of soybean nodule xanthine dehydrogenase. Arch Biochem Biophys 219:39–46
Truglio JJ, Theis K, Leimkühler S et al (2002) Crystal structures of the active and alloxanthine-inhibited forms of xanthine dehydrogenase from *Rhodobacter capsulatus*. Structure 10:115–125
Vitória AP, Mazzafera P (1999) Xanthine degradation and related enzyme activities in leaves and fruits of two coffea species differing in caffeine catabolism. J Agric Food Chem 47:1851–1855
Vorbach C, Scriven A, Capecchi MR (2002) The housekeeping gene xanthine oxidoreductase is necessary for milk fat droplet enveloping and secretion: gene sharing in the lactating mammary gland. Genes Dev 16:3223–3235

Wahl RC, Rajagopalan KV (1982) Evidence for the inorganic nature of the cyanolyzable sulfur of molybdenum hydroxylases. J Biol Chem 257:1354–1359

Wang BL, Tang XY, Cheng LY et al (2010) Nitric oxide is involved in phosphorus deficiency-induced cluster-root development and citrate exudation in white lupin. New Phytol 187:1112–1123

Watanabe S, Nakagawa A, Izumi S et al (2010) RNA interference-mediated suppression of xanthine dehydrogenase reveals the role of purine metabolism in drought tolerance in *Arabidopsis*. FEBS Lett 584:1181–1186

Werner AK, Witte CP (2011) The biochemistry of nitrogen mobilization: purine ring catabolism. Trends Plant Sci 16:381–387

Werner AK, Medina-Escobar N, Zulawski M et al (2013) The ureide-degrading reactions of purine ring catabolism employ three amidohydrolases and one aminohydrolase in Arabidopsis, soybean, and rice. Plant Physiol 163:672–681

Willems L, Garnham B, Headrick JP (2003) Aging-related changes in myocardial purine metabolism and ischemic tolerance. Exp Gerontol 38:1169–1177

Wollers S, Heidenreich T, Zarepour M et al (2008) Binding of sulfurated molybdenum cofactor to the C-terminal domain of ABA3 from *Arabidopsis thaliana* provides insight into the mechanism of molybdenum cofactor sulfuration. J Biol Chem 283:9642–9650

Woo KC, Atkins CA, Pate JS (1980) Biosynthesis of ureides from purines in a cell-free system from nodule extracts of cowpea [*Vigna unguiculata* (L) Walp.]. Plant Physiol 66:735–739

Woolfolk CA (1985) Purification and properties of a novel ferricyanide-linked xanthine dehydrogenase from *Pseudomonas putida* 40. J Bacteriol 163:600–609

Xi H, Schneider BL, Reitzer L (2000) Purine catabolism in *Escherichia coli* and function of xanthine dehydrogenase in purine salvage. J Bacteriol 182:5332–5341

Xiang Q, Edmondson DE (1996) Purification and characterization of a prokaryotic xanthine dehydrogenase from *Comamonas acidovorans*. Biochemistry 35:5441–5450

Xiong L, Gong Z, Rock CD et al (2001) Modulation of abscisic acid signal transduction and biosynthesis by an Sm-like protein in *Arabidopsis*. Dev Cell 1:771–781

Xu H, Zhai J, Liu Y et al (2012) The response of Mo-hydroxylases and abscisic acid to salinity in wheat genotypes with differing salt tolerances. Acta Physiol Plant 34:1767–1778

Yamaguchi Y, Matsumura T, Ichida K et al (2007) Human xanthine oxidase changes its substrate specificity to aldehyde oxidase type upon mutation of amino acid residues in the active site: roles of active site residues in binding and activation of purine substrate. J Biochem 141:513–524

Yesbergenova Z, Yang G, Oron E et al (2005) The plant Mo-hydroxylases aldehyde oxidase and xanthine dehydrogenase have distinct reactive oxygen species signatures and are induced by drought and abscisic acid. Plant J 42:862–876

Yoda H, Yamaguchi Y, Sano H (2003) Induction of hypersensitive cell death by hydrogen peroxide produced through polyamine degradation in tobacco plants. Plant Physiol 132:1973–1981

Zarepour M, Kaspari K, Stagge S et al (2010) Xanthine dehydrogenase AtXDH1 from *Arabidopsis thaliana* is a potent producer of superoxide anions via its NADH oxidase activity. Plant Mol Biol 72:301–310

Zdunek-Zastocka E, Lips HS (2003) Is xanthine dehydrogenase involved in response of pea plants (*Pisum sativum* L.) to salinity or ammonium treatment? Acta Physiol Plant 25:395–401

Zemojtel T, Fröhlich A, Palmieri MC et al (2006) Plant nitric oxide synthase: a never-ending story? Trends Plant Sci 11:524–525

Zhang H, Forde BG (2000) Regulation of *Arabidopsis* root development by nitrate availability. J Exp Bot 51:51–59

Zhang Y, Rump S, Gladyshev VN (2011) Comparative genomics and evolution of molybdenum utilization. Coord Chem Rev 255:1206–1217

Zrenner R, Stitt M, Sonnewald U, Boldt R (2006) Pyrimidine and purine biosynthesis and degradation in plants. Annu Rev Plant Biol 57:805–836

The Roles of Plant Peroxidases in the Metabolism of Reactive Nitrogen Species and Other Nitrogenous Compounds

Farida Minibayeva and Richard P. Beckett

1 Introduction: Interactions of Nitrogen and Peroxidases

Peroxidases (EC 1.11.1.x) are heme-containing proteins that catalyze one-electron oxidation of a variety of compounds in the presence of H_2O_2 or organic hydroperoxides (R-OOH) (Welinder et al. 1992). The products are free-radical cation intermediates and H_2O. The heme-containing peroxidases are classified into two main superfamilies, the peroxidase-cyclooxygenases (former mammalian peroxidases) and the non-animal peroxidases. The latter comprises class I or intracellular peroxidases, class II or secretory fungal peroxidases, and class III or higher plant secreted peroxidases. Peroxidases have broad substrate specificity and can metabolize nitrogenous compounds with various degrees of oxidation, such as nitrates, nitrites, and reactive nitrogen species (RNS) (Sakihama et al. 2003; Zhang et al. 2012). To date, the role of peroxidase in the metabolism of nitrogenous compounds in plants is poorly understood.

The participation of peroxidases in nitrogen metabolism can be considered as reduction/oxidation cycling of nitrogen in different states of oxidation. But on the other hand, it is also accompanied by the nitrogen-mediated formation of peroxidase intermediate compounds (Compounds I and II) with different heme redox status. The interactions of RNS with peroxidases are of a biological significance for several reasons. First of all, these interactions are implicated in signaling by producing or binding RNS and thus effecting the activities of components of signal

F. Minibayeva (✉)
Kazan Institute of Biochemistry and Biophysics, Kazan Science Centre, Russian Academy of Sciences, P.O. Box 30, Kazan 420111, Russian Federation
e-mail: minibayeva@mail.knc.ru; fminibayeva@gmail.com

R.P. Beckett
School of Life Sciences, University of KwaZulu-Natal, Pietermaritzburg, South Africa
e-mail: rpbeckett@gmail.com

K.J. Gupta, A.U. Igamberdiev (eds.), *Reactive Oxygen and Nitrogen Species Signaling and Communication in Plants*, Signaling and Communication in Plants 23,
DOI 10.1007/978-3-319-10079-1_3

transduction pathways. For example, RNS are stress signals involved in plant responses to stresses (Groβ et al. 2013). Second, the interactions of RNS with peroxidases also provide a key mechanism of posttranslational protein modifications as a consequence of the peroxidase-catalyzed nitration of tyrosine and tryptophan derivatives in proteins in the presence of nitrite and hydrogen peroxide. Activities of a large number of proteins and enzymes depend on tyrosine residues. At present, the nitrating pathways operating in vivo have not been clearly elucidated. Third, the interactions of RNS and other nitrogenous compounds with peroxidases may lead to the formation of nitrophenols, which have potentially defense and toxic effects. Forth, peroxidases may metabolize synthetic nitrophenol derivates, for example, by hydroxylating them to catechols and thus provide an essential mechanism for removal of toxic compounds including xenobiotics. In this review, we will critically assess the current knowledge on the unique roles of plant and fungal peroxidases in the metabolism of nitrogenous compounds.

2 Peroxidase-Mediated Signaling of Reactive Nitrogen Species

Similarly to reactive oxygen species (ROS), RNS are important messengers in many physiological processes. They are stress signals involved in plant responses to high salt, excess light, cold, heat, ozone, UV-B and various pathogens, and developmental processes in plants. Several excellent reviews cover these topics [for example, see Mur et al. (2013) and Groβ et al. (2013)]. It is now clear that nitric oxide ($NO^•$) and, in general, most of the RNS ($NO^•$, NO^+, NO^-, $NO_2^•$, and $ONOO^-$), are major signaling molecules (Durner and Klessig 1999), which can be synthesized during stress responses in plants at the same time as ROS (Almagro et al. 2009).

Reactive nitrogen species, such as NO, can bind to most transition metals and metalloproteins, including heme-containing peroxidases (Shapiro 2005). Nitric oxide is a relatively stable paramagnetic free-radical molecule (Ros Barceló et al. 2002), which is considered as a synchronizing chemical messenger involved in cytotoxicity and programmed cell death (Durner and Klessig 1999). Both activating and inhibitory effects of NO on heme proteins have been described. For example, in the presence of poor electron donors, such as catechols (Nappi and Vass 2001) or guaiacol (Huang et al. 2002; Uchida et al. 2002), NO may enhance peroxidase activity by promoting the formation of the native enzyme, FeIII, from the intermediate Compound II. In contrast, reversible binding of NO to the heme prosthetic group of peroxidases can also inhibit peroxidase-catalyzed reactions with a Ki in the μM range (Ferrer and Ros 1999). Inhibition of ascorbate peroxidase and catalase by NO donors was confirmed in isolated pea mitochondria, leaves of *Pelargonium peltatum* and suspension cultured cells of *Arabidopsis thaliana* and *Nicotiana tabacum* (Murgia et al. 2004; Arasimowicz-Jelonek et al. 2011; Martí et al. 2013). Thus, both activating and inhibitory effects of NO on peroxidase could be observed, depending on the redox state of the enzyme (Almagro et al. 2009).

Strong evidence has been presented that apoplastic ascorbate peroxidases and catalases are the targets of NO produced during the hypersensitive response in Arabidopsis. Nitric oxide can inhibit these enzymes and decrease the rate of H_2O_2 degradation; preventing NO accumulation will have the reverse effect (Zeier et al. 2004). However, the opposite effects were seen with the hypersensitive response in tobacco (Seregélyes et al. 2003). The thylakoid isoform of ascorbate peroxidase may also be a target for NO. The cell death elicited by treatment of Arabidopsis plants with 5 mM sodium nitroprusside, a donor of NO, was prevented by constitutive overexpression of the gene encoding ascorbate peroxidase. In wild-type plants, treatment with nitroprusside partially reduced the level of enzyme activity and strongly reduced RNA levels. Sodium nitroprusside in concentrations lower than 5 mM did not have any effect (Murgia et al. 2004). These observations indicate that the roles of NO in the hypersensitive response are facilitated by ascorbate peroxidases and catalases and that the effects of NO may vary between species.

Similar to ROS, the roles of RNS in signaling exert not only beneficial but also harmful effects. Peroxynitrite ($ONOO^-$), a reaction product between NO and superoxide ($O_2^{\bullet -}$), is considered to be the most toxic RNS that potentially oxidizes biomolecules including amino acids, lipids, and DNA (Groβ et al. 2013). In addition to the oxidative destruction of biomolecules, $ONOO^-$ is unique because it can nitrate tyrosine residues of proteins to form nitrotyrosine (van der Vliet et al. 1997). Interestingly, another mechanism has been found for nitrotyrosine formation that is independent of $ONOO^-$ production. This mechanism is mediated by heme-peroxidases, including horseradish peroxidase, myeloperoxidase, and lactoperoxidase (Sakihama et al. 2003). Peroxidases can oxidize the tyrosine contained in proteins to tyrosyl radicals ($Tyr^{\bullet}$), which combine with another $Tyr^{\bullet}$ to form dityrosine. When the oxidation of tyrosine by plant peroxidases is carried out in the presence of NO_2^-, a mixture of $Tyr^{\bullet}$ and NO_2 is formed in the reaction medium, species that couple to yield 3-nitrotyrosine (van der Vliet et al. 1997). In animals, this modification affects a small number of proteins and a few tyrosines within each protein (Souza et al. 2008). The addition of the nitro group causes a marked shift in the local pK_a of the hydroxyl group from 10.07 in tyrosine to 7.50 in 3-nitrotyrosine and enhances tyrosine hydrophobicity. These changes can lead to either a loss of function, such as with *S*-nitrosoglutathione reductase (GSNOR) (Savvides et al. 2002) or a gain of function, as with cytochrome c (Cassina et al. 2000). The pK_a shift also results in signal modulation, as it decreases the extent of tyrosine phosphorylation (Gow et al. 1996). The presence of 3-nitrotyrosine has been associated with a wide range of human and animal diseases (Casella et al. 2002). In plants, a strong evidence now exists that tyrosine nitration is involved in metabolic processes such as microtubule organization (Blume et al. 2013) and also in the response to abiotic and biotic stresses (Saito et al. 2006; Romero-Puertas et al. 2007; Valderrama et al. 2007; Corpas et al. 2008, Chaki et al. 2009, 2011a, b). Recently, ascorbate peroxidase has been identified as a potential target of tyrosine nitration in Arabidopsis (Lozano-Juste et al. 2011), *Citrus aurantium* (Tanou et al. 2012), and pea (Begara-Morales et al. 2014). The nitration-promoting activity of peroxidases is a widespread phenomenon. It was demonstrated that *A. thaliana* and *Medicago sativa* hemoglobins (Hb) (see Sect. 4) are capable of mediating protein nitration via NO_2^- oxidation to NO_2 by a H_2O_2-dependent peroxidase activity (Sakamoto et al. 2004; Maassen and Hennig 2011). However, when interpreting these data caution should be used, as the results on Hb acting as nitrite reductase and enzymatic nitration by peroxidases were obtained in vitro, and therefore more work is needed to uncover the exact molecular mechanisms governing this event in vivo.

S-Nitrosylation is another critical RNS-related posttranslational modification (Moreau et al. 2010). This process involves the reversible coupling of an NO moiety to a reactive cysteine thiol, leading to the formation of *S*-nitrosothiol (SNO). One-electron oxidation of NO yields NO_2, which can oxidize thiols to thiyl radicals (Keszler et al. 2010). In addition, $ONOO^-$ can form *S*-nitrosothiols either directly with thiolate anion (RS^-; van der Vliet et al. 1997) or through thiyl radicals (Keszler et al. 2010). Thus, any mechanism or cellular processes that could enhance RS^- formation, such as increased $O_2^{\bullet-}$ formation or the activity of peroxidases, also has the potential to generate *S*-nitrosothiols (Kovacs and Lindermayr 2013). S-Nitrosylation regulates a wide array of proteins involved in major cellular activities, such as photosynthesis, formation of secondary metabolites, genetic information processing, cellular architecture, and responses to biotic and abiotic stresses (Lindermayr et al. 2005). Therefore, understanding the impact of this posttranslational modification can provide insights into many physiological processes. In some plants, e.g., *Kalanchoe epinnata* (Abat et al. 2008), *Brassica juncea* (Abat and Deswal 2009), and *Citrus aurantium* L. (Tanou et al. 2009), about a thousand S-nitrosylated proteins were identified from leaf protein extracts exposed to GSNO. The proteins included enzymes involved in protection against oxidative stress such as glutaredoxin, ascorbate peroxidase, and Mn/Fe superoxide dismutase. Based on the analysis of the sites of NO production and the abundance of S-nitrosylated proteins, it has been suggested that the colocalization and spatial arrangement of the NO sources and NO targets could represent a major determinant of S-nitrosylation specificity in plants (Astier et al. 2012).

Although numerous plant proteins undergoing in vitro and/or in vivo S-nitrosylation have been identified, much regarding the impact of NO on their activity, structure, and function remains to be elucidated. To date, about 16 S-nitrosylated proteins, including peroxiredoxin II E (PrxII E), have been characterized in detail (Abat et al. 2008; Tada et al. 2008; Yun et al. 2011). Peroxiredoxins are a family of thiol-dependent peroxidases, which detoxify a broad set of peroxide substrates. Five major classes of peroxiredoxins are known in plants (Dietz 2011). These enzymes modulate redox signaling during photosynthesis, respiration, metabolism, and development. In addition, they serve in combating ROS generated in plants exposed to biotic and abiotic stresses. Interestingly, some peroxiredoxins including PrxII E also possess an $ONOO^-$ reductase activity (Romero-Puertas et al. 2007; Dietz 2011). PrxII E was shown to undergo

S-nitrosylation in *A. thaliana* plants exposed to the avirulent bacterial pathogen *Pseudomonas syringae* pv. *tomato* (Pst) (Romero-Puertas et al. 2007, 2008). Mechanistic studies demonstrated that both PrxII E peroxidase and $ONOO^-$ reductase activities were inhibited by NO through S-nitrosylation of Cys-121 in the active site. In planta, the S-nitrosylation of PrxII E led to a significant increase in the level of nitrated proteins, providing evidence that NO inhibits PrxII E $ONOO^-$ detoxification activity in vivo. It has been reported that other antioxidative enzymes such as catalase and ascorbate peroxidase are also regulated through S-nitrosylation by inhibiting their H_2O_2-degrading activity (Clark et al. 2000). Interestingly, as demonstrated by Begara-Morales et al. (2014), two RNS-mediated posttranslational modifications, tyrosine nitration and S-nitrosylation, can have opposite regulatory effects on cytosolic ascorbate peroxidase in pea plants exposed to salinity stress. Modulation of pea ascorbate peroxidase by irreversible tyrosine nitration and reversible S-nitrosylation leads to antagonistic effects. Nitration of Tyr235 inhibits ascorbate peroxidase activity, while S-nitrosylation of Cys32 causes an increase in ascorbate peroxidase activity (Begara-Morales et al. 2014). Therefore, one can suggest that an interplay between NO metabolism and antioxidative enzymes, such as peroxiredoxin, catalase, and ascorbate peroxidase, through their S-nitrosylation and tyrosine nitration, can provide fine tuning of the damaging and signaling effects of RNS and ROS.

Apart from proteins, many other molecules can be nitrated including lipids, fatty acids, amino acids, and nucleotides (Arasimowicz-Jelonek et al. 2011). Recently, 8-nitro-cGMP was identified as a down-stream signal of ABA, NO, and ROS in inducing stomatal closure at daytime, whereas cGMP regulated stomatal opening at night (Joudoi et al. 2013). 8-nitro-cGMP is a prime example of how NO, ROS, and cGMP can be integrated in one signaling cascade triggering a physical response.

3 RNS Production in Plants

RNS have chemical properties different from their precursors and may trigger specific physiological responses. For example, in the apoplast at an acidic pH exogenous NO_2^- can be nonenzymatically reduced to NO, a process that is accelerated by ascorbic acid and phenolics (Bethke et al. 2004). The pathway has been investigated in the barley aleuron layer but might occur also in other tissues. Among the possible enzymatic sources of production and metabolism of RNS, peroxidases are potential candidates. With broad substrate specificity, in addition to hydrogen peroxide, phenolic compounds and toxic compounds of aromatic nature, peroxidases can also utilize nitrogenous compounds with various degrees of oxidation (Sakihama et al. 2003; Herold 2004). Interaction of peroxidase with nitrogenous compounds may be accompanied by the formation of charged byproducts with an unpaired electron. Potentially, these interactions can be accompanied by the formation of RNS, such as $NO_2^{\bullet}$, $NO^{\bullet}$, and $ONOO^-$ (van der Vliet et al. 1997; Herold 2004). This was demonstrated to occur with some animal peroxidases, e.g., salivary peroxidase and bovine lactoperoxidase (Palmerini et al. 2012) and plant peroxidases. For example, NO formation was detected as a result of activity of horseradish peroxidase metabolizing hydroxyurea and H_2O_2 as substrates (Huang et al. 2002). Indeed, NO reacts with Compound I of peroxidase yielding the nitrosyl species, NO^+. Nitric oxide also reacts with Compound II of peroxidase leading probably to the formation of the nitrite anion, NO_2^-. Interestingly, the rate of reaction of Compound II with $NO^•$ is unusually high in comparison to that of Compound I, which reacts more quickly than Compound II with most substrates. The product of $NO^•$ oxidation by Compound II of peroxidase is NO_2^-, which can be further oxidized by plant peroxidases to form $NO_2^•$ (Almagro et al. 2009). NO_2^- can also be reduced to NO by a membrane-associated nitrite:NO reductase (NiNOR) as described for tobacco (Stöhr et al. 2001). However, NiNOR cannot be considered a major player in NO signaling because it is exclusively present in roots, functioning in the regulation of NO_3^- uptake. Copper amine oxidase 1 (CuAO1) is another candidate enzyme involved in NO synthesis (Wimalasekera et al. 2011). The *A. thaliana* cuao1 mutant is impaired in polyamine- and abscisic acid-induced NO production. The molecular background underlying this interesting phenotype is still unknown.

While chloroplasts and mitochondria are major sources of ROS from photosynthetic and respiratory electron-transport chains (ETCs), these organelles are also capable of NO synthesis. One proposed mechanism involves the transfer of electrons from the ETCs to NO_2^- by a nitrite:NO-reductase activity. Such ETC-dependent NO formation was observed in isolated chloroplasts from tobacco supplied with 25–100 μM NO_2^- and in mitochondria of tobacco suspension cells under anoxia (Planchet et al. 2005; Jasid et al. 2006). More work is needed for investigating if this pathway is active also in stress responses under normoxic conditions. Mammalian NOS oxidizes arginine to citrulline and NO. Although NOS-like activity is considered the most important source of NO accumulation in plant reactions to various stresses, the identity of corresponding plant NOS still remains disputable (Leitner et al. 2009; Mur et al. 2013). The first NOS sequence from plant kingdom has been reported for green algae species of the *Ostreococcus* genus by Foresi et al. (2010). Recently there have been a few reports showing a NOS-like activity in chloroplasts (Jasid et al. 2006; Tewari et al. 2013). To date, there is no convincing proof of NOS-like activity in plant mitochondria (Gupta et al. 2011a). In contrast, peroxisomes, which are active sites of ROS scavenging and formation, are also a source of NO both during salt stress as well as developmental processes such as lateral root growth (Schlicht et al. 2013). Corpas et al. (2009) using *A. thaliana* transgenic lines expressing green fluorescence protein (GFP) linked to peroxisomal targeting signal 1 (PTS1), showed that the fluorescence of the NO-specific dye diaminorhodamine colocalized with GFP fluorescence in the peroxisomes. Isolated peroxisomes displayed NOS-like activity, which was calcium dependent and could be inhibited by NOS inhibitors. High concentrations of NaCl stimulated NO synthesis in peroxisomes, which spread into the cytosol, where it probably contributed to $ONOO^-$ formation and protein tyrosine nitration (Corpas et al. 2009).

Available evidence suggests that in plant cells RNS can originate from various enzymatic sources and from different cellular locations. While in vitro experiments have shown that peroxidases can contribute to RNS production, whether this occurs in planta and the precise magnitude of such a contribution remain to be elucidated.

4 RNS Binding by Heme Proteins

The level of RNS is controlled by not only their production but also their reversible binding to various proteins, especially heme-containing proteins. Although hemoglobins (Hbs) are classically associated with O_2 binding and transport (Cooper 1999), they also are well known to play important roles in NO metabolism (Vázquez-Limón et al. 2012). The reactions with RNS are common to all hemeproteins and likely represent one of the earliest and most important Hb functions. The most significant advances in our understanding of the reactions with NO have come in two areas: vertebrate Hbs and myoglobins (Mbs) on one hand and bacterial and fungal flavohemoglobins (FHbs) on the other (Vinogradov and Moens 2008). Less is known about the physiological importance of RNS reactions with plant hemoglobins. The detailed characteristics and functions of plant hemoglobins and their interactions with RNS have been recently described in excellent papers and reviews (see Dordas et al. 2003; Dordas 2009; Gupta et al. 2011b; Vázquez-Limón et al. 2012; Hebelstrup et al. 2013). Importantly, being both heme proteins, peroxidases and hemoglobins share several fundamental characteristics. On the one hand, it is now understood that hemogloblins can display peroxidase activity (e.g., Reeder and Wilson 2005). Recombinant Arabidopsis globins were shown to possess peroxidase activity with diverse substrates including nitrite (Sakamoto et al. 2004). On the other, reactions with NO are common to all hemeproteins (Vinogradov and Moens 2008), and some peroxidases may bind NO even more readily than hemoglobins (Cooper 1999). Some Fe^{3+}–NO complexes are quite stable (e.g., catalase, horseradish peroxidase, cytochrome c peroxidase), whereas others (e.g., metmyoglobin, methemoglobin, cytochrome c) react further (Cooper 1999). In addition to simply binding NO, Hbs, similarly to peroxidases as described in Sect. 3, can catalyze the conversion of NO to NO_3^- (Hill et al. 2010; Hebelstrup et al. 2012). Another RNS, peroxynitrite, can act as a one-electron or two-electron oxidant for heme proteins. It can thus convert ferric iron to the ferryl intermediates with or without formation of a cation radical (Cooper 1999). Many heme proteins can stabilize ferryl intermediates $[(Fe^{4+} = O^{2-})^{2+}]$ formed as a result of the interactions of peroxynitrite, and this provides an alternative pathway for the species to dissociate. These reactions have clear physiological significance, as the ferryl iron and the associated free radical species (e.g., ferryl hemoglobin) are associated with oxidative stress damage to cells. It seems likely that binding of RNS by plant peroxidases and Hbs can modulate cytosolic levels of NO and hence regulate important signaling systems.

5 Peroxidase-Mediated Formation and Metabolism of Nitrophenols

5.1 Co-oxidation of Phenols and Natural Nitrogenous Compounds by Peroxidases

Nitration of natural phenols by peroxidases is a well-known process, although the significance of these reactions in plants remains somewhat speculative. In in vitro experiments Sakihama et al. (2003) demonstrated that the enzymatic nitration of *p*-coumaric acid by horseradish peroxidase in the presence of NO_2^- and H_2O_2 and the formation of phenolic nitro derivates go through the formation of intermediate phenoxyl radicals. Exogenous antioxidants, such as ascorbate and glutathione, inhibited peroxidase-mediated nitration of *p*-coumaric acid, by preferentially reacting with H_2O_2. Our recent data with native peroxidase isolated from wheat roots confirmed the nitration of *p*-coumaric acid in the presence of NO_2 and H_2O_2 resulting in the formation of corresponding nitrophenol (unpublished). This process was mediated by the generation of primary and secondary phenoxyl radicals, which were detected by electron spin resonance (ESR) and strongly depended on the pH of the reaction medium. Plant monohydroxyphenolics (monophenols such as *p*-coumaric acid) and polyhydroxyphenolics (polyphenols such as caffeic acid or chlorogenic acid) display different peroxidase-mediated scavenging of RNS and ROS. Polyphenols are more efficient antioxidants than monophenols, while monophenols are more efficient in removing $ONOO^-$ (or $NO_2^{\bullet}$) (Pannala et al. 1998). It has been speculated that preferential nitration of monophenols preserves the antioxidant activity of polyphenols during oxidative stress (Sakihama et al. 2003).

The physiological impact of natural nitrophenol synthesis on plant metabolism may depend on their concentration and cellular environment. In the apoplast, where hydroxycinnamic acids and class III peroxidases are present (Roach et al. 2010), phenoxyl radicals could be produced as intermediates for lignin biosynthesis. Negatively however, the presence of a combination of peroxidases and nitrates can have harmful consequences for plant cells. Normally, nitrite is produced through the one electron reduction of nitrate by nitrate reductase in the cytosol (Goshima et al. 1999). The concentration of nitrite in cells is very low due to its rapid transport into chloroplasts and subsequent reduction to ammonia by nitrite reductase. In the environment, however, nitrates can accumulate in soil due to the agricultural practices. After uptake by roots, they can be metabolized by plasma membrane-localized nitrate reductase to nitrites (Stöhr and Ullrich 2002). Theoretically, the potentially harmful peroxidase-mediated nitration of tyrosine amino acids in proteins can occur. Nitrotyrosine is a well-established marker of oxidative protein damage in mammals (van der Vliet et al. 1997), and evidence for nitrotyrosine formation in plant tissues has been reported (Romero-Puertas et al. 2007). As described in Sect. 2, the nitration of tyrosine and to a lesser degree tryptophan may change physicochemical properties and functions of proteins

(Abello et al. 2009). Furthermore, tyrosine and tryptophan nitration can increase the protein degradation in proteasomes (Nuriel et al. 2011). Preferential nitration of appropriate phenols, such as hydroxycinnamic acids, which are structurally related to tyrosine, can prevent the nitration of amino acids (Pannala et al. 1998).

Therefore, the formation of nitrophenols as a result of co-oxidation of phenols and nitrogen-containing compounds by peroxidases may play an important defense role by preventing proteins from deleterious nitration or synthesizing protective macromolecules. One further interaction of peroxidases with nitrogen-containing compounds that needs further investigation is that of eumelanin synthesis. Eumelanins are polymers of L-3,4-dihydroxyphenylalanine (DOPA), and typically contain c. 10 % nitrogen (Loganathan and Kalyanasundaram 1999). Melanin synthesis is a common response of organisms to stress. For example, in higher plants, the intensity of melanin formation is correlated with pathogen resistance (Bell 1981), presumably by forming a physical barrier preventing penetration. While tyrosinases are usually considered to be the main enzymes involved in melanin biosynthesis (Yoruk and Marshall 2003), some workers also consider that the importance of peroxidases in melanization has been underestimated (Takahama 2004; Mastore et al. 2005). Recent work has provided clear evidence that lichen and bryophyte peroxidases can readily polymerize DOPA in the presence of H_2O_2 (Li et al. 2010; Liers et al. 2011).

5.2 *Metabolism of N-Containing Pollutants by Class II Fungal Peroxidases*

Synthetic nitrophenols, which are nitrated aromatic ring structures consisting of benzene rings, hydroxyl (–OH) and nitro (–NO_2) groups, are widely used for the manufacture of dyes, drugs, pesticides, herbicides, fungicides, paints, and explosives (Arora et al. 2014). Nitroaromatic compounds are among the largest and most important groups of industrial chemicals in use today (Ju and Parales 2010). Being discharged into the environment in large quantities, nitrophenols represent a toxic threat for plants, animals, and humans. For example, they can react with heme-containing proteins including mitochondrial proteins, which may result in the uncoupling of mitochondrial oxidative phosphorylation and a reduction in ATP production. Furthermore, nitrophenols may induce mutagenesis by causing mis-incorporation of nucleotides during DNA synthesis (Arora et al. 2014). Therefore, the degradation of nitrophenols has a huge ecological significance.

Fungal class II peroxidases play important roles in the breakdown of N-containing pollutants. The best studied example of degradation of synthetic nitrophenols by peroxidases is that of 2,4,6-trinitrotoluene (TNT). TNT is a common nitroaromatic compound that has dominated the military production industry since 1902, reaching its peak during the World Wars. Coupled with this massive production and use, accidental and intentional release of TNT into the environment has also increased. In addition to its explosion hazard, TNT causes both ecotoxicological problems, including adverse effects on human health. While horseradish peroxidase, as demonstrated in vitro, can metabolize TNT (Lee et al. 2003), in the field degradation probably takes place by fungal nitroreductases (Nyanhongo et al. 2005). These enzymes reduce TNT to hydroxyl-amino-dinitrotoluenes and amino-dinitrotoluenes, while further degradation of these products is catalyzed by fungal peroxidases. Lignin peroxidase can transform 2,4-diamino-6-nitrotoluene and 2-amino-4,6-dinitrotoluene to nitroso-dinitro-toluenes. Manganese peroxidases are even more efficient, and can, for example, catalyze the breakdown of 2-amino-2,6-dinitrotoluene directly to CO_2 (Nyanhongo et al. 2005). Aromatic amines are another group of N-containing pollutants, and comprise amines attached to an aromatic hydrocarbon, e.g., aniline. These compounds can be readily metabolized by lignin peroxidase (Vazquez-Duhalt et al. 1995). Recently, immobilised peroxidases from plant and fungal sources have been used for the remediation of these industrial pollutants on a large scale (Husain et al. 2009).

5.3 Metabolism of N-Compounds by Non-class II Peroxidases

In addition to class II peroxidases, some fungi produce heme-thiolate and "dye-decolorizing" peroxidases (DyP-type or DyP) (Hofrichter et al. 2010). Heme-thiolate peroxidases are genuine "peroxygenases," in that they can transfer peroxide-oxygen from H_2O_2 to a substrate molecules. These enzymes include the chloroperoxidase from the ascomycete *Caldariomyces fumago*. In 2004, another group of heme-thiolate peroxidases was discovered in the mushroom *Agrocybe aegerita*, which later turned out to be a true aromatic peroxygenase (Ullrich et al. 2004). This peroxygenase can readily transfer oxygen to organic heteroatoms such as sulphur and nitrogen. For example, it can oxidize pyridine at the heterocyclic nitrogen to form pyridine N-oxide, while methylated pyridines are oxygenated both at the methyl group and at the ring nitrogen (Ullrich et al. 2008). Aromatic peroxygenase can hydroxylate 4-nitrophenol into 4-nitrocatechol (Peng et al. 2010), a reaction that is also catalyzed by the new class of peroxidases discovered in lichens (Liers et al. 2011).

DyP-type peroxidases represent a second new superfamily of heme peroxidases identified in fungi and bacteria. Fungal DyPs oxidize various N-containing dyes, as well as typical peroxidase substrates (Sugano 2009; Sugano et al. 2009). Structural comparison and sequence alignments of DyP-type peroxidases with representative members of all classes of the plant, bacterial, and fungal peroxidases demonstrated that DyPs cannot be classified as Class I, II or III or heme-thiolate peroxidases (Hofrichter et al. 2010). Peroxidases found in higher plants and fungi today had their origin in Class I enzymes in bacteria (Passardi et al. 2007). Although not widely studied, it is likely that bacterial peroxidases can metabolize N-containing compounds in the same ways as those of their higher plant and fungal descendants. In addition, almost five thousand bacterial DyP-type peroxidases are known (Colpa et al. 2014), and these enzymes probably catalyze similar reactions to those of fungal DyP-type peroxidases. For example, bacterial DyPs can catalyze the breakdown of N-containing dyes such as reactive blue (Santos et al. 2014). Furthermore, bacteria such as *Bacillus* that contain DyPs (Colpa et al. 2014) can readily degrade azobenzenes (Wackerow-Kouzova 2004), similarly to that in fungi (Kim and Shoda 1999). Further study is likely to increase our knowledge of the range of N-containing compounds that can be metabolized by bacterial peroxidases.

In conclusion, class II peroxidases, heme-thiolate peroxidases and DyP-type peroxidases may all detoxify xenobiotics and furthermore have all been detected in soils (Kellner et al. 2014). Therefore these enzymes have important implications in remediation of natural soils and waters and play roles in the preservation of ecosystems.

Conclusions

Table 1 and Fig. 1 summarize the interactions of peroxidases with nitrogenous compounds in various organisms. The most important role of peroxidases in plant N metabolism is probably their interactions with RNS that are involved as second messengers in signal transduction pathways. First, class I and class III peroxidases may be the actual targets of RNS, as their activity may be RNS regulated, for example to control ROS production during the stress-induced oxidative burst. Second, and potentially more importantly, peroxidase-mediated nitration of protein tyrosine can cause loss or gain in function of a variety of proteins. While there may be times when high levels of nitrites in the presence of peroxidase could cause excessive protein tyrosine nitration, it seems likely that proteins can be protected by preferential nitration of free (non-protein) phenolics. Third, peroxidases can regulate RNS concentrations, particularly NO. While in vitro evidence suggests that peroxidases can produce NO from nitrites, confirmation that this occurs in planta is lacking. Perhaps more likely, peroxidases bind NO, thereby reducing its cytosolic concentrations. Although NO binding is more often associated with Hb, the ability to bind NO is a common property of all heme containing proteins. While excellent in vitro evidence exists for a role in peroxidases in regulating cytosolic RNS levels, future work should be directed to obtaining convincing evidence for this in planta. Finally, peroxidases belonging to class II, hemethiolate, and DyP-type families are readily secreted into the soil and here play vital roles in the detoxification of nitrogenous xenobiotics. The considerable biotechnological significance of these fungal and bacterial enzymes has only begun to be exploited.

Table 1 Interactions of peroxidases with nitrogenous compounds in various organisms

Enzyme	Organism	Species	Nitrogenous compound	Reaction	Product	References
Guaiacol peroxidase Catechol peroxidase	Plants Mammals	*Armoracia rusticana* Rat	NO	Stimulation of guaiacol or catechol oxidation	Tetraguaiacol *o*-Benzoquinone	Huang et al. (2002) Uchida et al. (2002) Nappi and Vass (2001)
Ascorbate peroxidase	Plants	*Pisum sativum* *Pelargonium peltatum* *Arabidopsis thaliana* *Nicotiana tabacum*	NO donors	Inhibition of ascorbate oxidation	Dehydroascorbate	Murgia et al. (2004) Arasimowicz-Jelonek et al. (2011) Martí et al. (2013)
Apoplastic ascorbate peroxidase	Plants	*Arabidopsis thaliana*	NO	Inhibition of ascorbate oxidation	Dehydroascorbate	Zeier et al. (2004)
Apoplastic ascorbate peroxidases	Plants	*Nicotiana tabacum*	NO	Stimulation of ascorbate oxidation	Dehydroascorbate	Shapiro 2005
Horseradish peroxidase Myeloperoxidase Lactoperoxidase	Plants Mammals	*Armoracia rusticana* Human Bovine	NO_2^-	Tyrosine nitration	3-Nitrotyrosine	Sakihama et al. (2003) van der Vliet et al. (1997)
Ascorbate peroxidase	Plants	*Arabidopsis thaliana* *Citrus aurantium* *Pisum sativum*	NO_2^-	Tyrosine nitration	3-Nitrotyrosine	Lozano-Juste et al. (2011) Tanou et al. (2012) Begara-Morales et al. (2014)
Peroxiredoxin	Plants	*Arabidopsis thaliana*	GSNO	S-nitrosylation	S-nitrosylated proteins	Romero-Puertas et al. (2007, 2008)
Ascorbate peroxidase Catalase	Plants	*Nicotiana tabacum*	GSNO	S-nitrosylation	S-nitrosylated proteins Inhibition of H_2O_2-degrading activity	Clark et al. (2000)

Ascorbate peroxidase	Plants	*Pisum sativum*	GSNO	S-nitrosylation Tyrosine nitration	S-nitrosylated proteins 3-Nitrotyrosine Stimulation of H_2O_2-degrading activity	Begara-Morales et al. (2014)
Salivary peroxidase Lactoperoxidase	Mammals	Human Bovine	$NO^-{}_2$	Nitrite oxidation	NO	Palmerini et al. (2012)
Horseradish peroxidase	Plants	*Armoracia rusticana*	Hydroxyurea and H_2O_2	NO with Compound I NO with Compound II $NO^•$ with Compound II	NO^+ NO_2^- $NO_2^•$	Huang et al. (2002) Almagro et al. (2009)
Horseradish peroxidase Catalase Cytochrome c peroxidase	Plants Fungi	*Armoracia rusticana* Yeasts	RNS	Binding	Fe^{3+}-NO complexes	Cooper (1999)
Horseradish peroxidase	Plants	*Armoracia rusticana*	NO_2^-	Reaction with H_2O_2 goes via intermediate phenoxyl radicals	Phenolic nitroderivates Can prevent nitration of amino acids	Pannala et al. (1998) Sakihama et al. (2003)
Various Peroxidases	Plants Fungi	Various species	DOPA	Reaction with H_2O_2 to form complex polymers	Melanin	Takahama (2004), Mastore et al. (2005) Li et al. (2010) Liers et al. (2011)
Horseradish peroxidase Manganese peroxidase	Plants Fungi	*Armoracia rusticana Phlebia radiata*	TNT	Metabolic conversion	Not known Complete breakdown	Lee et al. (2003) Nyanhongo et al. (2005)
Lignin peroxidase	Fungi	*Phanerochaete chrysosporium*	TNT breakdown products (2,4-diamino-6-nitrotoluene and 2-amino-4,6-dinitro-toluene)	Metabolization	Nitroso-dinitro-toluenes	Nyanhongo et al. (2005)

(continued)

Table 1 (continued)

Enzyme	Organism	Species	Nitrogenous compound	Reaction	Product	References
Horseradish peroxidase Lignin peroxidase	Plants Fungi	*Armoracia rusticana* *Phanerochaete chrysosporium*	Aniline	Oxidation	Polymerization	Mann and Saunders (1935) Vazquez-Duhalt et al. (1995)
Aromatic peroxygenase Peroxidase	Fungi Lichens	*Agrocybe aegerita* *Leptogium saturninum*	4-Nitrophenol	Hydroxylation	4-Nitrocatechol	Peng et al. (2010) Liers et al. (2011)
DyP-type peroxidase	Fungi	*Geotrichum candidum*	Azobenzene	Oxidative cleavage	Aniline	Kim and Shoda (1999)
DyP-type peroxidase	Fungi Bacteria	*Thanatephorus cucumeris* *Bacillus subtilis* *Pseudomonas putida*	Reactive blue	Metabolic conversion	Phthalate + 2,2'-disulfonyl azobenzene + 3-[(4-amino-6-chloro-1,3,5-triazin-2-yl)amino]benzene-sulfonate	Sugano et al. (2009) Santos et al. (2014)
DyP-type peroxidase	Bacteria	*Bacillus*	Azobenzene	Metabolic conversion	Complete break-down, utilization as carbon source	Wackerow-Kouzova (2004)
Heme-thiolate peroxidase	Fungi	*Agrocybe aegerita*	Pyridine	Oxidation	Pyridine N-oxide	Ullrich et al. (2008)

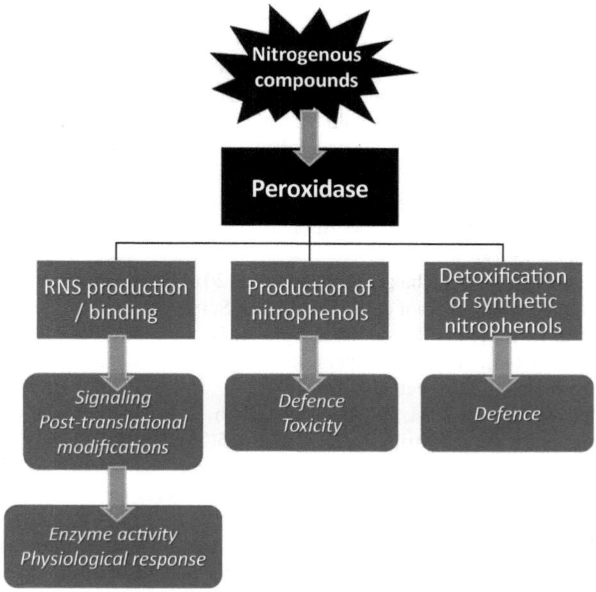

Fig. 1 Hypothetical scheme of interactions of peroxidases with RNS and other nitrogenous compounds. The interactions of nitrogenous compounds with peroxidases are implicated in signaling by producing or binding RNS and thus effecting the activities of components of signal transduction pathways. The interactions of RNS with peroxidases also provide a key mechanism of post-translational protein modifications as a consequence of the peroxidase-catalyzed nitration of tyrosine and tryptophan derivatives in proteins in the presence of nitrite and hydrogen peroxide. Activities of a large number of proteins and enzymes depend on tyrosine residues. The interactions of RNS and other nitrogenous compounds with peroxidases may lead to the formation of nitrophenols, which exhibit regulatory effects. Peroxidases may metabolize synthetic nitrophenol derivatives and thus provide an essential mechanism for removal of toxic compounds including xenobiotics

References

Abat JK, Deswal R (2009) Differential modulation of S-nitrosoproteome of *Brassica juncea* by low temperature: change in S-nitrosylation of Rubisco is responsible for the inactivation of its carboxylase activity. Proteomics 9(18):4368–4380

Abat JK, Mattoo AK, Deswal R (2008) S-nitrosylated proteins of a medicinal CAM plant *Kalanchoe pinnata* - ribulose-1,5-bisphosphate carboxylase/oxygenase activity targeted for inhibition. FEBS J 275(11):2862–2872

Abello N, Kerstjens HAM, Postma DS, Bischoff R (2009) Protein tyrosine nitration: selectivity, physicochemical and biological consequences, denitration, and proteomics methods for the identification of tyrosine-nitrated proteins. J Proteome Res 8(7):3222–3238

Almagro L, Ros LVG, Belchi-Navarro S, Bru R, Ros Barceló A, Pedreño MA (2009) Class III peroxidases in plant defence reactions. J Exp Bot 60(2):377–390

Arasimowicz-Jelonek M, Floryszak-Wieczorek J, Kosmala A (2011) Are nitric oxide donors a valuable tool to study the functional role of nitric oxide in plant metabolism? Plant Biol 13(5): 747–756

Arora PK, Srivastava A, Singh VP (2014) Bacterial degradation of nitrophenols and their derivatives. J Hazard Mater 266:42–59

Astier J, Kulik A, Koen E, Besson-Bard A, Bourque S, Jeandroz S et al (2012) Protein S-nitrosylation: what's going on in plants? Free Radic Biol Med 53(5):1101–1110

Begara-Morales JC, Sanchez-Calvo B, Chaki M, Valderrama R, Mata-Perez C, Lopez-Jaramillo J et al (2014) Dual regulation of cytosolic ascorbate peroxidase (APX) by tyrosine nitration and S-nitrosylation. J Exp Bot 65(2):527–538

Bell AA (1981) Biochemical mechanisms of disease resistance. Annu Rev Plant Physiol 32:21–81

Bethke PC, Badger MR, Jones RL (2004) Apoplastic synthesis of nitric oxide by plant tissues. Plant Cell 16(2):332–341

Blume YB, Krasylenko YA, Demchuk OM, Yemets AI (2013) Tubulin tyrosine nitration regulates microtubule organization in plant cells. Front Plant Sci 4:530

Casella L, Monzani E, Roncone R, Nicolis S, Sala A, De Riso A (2002) Formation of reactive nitrogen species at biologic heme centers: a potential mechanism of nitric oxide-dependent toxicity. Environ Health Perspect 110:709–711

Cassina AM, Hodara R, Souza JM, Thomson L, Castro L, Ischiropoulos H et al (2000) Cytochrome c nitration by peroxynitrite. J Biol Chem 275(28):21409–21415

Chaki M, Valderrama R, Fernandez-Ocana AM, Carreras A, Lopez-Jaramillo J, Luque F et al (2009) Protein targets of tyrosine nitration in sunflower (*Helianthus annuus* L.) hypocotyls. J Exp Bot 60(15):4221–4234

Chaki M, Valderrama R, Fernandez-Ocana AM, Carreras A, Gomez-Rodriguez MV, Lopez-Jaramillo J et al (2011a) High temperature triggers the metabolism of S-nitrosothiols in sunflower mediating a process of nitrosative stress which provokes the inhibition of ferredoxin-NADP reductase by tyrosine nitration. Plant Cell Environ 34(11):1803–1818

Chaki M, Valderrama R, Fernandez-Ocana AM, Carreras A, Gomez-Rodriguez MV, Pedrajas JR et al (2011b) Mechanical wounding induces a nitrosative stress by down-regulation of GSNO reductase and an increase in S-nitrosothiols in sunflower (*Helianthus annuus*) seedlings. J Exp Bot 62(6):1803–1813

Clark D, Durner J, Navarre DA, Klessig DF (2000) Nitric oxide inhibition of tobacco catalase and ascorbate peroxidase. Mol Plant Microbe Interact 13(12):1380–1384

Colpa DI, Fraaije MW, van Bloois E (2014) DyP-type peroxidases: a promising and versatile class of enzymes. J Ind Microbiol Biotechnol 41(1):1–7

Cooper CE (1999) Nitric oxide and iron proteins. Biochim Biophys Acta 1411(2–3):290–309

Corpas FJ, Chaki M, Fernandez-Ocana A, Valderrama R, Palma JM, Carreras A et al (2008) Metabolism of reactive nitrogen species in pea plants under abiotic stress conditions. Plant Cell Physiol 49(11):1711–1722

Corpas FJ, Hayashi M, Mano S, Nishimura M, Barroso JB (2009) Peroxisomes are required for in vivo nitric oxide accumulation in the cytosol following salinity stress of Arabidopsis plants. Plant Physiol 151(4):2083–2094

Dietz KJ (2011) Peroxiredoxins in plants and cyanobacteria. Antioxid Redox Signal 15(4): 1129–1159

Dordas C (2009) Non-symbiotic hemoglobins and stress tolerance in plants. Plant Sci 176(4): 433–440

Dordas C, Hasinoff BB, Igamberdiev AU, Manac'h N, Rivoal J, Hill RD (2003) Expression of a stress-induced hemoglobin affects NO levels produced by alfalfa root cultures under hypoxic stress. Plant J 35(6):763–770

Durner J, Klessig DF (1999) Nitric oxide as a signal in plants. Curr Opin Plant Biol 2(5):369–374

Ferrer MA, Ros BA (1999) Differential effects of nitric oxide on peroxidase and H_2O_2 production by the xylem of *Zinnia elegans*. Plant Cell Environ 22(7):891–897

Foresi N, Correa-Aragunde N, Parisi G, Caló G, Salerno G, Lamattina L (2010) Characterization of a nitric oxide synthase from the plant kingdom: NO generation from the green alga Ostreococcustauri is light irradiance and growth phase dependent. Plant Cell 22(11): 3816–3830

Goshima N, Mukai T, Suemori M, Takahashi M, Caboche M, Morikawa H (1999) Emission of nitrous oxide (N_2O) from transgenic tobacco expressing antisense NiR mRNA. Plant J 19(1): 75–80

Gow AJ, Duran D, Malcolm S, Ischiropoulos H (1996) Effects of peroxynitrite-induced protein modifications on tyrosine phosphorylation and degradation. FEBS Lett 385(1–2):63–66

Groβ F, Durner J, Gauples F (2013) Nitric oxide, antioxidants and prooxidants in plant defence responses. Front Plant Sci 4:419

Gupta KJ, Fernie AR, Kaiser WM, van Dongen JT (2011a) On the origins of nitric oxide. Trends Plant Sci 16(3):160–168

Gupta KJ, Hebelstrup KH, Mur LAJ, Igamberdiev AU (2011b) Plant hemoglobins: important players at the crossroads between oxygen and nitric oxide. FEBS Lett 585(24):3843–3849

Hebelstrup KH, van Zanten M, Mandon J, Voesenek LACJ, Harren FJM, Cristescu SM et al (2012) Haemoglobin modulates NO emission and hyponasty under hypoxia-related stress in Arabidopsis thaliana. J Exp Bot 63(15):5581–5591

Hebelstrup KH, Shahb JK, Igamberdiev AU (2013) The role of nitric oxide and hemoglobin in plant development and morphogenesis. Physiol Plant 148(4):457–469

Herold S (2004) Nitrotyrosine, dityrosine, and nitrotryptophan formation from metmyoglobin, hydrogen peroxide, and nitrite. Free Radic Biol Med 36(5):565–579

Hill BG, Dranka BP, Bailey SM, Lancaster JR Jr, Darley-Usmar VM (2010) What part of NO don't you understand? Some answers to the cardinal questions in nitric oxide biology. J Biol Chem 285(26):19699–19704

Hofrichter M, Ullrich R, Pecyna MJ, Liers C, Lundell T (2010) New and classic families of secreted fungal heme peroxidases. Appl Microbiol Biotechnol 87(3):871–897

Huang JM, Sommers EM, Kim-Shapiro DB, King SB (2002) Horseradish peroxidase catalyzed nitric oxide formation from hydroxyurea. J Am Chem Soc 124(13):3473–3480

Husain Q, Husain M, Kulshrestha Y (2009) Remediation and treatment of organopollutants mediated by peroxidases: a review. Crit Rev Biotechnol 29(2):94–119

Jasid S, Simontacchi M, Bartoli CG, Puntarulo S (2006) Chloroplasts as a nitric oxide cellular source. Effect of reactive nitrogen species on chloroplastic lipids and proteins. Plant Physiol 142(3):1246–1255

Joudoi T, Shichiri Y, Kamizono N, Akaike T, Sawa T, Yoshitake J et al (2013) Nitrated cyclic GMP modulates guard cell signaling in Arabidopsis. Plant Cell 25(2):558–571

Ju K-S, Parales RE (2010) Nitroaromatic compounds, from synthesis to biodegradation. Microbiol Mol Biol Rev 74(2):250–272

Kellner H, Luis P, Pecyna MJ, Barbi F, Kapturska D, Kruger D et al (2014) Widespread occurrence of expressed fungal secretory peroxidases in forest soils. PLoS One 9(4):e95557

Keszler A, Zhang Y, Hogg N (2010) Reaction between nitric oxide, glutathione, and oxygen in the presence and absence of protein: how are S-nitrosothiols formed? Free Radic Biol Med 48(1): 55–64

Kim SJ, Shoda M (1999) Purification and characterization of a novel peroxidase from *Geotrichum-candidum* Dec 1 involved in decolorization of dyes. Appl Environ Microbiol 65(3):1029–1035

Kovacs I, Lindermayr C (2013) Nitric oxide-based protein modification: formation and site-specificity of protein S-nitrosylation. Front Plant Sci 4:137

Lee KB, Gu MB, Moon SH (2003) Degradation of 2,4,6-trinitrotoluene by immobilized horse-radish peroxidase and electrogenerated peroxide. Water Res 37(5):983–992

Leitner M, Vandelle E, Gaupels F, Bellin D, Delledonne M (2009) NO signals in the haze nitric oxide signalling in plant defence. Curr Opin Plant Biol 12(4):451–458

Li JL, Sulaiman M, Beckett RP, Minibayeva FV (2010) Cell wall peroxidases in the liverwort *Dumortierahirsuta* are responsible for extracellular superoxide production, and can display tyrosinase activity. Physiol Plant 138(4):474–484

Liers C, Ullrich R, Hofrichter M, Minibayeva FV, Beckett RP (2011) A heme peroxidase of the ascomyceteous lichen Leptogiumsaturninum oxidizes high-redox potential substrates. Fungal Genet Biol 48(12):1139–1145

Lindermayr C, Saalbach G, Durner J (2005) Proteomic identification of S-nitrosylated proteins in Arabidopsis. Plant Physiol 137(3):921–930

Loganathan P, Kalyanasundaram I (1999) The melanin of the myxomycete *Stemonitisherbatica*. Acta Protozool 38:97–103

Lozano-Juste J, Colom-Moreno R, Leon J (2011) In vivo protein tyrosine nitration in *Arabidopsis thaliana*. J Exp Bot 62(10):3501–3517

Maassen A, Hennig J (2011) Effect of *Medicago sativa* Mhb1 gene expression on defense response of *Arabidopsis thaliana* plants. Acta Biochim Pol 58(3):427–432

Mann PJC, Saunders BC (1935) Studies in peroxidase action. I—The oxidation of aniline. Proc R Soc B Biol Sci 119:47–60

Martí MC, Florez-Sarasa I, Camejo D, Pallol B, Ortiz A, Ribas-Carbó M et al (2013) Response of mitochondrial antioxidant system and respiratory pathways to reactive nitrogen species in pea leaves. Physiol Plant 147(2):194–206

Mastore M, Kohler L, Nappi AJ (2005) Production and utilization of hydrogen peroxide associated with melanogenesis and tyrosinase-mediated oxidations of DOPA and dopamine. FEBS J 272 (10):2407–2415

Moreau M, Lindermayr C, Durner J, Klessig DF (2010) NO synthesis and signaling in plants—where do we stand? Physiol Plant 138(4):372–383

Mur LAJ, Mandon J, Persijn S, Cristescu SM, Moshkov IE, Novikova GV et al (2013) Nitric oxide in plants: an assessment of the current state of knowledge. AoB Plants 5:Pls052

Murgia I, de Pinto MC, Delledonne M, Soave C, De Gara L (2004) Comparative effects of various nitric oxide donors on ferritin regulation, programmed cell death, and cell redox state in plant cells. J Plant Physiol 161(7):777–783

Nappi AJ, Vass E (2001) The effects of nitric oxide on the oxidations of L-dopa and dopamine mediated by tyrosinase and peroxidase. J Biol Chem 276(14):11214–11222

Nuriel T, Hansler A, Gross SS (2011) Protein nitrotryptophan: formation, significance and identification. J Proteomics 74(11):2300–2312

Nyanhongo GS, Schroeder M, Steiner W, Gubitz GM (2005) Biodegradation of 2,4,6-trinitrotoluene (TNT): an enzymatic perspective. Biocatal Biotransform 23(2):53–69

Palmerini CA, Marmottini F, Arienti G (2012) Production of nitric oxide by human salivary peroxidase and by bovine lactoperoxidase. J Biochem Mol Toxicol 26(3):87–93

Pannala A, Razaq R, Halliwell B, Singh S, Rice-Evans CA (1998) Inhibition of peroxynitrite dependent tyrosine nitration by hydroxycinnamates: nitration or electron donation? Free Radic Biol Med 24(4):594–606

Passardi F, Bakalovic N, Teixeira FK, Margis-Pinheiro M, Penel C, Dunand C (2007) Prokaryotic origins of the non-animal peroxidase superfamily and organelle-mediated transmission to eukaryotes. Genomics 89(5):567–579

Peng L, Wollenberger U, Hofrichter M, Ullrich R, Scheibner K, Scheller FW (2010) Bioelectro-catalytic properties of *Agrocybeaegerita* peroxygenase. Electrochim Acta 55(27):7809–7813

Planchet E, Gupta KJ, Sonoda M, Kaiser WM (2005) Nitric oxide emission from tobacco leaves and cell suspensions: rate limiting factors and evidence for the involvement of mitochondrial electron transport. Plant J 41(5):732–743

Reeder BJ, Wilson MT (2005) Hemoglobin and myoglobin associated oxidative stress: from molecular mechanisms to disease states. Curr Med Chem 12(23):2741–2751

Roach T, Beckett RP, Minibayeva FV, Colville L, Whitaker C, Chen HY, Bailly C, Kranner I (2010) Extracellular superoxide production, viability and redox poise in response to desiccation in recalcitrant *Castanea sativa* seeds. Plant Cell Environ 33(1):59–75

Romero-Puertas MC, Laxa M, Matte A, Zaninotto F, Finkemeier I, Jones AME et al (2007) S-nitrosylation of peroxiredoxin II E promotes peroxynitrite-mediated tyrosine nitration. Plant Cell 19(12):4120–4130

Romero-Puertas MC, Campostrini N, Matte A, Righetti PG, Perazzolli M, Zolla L et al (2008) Proteomic analysis of S-nitrosylated proteins in Arabidopsis thaliana undergoing hypersensitive response. Proteomics 8(7):1459–1469

Ros Barceló A, Pomar F, Ferrer MA, Martínez P, Ballesta MC, Pedreño MA (2002) In situ characterization of a NO-sensitive peroxidase in the lignifying xylem of *Zinnia elegans*. Physiol Plant 114(1):33–40

Saito S, Yamamoto-Katou A, Yoshioka H, Doke N, Kawakita K (2006) Peroxynitrite generation and tyrosine nitration in defense responses in tobacco BY-2 cells. Plant Cell Physiol 47(6): 689–697

Sakamoto A, Sakurao S, Fukunaga K, Matsubara T, Ueda-Hashimoto M, Tsukamoto S et al (2004) Three distinct Arabidopsis hemoglobins exhibit peroxidase-like activity and differentially mediate nitrite-dependent protein nitration. FEBS Lett 572(1–3):27–32

Sakihama Y, Tamaki R, Shimoji H, Ichiba T, Fukushi Y, Tahara S et al (2003) Enzymatic nitration of phytophenolics: evidence for peroxynitrite-independent nitration of plant secondary metabolites. FEBS Lett 553(3):377–380

Santos A, Mendes S, Brissos V, Martins LO (2014) New dye-decolorizing peroxidases from *Bacillus subtilis* and *Pseudomonas putida* MET94: towards biotechnological applications. Appl Microbiol Biotechnol 98(5):2053–2065

Savvides SN, Scheiwein M, Bohme CC, Arteel GE, Karplus PA, Becker K et al (2002) Crystal structure of the antioxidant enzyme glutathione reductase inactivated by peroxynitrite. J Biol Chem 277(4):2779–2784

Schlicht M, Ludwig-Müller J, Burbach C, Volkmann D, Baluska F (2013) Indole-3-butyric acid induces lateral root formation via peroxisome-derived indole-3-acetic acid and nitric oxide. New Phytol 200(2):473–482

Seregélyes C, Barna B, Hennig J, Konopka D, Pasternak TP, Lukács N et al (2003) Phytoglobins can interfere with nitric oxide functions during plant growth and pathogenic responses: a transgenic approach. Plant Sci 165(3):541–550

Shapiro AD (2005) Nitric oxide signaling in plants. Vitam Horm 72:339–398 (Litwack G, editor, Plant hormones)

Souza JM, Peluffo G, Radi R (2008) Protein tyrosine nitration—functional alteration or just a biomarker? Free Radic Biol Med 45(4):357–366

Stöhr C, Ullrich WR (2002) Generation and possible roles of NO in plant roots and their apoplastic space. J Exp Bot 53(379):2293–2303

Stöhr C, Strube F, Marx G, Ullrich WR, Rockel P (2001) A plasma membrane-bound enzyme of tobacco roots catalyses the formation of nitric oxide from nitrite. Planta 212(5–6):835–841

Sugano Y (2009) DyP-type peroxidases comprise a novel heme peroxidase family. Cell Mol Life Sci 66(8):1387–1403

Sugano Y, Matsushima Y, Tsuchiya K, Aoki H, Hirai M, Shoda M (2009) Degradation pathway of an anthraquinone dye catalyzed by a unique peroxidase DyP from *Thanatephoruscucumeris* Dec 1. Biodegradation 20(3):433–440

Tada Y, Spoel SH, Pajerowska-Mukhtar K, Mou Z, Song J, Wang C et al (2008) Plant immunity requires conformational changes of NPR1 via S-nitrosylation and thioredoxins. Science 321:952–956

Takahama U (2004) Oxidation of vacuolar and apoplastic phenolic substrates by peroxidase: physiological significance of the oxidation reactions. Phytochem Rev 3:207–219

Tanou G, Job C, Rajjou L, Arc E, Belghazi M, Diamantidis G et al (2009) Proteomics reveals the overlapping roles of hydrogen peroxide and nitric oxide in the acclimation of citrus plants to salinity. Plant J 60:795–804

Tanou G, Filippou P, Belghazi M, Job D, Diamantidis G, Fotopoulos V et al (2012) Oxidative and nitrosative-based signaling and associated posttranslational modifications orchestrate the acclimation of citrus plants to salinity stress. Plant J 72(4):585–599

Tewari RK, Prommer J, Watanabe M (2013) Endogenous nitric oxide generation in protoplast chloroplasts. Plant Cell Rep 32(1):31–44

Uchida T, Okumura K, Ito T, Kamiya H, Nishimoto Y, Yamada M et al (2002) Quinapril treatment restores the vasodilator action of insulin in fructose-hypertensive rats. Clin Exp Pharmacol Physiol 29(5–6):381–385

Ullrich R, Nueske J, Scheibner K, Spantzel J, Hofrichter M (2004) Novel haloperoxidase from the agaric basidiomycete *Agrocybeaegerita* oxidizes aryl alcohols and aldehydes. Appl Environ Microbiol 70(8):4575–4581

Ullrich R, Dolge C, Kluge M, Hofrichter M (2008) Pyridine as novel substrate for regioselective oxygenation with aromatic peroxygenase from *Agrocybeaegerita*. FEBS Lett 582(29): 4100–4106

Valderrama R, Corpas FJ, Carreras A, Fernandez-Ocana A, Chaki M, Luque F et al (2007) Nitrosative stress in plants. FEBS Lett 581(3):453–461

van der Vliet A, Eiserich JP, Halliwell B, Cross CE (1997) Formation of reactive nitrogen species during peroxidase-catalyzed oxidation of nitrite—a potential additional, mechanism of nitric oxide-dependent toxicity. J Biol Chem 272(12):7617–7625

Vazquez-Duhalt R, Westlake DWS, Fedorak PM (1995) Kinetics of chemically-modified lignin peroxidase and enzymatic oxidation of aromatic nitrogen-containing compounds. Appl Microbiol Biotechnol 42(5):675–681

Vázquez-Limón C, Hoogewijs D, Vinogradov SN, Arredondo-Peter R (2012) The evolution of land plant hemoglobins. Plant Sci 191:71–81

Vinogradov SN, Moens L (2008) Diversity of globin function: enzymatic, transport, storage, and sensing. J Biol Chem 283(14):8773–8777

Wackerow-Kouzova ND (2004) Isolation and study of azobenzene-transforming soil bacteria. Appl Biochem Microbiol 41:62–64

Welinder KG, Mauro JM, Norskovlauritsen L (1992) Structure of plant and fungal peroxidases. Biochem Soc Trans 20(2):337–340

Wimalasekera R, Villar C, Begum T, Scherer GFE (2011) COPPER AMINE OXIDASE1 (CuAO1) of *Arabidopsis thaliana* contributes to abscisic acid- and polyamine-induced nitric oxide biosynthesis and abscisic acid signal transduction. Mol Plant 4(4):663–678

Yoruk R, Marshall MR (2003) Physicochemical properties and function of plant polyphenol oxidase: a review. J Food Biochem 27(5):361–422

Yun BW, Feechan A, Yin M, Saidi NB, Le Bihan T, Yu M et al (2011) S-nitrosylation of NADPH oxidase regulates cell death in plant immunity. Nature 478(7368):264–268

Zeier J, Delledonne M, Mishina T, Severi E, Sonoda M, Lamb C (2004) Genetic elucidation of nitric oxide signaling in incompatible plant-pathogen interactions. Plant Physiol 136(1): 2875–2886

Zhang Y, Liu H, Hu N (2012) Electrochemical detection of natural DNA damage induced by in situ peroxidase-generated reactive nitrogen species in DNA layer-by-layer films. Bioelectrochemistry 86:67–71

Mitochondrial Signaling in Plants Under Hypoxia: Use of Reactive Oxygen Species (ROS) and Reactive Nitrogen Species (RNS)

Kim H. Hebelstrup and Ian M. Møller

Abbreviations

AOX	Alternative oxidase
CCO	Cytochrome c oxidase
Complex I	NADH dehydrogenase complex
Complex III	Cytochrome bc_1 complex
Complex IV	Cytochrome c oxidase complex
ETC	Electron transport chain
HNE	4-Hydroxy-2-nonenal
IMM	Inner mitochondrial membrane
OMM	Outer mitochondrial membrane
PTM	Posttranslational modification
RNS	Reactive nitrogen species
ROS	Reactive oxygen species
SOD	Superoxide dismutase

1 Introduction

The plant cell depends on its mitochondria for a wide variety of services. The mitochondria provide energy in the form of reducing equivalents and ATP produced mainly in the Krebs cycle and in oxidative phosphorylation, respectively. They also provide metabolic building blocks in the form of carbon skeletons, coenzymes (vitamins), and FeS centers, etc. (Millar et al. 2011; Salvato et al. 2014).

K.H. Hebelstrup (✉) • I.M. Møller
Department of Molecular Biology and Genetics, Aarhus University, 4200 Slagelse, Denmark
e-mail: kim.hebelstrup@agrsci.dk

K.J. Gupta, A.U. Igamberdiev (eds.), *Reactive Oxygen and Nitrogen Species Signaling and Communication in Plants*, Signaling and Communication in Plants 23,
DOI 10.1007/978-3-319-10079-1_4

Finally, plant mitochondria participate in pathogen defense reactions including programmed cell death (Colombatti et al. 2014).

Plant mitochondria are semiautonomous, which means that they grow and divide to keep pace with cell division and the need to replace damaged units (Welchen et al. 2014). They can also change their metabolism to suit changing cellular function as happens during cell differentiation or in response to changing environmental conditions.

To do this, mitochondrial metabolism needs to be coordinated with that of the rest of the cell. The vast majority of the proteins making up the mitochondria have to be imported from the cytosol and metabolites and signaling molecules have to be received and exported to the rest of the cell via the cytosol. In some cases—e.g., citrate (Finkemeier et al. 2013)—a metabolic intermediate can be the signal, but there are also a variety of specialized signaling molecules, including the reactive oxygen species (ROS) and reactive nitrogen species (RNS).

In this chapter we intend to review our knowledge about ROS and RNS signaling from the mitochondria to the rest of the cell under hypoxic conditions. We start by briefly describing the situations where hypoxia occurs in plant tissues. We then present an overview of ROS and RNS turnover in plant mitochondria, followed by a more detailed description of cellular NO turnover including the role of hemoglobin/NO cycle. In the final sections we describe different ways in which ROS and NO can modify proteins and lipids and discuss how this might contribute to intracellular and intercellular signaling.

2 Hypoxia in Plant Tissues: Roots and Seeds

Oxygen supply to non-photosynthetic plant tissues is restricted by diffusion pathways. This means that the innermost tissues of bulky organs, such as storage tissues, often experience some degree of hypoxia (Geigenberger et al. 2000). Plant roots are particularly at risk, because the soil can become water-logged and this generates steep O_2 gradients caused by two factors: (1) the oxygen concentration in air-equilibrated water is 30–40 times lower than in air; (2) The diffusion rate of O_2 10,000 times lower in water than in air. The presence of aerenchyma (porous gas-filled spaces) in stems and roots is therefore a pre-requisite for enhanced tolerance to long-term flooded soils. Oxygen transport through these porous air-filled tissues may be both diffusive and convective (Armstrong and Armstrong 2014). Non-aerenchymous plants are dependent on O_2 diffusion through the soil, and roots will be subjected to a depth gradient as well as radial gradients where local O_2 concentrations decline to below the apparent Km of the terminal oxidases in the mitochondria, so that aerobic metabolism becomes limited (Darwent et al. 2003; Armstrong and Armstrong 2014), in particularly when the plant is in darkness so that photosynthesis is not active (Vashisht et al. 2011). The formation of aerenchyma is induced in some plants when subjected to hypoxic stress. This response is often triggered by ethylene and involves the onset of programmed cell death (PCD) in certain lysogenic cells, which become destined for aerenchyma formation. This PCD is promoted by H_2O_2 and $O_2^{\bullet-}$ (Steffens et al. 2011), which is induced by a genetic program involving increased expression of ROS-forming enzymes, such as Respiratory Burst Oxidase Homolog (RBOH), and decreased expression of proteins restricting ROS formation, such as metallothioenin (binds any free metal ions thereby preventing the Fenton reaction), in the tissues destined for PCD (Rajhi et al. 2011). A significant amount of NO is generated in plant tissues in response to severe hypoxia, where O_2 concentrations are below the apparent K_m of the mitochondria (Hebelstrup et al. 2012, 2014), and experiments in hypoxic *Medicago sativa* roots indicate that NO stimulates aerenchyma formation in hypoxic roots (Dordas et al. 2003). NO and ROS therefore seem to be important signals in the adaptation to hypoxia in roots, by inducing formation of aerenchyma.

Nitrogen-fixing nodule of legume roots represents a special case. Here the concentration of free oxygen must be extremely low to maintain the nitrogenase responsible for nitrogen reduction in a functional state. A diffusion barrier to oxygen in the nodule parenchyma, a very high concentration of the oxygen-binding leghemoglobin (different from the other hemoglobins treated later in this chapter) as well as oxygen-consuming reactions all contribute to limiting the oxygen concentration. The result is that there is a high rate of ROS production in the nodules and a number of enzymatic mechanisms to remove the ROS (Becana et al. 2000).

Hypoxia in developing and germinating seeds is also a common occurrence caused by a diffusion permeability barrier in the seed coat as well as relatively high metabolic rates inside the seed (Borisjuk et al. 2007).

3 Turnover of ROS and RNS in Plant Mitochondria

Mitochondria are the centers of aerobic metabolism. Electrons enter the electron transport chain (ETC) through a number of different dehydrogenases in or associated with the inner mitochondrial membrane (IMM), two of which are Complex I, the NADH dehydrogenase complex, and Complex II, succinate dehydrogenase. The latter is also part of the Krebs cycle. Most of these dehydrogenases donate their electrons to the ubiquinone pool from where they can either pass through Complex III, the cytochrome bc_1 complex, and Complex IV, cytochrome c oxidase (CCO), or through the alternative oxidase (AOX) in both cases to reduce oxygen to water by the transfer of four electrons (Rasmusson and Møller 2011). Coupled to electron transport through Complexes I, III, and IV is proton pumping across the IMM (10 H^+ per $2e^-$ or 1 NADH) to set up an electrochemical proton gradient, which is used by the ATP synthase (often called Complex V) to synthesize ATP or to drive transport processes in and out of the mitochondria across the IMM. Since the synthesis and export of 1 ATP molecule costs 4 H^+ (Brand 1994), the ATP/NADH stoichiometry is a maximum of 2.5, and only 1.5 when external NAD(P)H dehydrogenases are used, as the electrons will bypass the proton-pumping Complex

I. Electron transport bypassing all three proton pumping sites, e.g., electrons donated by a dehydrogenase other than Complex I and going to AOX rather than to CCO, will not contribute to IMM energization and ATP synthesis (Rasmusson and Møller 2011).

The two terminal oxidases differ markedly in their affinity for oxygen in that CCO has a much lower K_m (O_2) (0.14 μM) than AOX (2–20 μM) (Millar et al. 1994; Ribas-Carbo et al. 1994; Affourtit et al. 2001). This means that as the oxygen concentration decreases below 10–50 μM or 4–20 % of that in water in equilibrium with normal air, AOX activity is restricted resulting in a higher ubiquinone reduction level and an increased back-pressure of electrons on Complexes III and IV. At oxygen concentrations below 1 μM (0.4 %) even CCO activity becomes restricted and the ETC components are very highly reduced.

As outlined above, when oxygen is reduced by one of the terminal oxidases, water is the product. However, oxygen can also be reduced at other sites in the electron transport chain, notably at Complexes I and III, as well as by other enzymes, accepting only one electron to give the superoxide anion ($O_2^{-\bullet}$). This typically occurs when the reduction level of the redox centers involved is high and induction of the AOX is one way to prevent or limit the rate of superoxide production (Maxwell et al. 1999; Møller 2001; Halliwell and Gutteridge 2007; Brand 2010). As an example, the AOX is specifically induced in hypoxic barley roots in response to reoxygenation (Szal et al. 2003). The superoxide anion is not particularly reactive, but does react with iron–sulfur centers, e.g., in aconitase, releasing free Fe^{3+}, which presents its own problems (see below).

The internal redox process converting superoxide to hydrogen peroxide (H_2O_2) and water is either spontaneous or catalyzed by the enzyme superoxide dismutase (SOD) (Halliwell and Gutteridge 2007). Hydrogen peroxide is not very reactive, it is uncharged and can pass membranes, e.g., the IMM, through aquaporins (Bienert et al. 2007). It is therefore generally considered to be the ROS molecule responsible for stress signaling (see later). Hydrogen peroxide can be removed by no fewer than five different enzymes/enzyme systems (Møller 2007), all of which are present in potato tuber mitochondria (Salvato et al. 2014). The efficient removal of hydrogen peroxide is important, because when it meets a free metal ion, primarily Fe^{3+}, Cu^{2+}, or Mn^{2+}, the hydroxyl radical, the most reactive of the ROS, can be formed through the Fenton reaction. For the same reason, free metal ions should also be removed, before they can react with hydrogen peroxide, by binding to protected sites on proteins (Møller et al. 2011). Although the total concentration of metal ions is very high in plant mitochondria (Tan et al. 2010), the concentration of free metal ions is kept very low by several dedicated carrier and storage proteins, such as frataxin and ferritin (Møller et al. 2011).

RNS are also produced in plant mitochondria. NO is considered the most important of RNS, because reversible nitrosylation of cysteines is a recognized regulatory posttranslational modification (PTM) (Lindermayr et al. 2005). The turnover of NO will be discussed in the following section. However, the presence of NO gives rise to a range of other RNS. NO can react with superoxide to give $ONOO^-$, peroxynitrite, a quite reactive RNS, which is able to add a nitrate group to tyrosine (and tryptophan and phenylalanine) in a process called nitration and oxidize methionine to methionine sulfoxide (Halliwell and Gutteridge 2007).

4 Formation of NO in Plant Mitochondria

In mammalian tissues, where NO is recognized as an important hormone, several different NO synthases (NOS), one of which is localized in the mitochondria (Zaobornyj and Ghafourifar 2012), oxidize arginine to NO using molecular oxygen and NAD(P)H. Under anoxic conditions, mammalian mitochondria can also reduce nitrite to NO at a site in Complex III probably involving the ubisemiquinone (Kozlov et al. 1999), or in CCO involving one of the cytochromes (Cooper 2002). The rate of NO production in mammalian mitochondria is 1–2 nmol min^{-1} (mg $protein)^{-1}$, highest under reducing conditions, which is sufficient to maintain millimolar concentrations of NO in the matrix. In spite of this, the steady-state NO concentration in mammalian mitochondria has been estimated to be in the nanomolar range indicating that very efficient removal mechanisms are at work (Zaobornyj and Ghafourifar 2012). NO is also an important signaling molecule in plants (Freschi 2013), but there is no reliable evidence for the presence of a mammalian-type NO synthase in plant cells. Instead plant cells appear to produce NO, mainly by nitrate reductase in the cytosol (Yamasaki et al. 1999; Gupta et al. 2005), but also by nitrite reduction in the mitochondria probably by mechanisms similar to those observed in mammalian mitochondria (Gupta et al. 2005; Stoimenova et al. 2007). However, it is likely that the balance between cytosolic and mitochondrial NO formation mechanisms varies depending on the physiological and environmental conditions (Gupta et al. 2005).

An open question is whether nitrite reduction in the mitochondria takes place on the inner or outer surface of the IMM. For nitrite to be reduced on the inner surface of the IMM would require the presence of a nitrite/$2H^+$ symporter in the IMM, a carrier not yet identified in mitochondria. However, since such a carrier is present in the inner chloroplast envelope (Sugiura et al. 2007), it is entirely possible that it is also found in the IMM, as the two membranes share a number of similarities.

The study of anaerobic nitrite reduction by mitochondria isolated from barley and rice roots showed that the stoichiometry of the reaction was 1 NAD(P)H oxidized to 1 nitrite consumed and this resulted in the production of about 1 ATP per 1 nitrite (Stoimenova et al. 2007). As we shall see in the following section, this is close to what one would expect from the operation of the mitochondrial ETC. Interestingly, Stoimenova et al. (2007) reported that the rate of NO production was only around 1 % of nitrite consumption. In combination, these observations are consistent with a reaction where nitrous oxide (N_2O), rather than NO is the main product of nitrite reduction under complete anoxia, a condition that is perhaps relatively rare in nature.

5 Hemoglobin and NO

Plant cells express hemoglobin molecules of three different types (type 1, 2 and 3), with type classification based on amino acid sequence cladistics (Hebelstrup et al. 2007). An inverse correlation between expression levels of these plant hemoglobins and NO over a broad range of O_2 concentrations has been demonstrated independently in several plant species using different methods for detection of NO. In *Arabidopsis thaliana* such correlations were confirmed using both fluorescence detection of NO within cells by cPTIO (Hebelstrup et al. 2008; Hebelstrup and Jensen 2008) and with quantum-laser or chemiluminescence-based detection of gaseous NO emission (Hebelstrup et al. 2012; Perazzolli et al. 2004). Similarly in *Hordeum vulgare* the correlation was observed also by using quantum-laser or chemiluminescence-based detection of gaseous NO emission (Hebelstrup et al. 2014; Gupta et al. 2014). In *Medicago sativa*, the correlation was determined using an NO electrode (Igamberdiev et al. 2004) or electron paramagnetic resonance (EPR) detection for NO measurements in vivo (Dordas et al. 2004). An explanation for this correlation may be that oxyhemoglobin directly reacts with NO. This reaction seems to be ubiquitous for all hemoglobins due to the capacity of the ferrous heme–O_2 complex to oxidize NO, also called the NO dioxygenase (NOD) mechanism. However, for the continuation of this reaction the heme complex must be constitutively reduced, and such reduction could be limiting in vivo (Smagghe et al. 2008). A cycle involving monodehydroascorbate reductase (MDHAR) and ascorbic acid has been demonstrated to fulfill the role of plant met-hemoglobin reduction in the NOD mechanism (Igamberdiev et al. 2006). However, this mechanism was not supported by in vitro experiments in another study, where it was suggested that the NOD mechanism in this form is an unlikely explanation of the inverse correlation between hemoglobin and NO levels in plant tissues (Wang and Hargrove 2013). Alternatively, free reduced flavins which have recently been demonstrated to directly reduce plant met-hemoglobins (Sainz et al. 2013), could be involved in maintaining continuation of the NOD mechanism in vivo. Finally, we cannot exclude that still other, as yet unidentified, mechanisms of met-hemoglobin reduction exist in plants.

NO is a signaling molecule in several developmental and physiological processes, and plant hemoglobin gene expression modulates the NO signal in such processes (reviewed in Hebelstrup et al. 2013; Hill 2012). In particular, over-expression of hemoglobin in plants contributes to the maintenance of energy charge during hypoxia in plant tissues (Sowa et al. 1998; Igamberdiev et al. 2005). The NOD mechanism of plant hemoglobins may be linked with nitrate and nitrite reduction in mitochondria through the reactions (1), (2) and (3) in Fig. 1 (Igamberdiev et al. 2005). This cycle, known as the hemoglobin/NO cycle, may contribute to production of ATP under hypoxic conditions through reaction (3) (Table 1), and it has been proposed to be the reason hemoglobin-overexpressing plants have as high ATP levels under hypoxia as WT plants (Dordas et al. 2003). However, it is important to note that operation of the cycle also leads to net oxygen

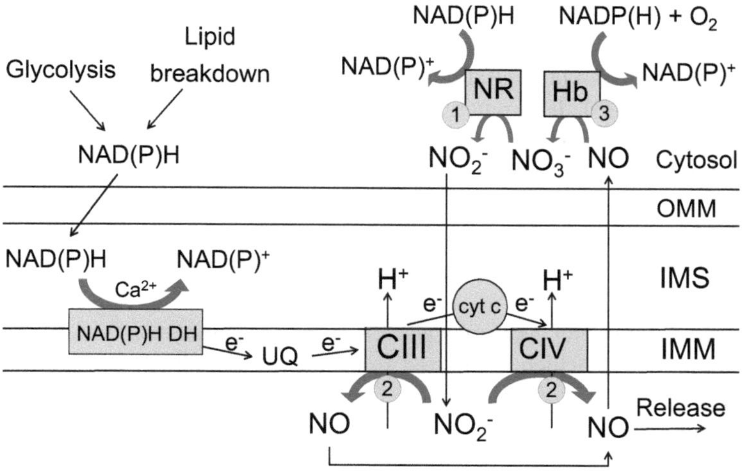

Fig. 1 Schematic drawing of the hemoglobin-NO cycle in plant cells. Under anoxia, reaction 3 is not operational and NO diffuses out of the cells and into the surrounding soil or air to be lost or act as a gaseous hormone affecting the above-ground parts of the plant. The numbered reactions (1–3) are shown as balanced equations in Table 1. The alternative oxidase, which receives its electrons from UQ and passes them to oxygen without any proton pumping, is not shown. CIII, CIV, respiratory complexes III and IV; DH, dehydrogenase; IMM, inner mitochondrial membrane; IMS, intermembrane space; NR, nitrate reductase; OMM, outer mitochondrial membrane

Table 1 Balanced equations and the net reaction of the hemoglobin–NO cycle

1. $2NO_3^- + 2NAD(P)H + 2H^+ \rightarrow 2NO_2^- + 2H_2O + 2NAD(P)^+$ 2. $2NO_2^- + NAD(P)H + 3H^+ \rightarrow 2NO + 2H_2O + NAD(P)^+$ 3. $2O_2 + 2NO + NAD(P)H \rightarrow 2NO_3^- + NAD(P)^+ + H^+$
Net reaction $2O_2 + 4NAD(P)H + 4\ H^+ \rightarrow 4H_2O + 4NAD(P)^+$ (for 2 NO_3^- molecules)
Net reaction for 1 molecule of nitrate: $O_2 + 2NAD(P)H + 2H^+ \rightarrow 2H_2O + 2NAD(P)^+$

The numbers refer to the reactions in Fig. 1. The NO formed in reaction 2 is either released or recycled through reaction 3. The net result of a full turn of the cycle is the transfer of electrons from NAD(P)H to oxygen while the nitrogen-containing compounds are recycled. In connection with the electron transport from NAD(P)H to oxygen (reaction 2) protons are pumped out of the mitochondrial matrix to give an electrochemical proton gradient (not the protons appearing in the equation). This gradient can be used to produce a maximum of 1.5 ATP/NAD(P)H as long as one of the external NAD(P)H dehydrogenases and COX is used. So the maximum ATP yield is 1.5 ATP per two molecules of $NO_3^- = 0.75$ ATP/O_2 (again only reaction 2 is relevant). Use of AOX instead of COX, e.g., when NO inhibits COX, will reduce the ATP yield significantly. The degree to which nitrite reduction by Complexes III and IV affects the number of protons pumped across the IMM is not known

consumption (Table 1), so that it can at the most contribute about 1 ATP per oxygen molecule. Thus, if the concentration of oxygen is about 10 μM, which appears to be typical for crop seeds (Borisjuk et al. 2007), the most the pathway can contribute is 10 μM ATP, which is only about 10 % of the tissue concentration, unless diffusion replenishes the oxygen at a rapid pace.

The above calculations indicate that it is unlikely that cycling NO through the hemoglobin/NO cycle is solely responsible for maintaining the ATP/ADP level observed under hypoxia in alfalfa roots in response to hemoglobin overexpression (Dordas et al. 2003). It is much more likely that the major part of the massive amount of NO produced by reaction 2 (Fig. 1) under hypoxic conditions (Hebelstrup et al. 2013) is released from the cells and the tissues into the surrounding soil and air and is lost to the plant. In that case, the mitochondrial ETC uses nitrite (from nitrate) as a reserve electron acceptor and hemoglobin (reaction 1, Fig. 1) only functions to keep the local concentration of NO down to manageable levels.

The hemoglobin/NO cycle may also contribute to conservation of fixed nitrogen in the form of NO, which is emitted from anoxic mitochondria, where nitrite is reduced (Gupta et al. 2005). At severe hypoxia or anoxia, plants with silencing of hemoglobin genes emit NO to the atmosphere at a rate of more than 500 nL h^{-1} gFW^{-1}, which corresponds to losing cellular N at a rate of around 0.5 μmol/gFW (0.5 mM) per day (Hebelstrup et al. 2012), or an amount comparable to that of free nitrate in plant cells. In the absence of a functional hemoglobin/NO cycle, long-term hypoxia would therefore interfere very significantly with the plant's nitrogen uptake and metabolism. An increased NO_x emission from agricultural soils is one of the deleterious effects of modern agriculture caused by the use of high doses of nitrogen fertilizer. For this reason the fate of NO emitted from hypoxic plants and the extent of contribution to the microbial nitrogen metabolism in the rhizosphere and to the surrounding environment should be explored further. It is likely that hemoglobin overexpression reduces such NO emissions (Hebelstrup et al. 2014).

6 Role of Hemoglobin in Plant Mitochondrial Respiration and ROS Generation

Hemo- and myoglobins are known to transport O_2 by two mechanisms: By active transport in erythrocytes, where O_2 is bound by hemoglobin in oxygen-rich environments and released in oxygen-poor environments. However, facilitation of diffusion by increasing O_2 flux in tissues provides another mechanism whereby both hemo- and myoglobin can contribute to the transport of O_2 (Wittenberg 1966; Wyman 1966). Such facilitation of diffusion has been demonstrated by both experimental evidence and extensive mathematical modeling (reviewed in Wittenberg 1970). Specific overexpression in seeds of type-2 hemoglobin (Glb2) in Arabidopsis results in enhanced seed yield and seed size. This effect was proposed to be the result of an increased oxygen supply (Vigeolas et al. 2011). However, facilitation of O_2 diffusion requires hemo-/myoglobin concentrations in the low millimolar range (Wittenberg 1970). Such concentrations will color the tissue red and are unlikely for non-symbiotic plant hemoglobins—even when these are overexpressed in plants by transgenic methods. In fact, overexpressing alfalfa only contains 12 nmol hemoglobin/g fresh weight (about 10 μM) (Dordas et al. 2003), which is at least two orders of magnitude below that required for facilitation of oxygen diffusion. It was recently demonstrated that overexpression of type-1 hemoglobin in barley roots increases respiration (as measured by O_2 uptake and CO_2 release), and simultaneously decreases the O_2 concentration internally in the roots (Gupta et al. 2014). This effect was linked to a decrease in internal NO, and an increase in ROS (as measured by fluorescence of 2′,7′-dichlorodihydro-fluorescein diacetate—DFC), and therefore interpreted to be caused by NO inhibition of CCO in the mitochondrial electron transport chain. This mechanism has also been suggested to control respiration in developing hypoxic seeds (Borisjuk et al. 2007). However, seed-specific overexpression of type-1 hemoglobins in either Arabidopsis (Thiel et al. 2011) or *Hordeum vulgare* (barley, Hebelstrup et al. 2014) failed to increase seed size, in contradiction to overexpression of type-2 hemoglobin in Arabidopsis. We can therefore not exclude the possibility that the effect of type-2 hemoglobin overexpression in seeds is caused by mechanisms other than those related to increased respiration.

7 ROS Signaling/Regulation of Metabolism

ROS can influence the mitochondrial and cellular metabolism either directly or indirectly. Direct interaction occurs via reactions with other metabolites or with proteins. The sulfur-containing amino acids, methionine and cysteine, can be oxidized by ROS in several steps where the first is reversible and therefore useful for regulation and perhaps signaling. The reversible oxidation of cysteine groups to cysteine is one of the fundamental regulatory mechanisms of protein function in cellular metabolism (Buchanan and Balmer 2005). Evidence is now appearing indicating that the reversible oxidation of methionine to methionine sulfoxide is also an important regulatory mechanism, which interacts with protein phosphorylation, a very well-known regulatory mechanism (Rao et al. 2014). It has been suggested that some peripheral methionine residues act as endogenous antioxidants protecting the active site and other sensitive domains in the protein while helping to remove ROS (Levine et al. 1996), particularly in mitochondria (Bender et al. 2008; Sweetlove and Møller 2009). Consistent with this, a very large number of proteins with oxidized methionine residues were as identified in potato tuber mitochondria (Salvato et al. 2014).

Protein carbonylation is probably the most common form of ROS-induced irreversible protein oxidation and it occurs on the side chains of four amino acids, arginine, lysine, proline, and threonine but also on tryptophan, although the latter appears to have different chemical properties. Carbonylation is thought to lead to changes in protein conformation, inhibition of protein function and degradation. In plant mitochondria, carbonylated proteins are found at high steady-state concentrations and on many different proteins indicating that it probably has an important effect on mitochondrial metabolism. In senescing legume nodules, the mitochondria are early targets of oxidative modifications and have been suggested to be a likely source of redox signals (Matamoros et al. 2013).

The protein carbonylation sites appear to be clustered and it is possible that this occurs near the binding of rogue metal ions, which interact with hydrogen peroxide and produce the hydroxyl radical via the Fenton reaction (Maisonneuve et al. 2009; Rao and Møller 2011; Møller et al. 2011). However, with the exception of aconitase, which is known to be inhibited by oxidation, the effect of protein oxidation on the properties of the proteins is not known (Kristensen et al. 2004; Møller and Kristensen 2006; Møller et al. 2007; Salvato et al. 2014). The carbonylated proteins are degraded by dedicated proteases, so protein oxidation is likely to contribute significantly to protein turnover (Smakowska et al. 2014).

Protein modification by ROS can also be indirect for instance via oxidation of polyunsaturated fatty acids, which are abundant in the phospholipids of plant mitochondria. This oxidation leads to the production of reactive aldehydes such as 2-hydroxynonenal, which can form conjugation products with proteins typically on lysine residues. A number of plant mitochondrial proteins have been observed to be modified in this way leading to inactivation (Winger et al. 2007).

Increased ROS production, the sign of oxidative stress, can also lead to retrograde signaling to the nucleus to regulate gene expression. Gadjev et al. (2006) observed that oxidative stress in specific subcellular compartments, e.g., mitochondria, led to the expression of a group of genes specific for that compartment. This indicated that oxidative stress in each compartment led to the release of a unique "signature" signal, which somehow reached the nucleus. Based on this observation, it was proposed that the specific signals, at least from chloroplasts and mitochondria, could be oxidized peptides deriving from the proteolytic degradation of irreversibly oxidized proteins rather than ROS molecules themselves (Møller and Sweetlove 2010). As the basis for this hypothesis, it was argued that hydrogen peroxide, the most likely ROS messenger, is a simple molecule unable to transmit information about its origin, unlike unique peptides deriving from mitochondrial or chloroplastic proteins. In a later modeling study (Vestergaard et al. 2012), it was shown that transmission of information by frequency- and amplitude-modulated waves of hydrogen peroxide, similar to that thought to occur for calcium ions, is possible given a number of assumptions. It is, however, unclear whether this actually occurs in plant cells as the typical average steady-state concentrations of hydrogen peroxide reported for plant tissues are so high as to make this signaling mode difficult.

8 NO Signaling/Regulation of Metabolism

Similar to ROS, NO can influence the mitochondrial and cellular metabolism either directly or indirectly. Direct interaction occurs via two mechanisms: The first is direct binding of NO to hemes competing with oxygen. In this way, NO inhibits

CCO as discussed above. NO also binds to Complex III, but here the binding site may be iron–sulfur centers (Zaobornyj and Ghafourifar 2012).

The other direct mechanism involves nitrosylation of cysteines in proteins or in small cysteine-containing metabolites such as glutathione. This occurs either by reaction with NO directly or via peroxinitrite formed by the reaction between NO and superoxide. The nitrosylation is reversible (Liu et al. 2001), which makes it ideal for regulation. In mammalian cells, protein nitrosylation regulates a number of enzymes including Complex I.

The amount of nitrosothiols (–SNO groups) in whole leaf extracts increased almost 1,000-fold when Arabidopsis rosettes were placed under hypoxia (0.1 % oxygen), which correlated quite well with the amount of NO released from the rosettes under the same conditions (Hebelstrup et al. 2012). In plant cells, a number of nitrosylated proteins have been identified (e.g., Lindermayr et al. 2005), some of which are mitochondrial. The effect of nitrosylation on the properties of the affected plant proteins has not yet been described for any mitochondrial protein.

NO not metabolized inside the mitochondria, leaves them by diffusion to be recycled through the hemoglobin-NO cycle (Fig. 1) or to regulate proteins and genes in the rest of the cell. For example, NO may interfere with hormone perception through the direct nitrosylation of hormone receptor or factors in the downstream signaling cascade of hormone receptors (Hebelstrup et al. 2012, 2013). However, at least under hypoxia, a large proportion diffuses out of the plant tissue. In that way it may act as a gaseous hormone, for instance signaling from hypoxic roots to the normoxic shoots perhaps by diffusion through the aerenchyma.

9 Perspectives

We are beginning to get a clearer picture of the turnover of ROS and RNS in plant mitochondria, while their signaling functions are still relatively unexplored. It is likely that the final picture will be very complex given that the ROS and RNS families have a number of family members with widely different reactivities and therefore different abilities to initiate signaling pathways. The effects of reversible and irreversible protein oxidation and nitrosylation need to be determined for key enzymes and key processes before we can fully understand how ROS and RNS contribute to the regulation of the mitochondrial and the cellular metabolism. We also need to know more about the way ROS and RNS and their signals move within the cell, between cells, and between plant parts such as roots and shoots.

References

Affourtit C, Krab K, Moore AL (2001) Control of plant mitochondrial respiration. Biochim Biophys Acta 1504:58–69

Armstrong W, Armstrong J (2014) Plant internal oxygen transport (diffusion and convection) and measuring and modelling oxygen gradients. In: van Dongen JT, Licausi F (eds) Low-oxygen stress in plants. Springer, Vienna, pp 267–297

Becana M, Dalton DA, Morana JF, Iturbe-Ormaetxea I, Matamorosa MA, Rubio MC (2000) Reactive oxygen species and antioxidants in legume nodules. Physiol Plant 109:372–381

Bender A, Hajieva P, Moosmann B (2008) Adaptive antioxidant methionine accumulation in respiratory chain complexes explains the use of a deviant genetic code in mitochondria. Proc Natl Acad Sci USA 105:16496–16501

Bienert GP, Møller ALB, Kristiansen KA, Schulz A, Møller IM, Schjoerring JK, Jahn JP (2007) Specific aquaporins facilitate the diffusion of hydrogen peroxide across membranes. J Biol Chem 282:1183–1192

Borisjuk L, Macherel D, Benamar A, Wobus U, Rolletschek H (2007) Low oxygen sensing and balancing in plant seeds: a role for nitric oxide. New Phytol 176:813–823

Brand MD (1994) The stoichiometry of proton pumping and ATP synthesis in mitochondria. Biochemist 16:20–24

Brand MD (2010) The sites and topology of mitochondrial superoxide production. Exp Gerontol 45:466–472

Buchanan BB, Balmer Y (2005) Redox regulation: a broadening horizon. Annu Rev Plant Biol 56: 187–220

Colombatti F, Gonzalez DH, Welchen E (2014) Plant mitochondria under pathogen attack: A sigh of relief or a last breath? Mitochondrion, in press

Cooper CE (2002) Nitric oxide and cytochrome oxidase: substrate, inhibitor or effector? Trends Biochem Sci 27:33–39

Darwent MJ, Armstrong W, Armstrong J, Beckett PM (2003) Exploring the radial and longitudinal aeration of primary maize roots by means of Clark-Type oxygen microelectrodes. Russ J Plant Physiol 50:722–732

Dordas C, Hasinoff BB, Igamberdiev AU, Manac'h N, Rivoal J, Hill RD (2003) Expression of a stress-induced hemoglobin affects NO levels produced by alfalfa root cultures under hypoxic stress. Plant J 35:763–770

Dordas C, Hasinoff BB, Rivoal J, Hill RD (2004) Class-1 hemoglobins, nitrate and NO levels in anoxic maize cell-suspension cultures. Planta 219:66–72

Finkemeier I, Konig AC, Heard W, Nunes-Nesi A, Pham PA, Leister D, Fernie AR, Sweetlove LJ (2013) Transcriptomic analysis of the role of carboxylic acids in metabolite signaling in arabidopsis leaves. Plant Physiol 162:239–253

Freschi L (2013) Nitric oxide and phytohormone interactions: current status and perspectives. Front Plant Sci 4:398. doi:10.3389/fpls.2013.00398

Gadjev I, Vanderauwera S, Gechev TS, Laloi C, Minkov IN, Shulaev V, Apel K, Inzé D, Mittler R, Van Breusegem F (2006) Transcriptomic footprints disclose specificity of reactive oxygen species signaling in Arabidopsis. Plant Physiol 141:436–445

Geigenberger P, Fernie AR, Gibon Y, Maja Christ M, Stitt M (2000) Metabolic activity decreases as an adaptive response to low internal oxygen in growing potato tubers. Biol Chem 381: 723–740

Gupta KJ, Stoimenova M, Kaiser WM (2005) In higher plants, only root mitochondria, but not leaf mitochondria reduce nitrite to NO, in vitro and in situ. J Exp Bot 56:2601–2609

Gupta KJ, Hebelstrup KH, Kruger NJ, Ratcliffe GR (2014) Nitric oxide is required for homeostasis of oxygen and reactive oxygen species in barley roots under aerobic conditions. Mol Plant 7:747–750

Halliwell B, Gutteridge JMC (2007) Free radicals in biology and medicine, 4th edn. Oxford University Press, Oxford

Hebelstrup KH, Jensen EO (2008) Expression of NO scavenging hemoglobin is involved in the timing of bolting in Arabidopsis thaliana. Planta 227:917–927

Hebelstrup KH, Igamberdiev AU, Hill RD (2007) Metabolic effects of hemoglobin gene expression in plants. Gene 398:86–93

Hebelstrup KH, Ostergaard-Jensen E, Hill RD (2008) Bioimaging techniques for subcellular localization of plant hemoglobins and measurement of hemoglobin-dependent nitric oxide scavenging in planta. Methods Enzymol 437:595–604

Hebelstrup KH, van Zanten M, Mandon J, Voesenek LACJ, Harren FJM, Cristescu SM, Møller IM, Mur LAJ (2012) Haemoglobin modulates NO emission and hyponasty under hypoxia-related stress in Arabidopsis thaliana. J Exp Bot 63:5581–5591

Hebelstrup KH, Shah JK, Igamberdiev AU (2013) The role of nitric oxide and hemoglobin in plant development and morphogenesis. Physiol Plant 148:457–469

Hebelstrup KH, Shah JK, Simpson C, Schjoerring JK, Mandon J, Cristescu SM, Harren FJM, Christiansen MW, Mur LAJ, Igamberdiev AU (2014) An assessment of the biotechnological use of hemoglobin modulation in cereals. Physiol Plant 150:593–603

Hill RD (2012) Non-symbiotic haemoglobins—what's happening beyond nitric oxide scavenging? AoB PLANTS 2012:pls004

Igamberdiev AU, Seregelyes C, Manac'h N, Hill RD (2004) NADH-dependent metabolism of nitric oxide in alfalfa root cultures expressing barley hemoglobin. Planta 219:95–102

Igamberdiev AU, Baron K, Manac'h-Little N, Stoimenova M, Hill RD (2005) The haemoglobin/nitric oxide cycle: involvement in flooding stress and effects on hormone signalling. Ann Bot 96:557–564

Igamberdiev AU, Bykova NV, Hill RD (2006) Nitric oxide scavenging by barley hemoglobin is facilitated by a monodehydroascorbate reductase-mediated ascorbate reduction of methemoglobin. Planta 223:1033–1040

Kozlov AV, Staniek K, Nohl H (1999) Nitrite reductase activity is a novel function of mammalian mitochondria. FEBS Lett 454:127–130

Kristensen BK, Askerlund P, Bykova NV, Egsgaard H, Møller IM (2004) Identification of oxidised proteins in the matrix of rice leaf mitochondria by immunoprecipitation and two-dimensional liquid chromatography-tandem mass spectrometry. Phytochemistry 65: 1839–1851

Levine RL, Mosoni L, Berlett BS, Stadtman ER (1996) Methionine residues as endogenous antioxidants in proteins. Proc Natl Acad Sci USA 93:15036–15040

Lindermayr C, Saalbach G, Durner J (2005) Proteomic identification of S-nitrosylated proteins in Arabidopsis. Plant Physiol 137:921–930

Liu L, Hausladen A, Zeng M, Que L, Heitman J, Stamler JS (2001) A metabolic enzyme for S-nitrosothiol conserved from bacteria to humans. Nature 410:490–494

Maisonneuve E, Ducret A, Khoueiry P, Lignon S, Longhi S, Talla E, Dukan S (2009) Rules governing selective protein carbonylation. PLoS One 4(10):e7269. doi:10.1371/journal.pone.0007269

Matamoros MA, Fernandez-Garcia N, Wienkoop S, Loscos J, Saiz A, Becana M (2013) Mitochondria are an early target of oxidative modifications in senescing legume nodules. New Phytol 197:873–885. doi:10.1111/nph.12049

Maxwell DP, Wang Y, McIntosh L (1999) The alternative oxidase lowers mitochondrial reactive oxygen production in plant cells. Proc Natl Acad Sci USA 96:8271–8276

Millar AH, Bergersen FJ, Day DA (1994) Oxygen-affinity of terminal oxidases in soybean mitochondria. Plant Physiol Biochem 32:847–852

Millar AH, Whelan J, Soole KL, Day DA (2011) Organization and regulation of mitochondrial respiration in plants. Annu Rev Plant Biol 62:79–104

Møller IM (2001) Plant mitochondria and oxidative stress. Electron transport, NADPH turnover and metabolism of reactive oxygen species. Annu Rev Plant Physiol Plant Mol Biol 52: 561–591

Møller IM (2007) Mitochondrial electron transport and oxidative stress. In: Logan DM (ed) Plant mitochondria, Annual plant reviews. Blackwell, Oxford, pp 185–211
Møller IM, Kristensen BK (2006) Protein oxidation in plant mitochondria detected as oxidized tryptophan. Free Radic Biol Med 40:430–435
Møller IM, Jensen PE, Hansson A (2007) Oxidative modifications to cellular components in plants. Annu Rev Plant Biol 58:459–481
Møller IM, Sweetlove LJ (2010) ROS signaling—specificity is required. Trends Plant Sci 15: 370–374
Møller IM, Rogowska-Wrzesinska A, Rao RSP (2011) Protein carbonylation and metal-catalyzed oxidation in a cellular perspective. J Proteomics 74:2228–2242
Perazzolli M, Dominici P, Romero-Puertas MC, Zago E, Zeier A, Sonoda M, Lamb C, Delledonne M (2004) Arabidopsis nonsymbiotic hemoglobin AHb1 modulates nitric oxide bioactivity. Plant Cell 16:2785–2794
Rajhi I, Yamauchi T, Takahashi H, Nishiuchi S, Shiono K, Watanabe R, Mliki A, Nagamura Y, Tsutsumi N, Nishizawa NK, Nakazono M (2011) Identification of genes expressed in maize root cortical cells during lysigenous aerenchyma formation using laser microdissection and microarray analyses. New Phytol 190:351–368
Rao RS, Møller IM (2011) Pattern of occurrence and occupancy of carbonylation sites in proteins. Proteomics 11(21):4166–4173
Rao RSP, Møller IM, Thelen JJ, Miernyk JA (2014) Methionine oxidation: Bridging ROS and protein phosphorylation signaling. Cell Stress and Chaperones, in press
Rasmusson AG, Møller IM (2011) Mitochondrial electron transport and plant stress. In: Kempken F (ed) Plant mitochondria. Springer, New York, pp 357–381
Ribas-Carbo M, Berry JA, Azconn-Bieto J, Siedow JN (1994) The reaction of the plant mitochondrial cyanide-resistant alternative oxidase with oxygen. Biochim Biophys Acta 1188:205–212
Sainz M, Pérez-Rontomé C, Ramos J, Mulet JM, James EK, Bhattacharjee U, Petrich JW, Becana M (2013) Plant hemoglobins may be maintained in functional form by reduced flavins in the nuclei, and confer differential tolerance to nitro-oxidative stress. Plant J 76:875–887
Salvato F, Havelund JF, Chen M, Rao RSP, Wrzesinska-Rogowska A, Jensen ON, Gang DR, Thelen JJ, Møller IM (2014) The potato tuber mitochondrial proteome. Plant Physiol 164: 637–653
Smagghe BJ, Trent JT, Hargrove MS (2008) NO dioxygenase activity in hemoglobins is ubiquitous in vitro, but limited by reduction in vivo. PLos One 3:e2039
Smakowska E, Czarna M, Janska H (2014) Mitochondrial ATP-dependent proteases in protection against accumulation of carbonylated proteins. Mitochondrion, in press
Sowa AW, Duff SMG, Guy PA, Hill RD (1998) Altering hemoglobin levels changes energy status in maize cells under hypoxia. Proc Natl Acad Sci USA 95:10317–10321
Steffens B, Geske T, Sauter M (2011) Aerenchyma formation in the rice stem and its promotion by H_2O_2. New Phytol 190:369–378
Stoimenova M, Igamberdiev AU, Gupta KJ, Hill RD (2007) Nitrite-driven anaerobic ATP synthesis in barley and rice root mitochondria. Planta 226:465–474
Sugiura M, Georgescu MN, Takahashi M (2007) A nitrite transporter associated with nitrite uptake by higher plant chloroplasts. Plant Cell Physiol 48:1022–1035
Sweetlove LJ, Møller IM (2009) Oxidation of proteins in plants—mechanisms and consequences. Adv Bot Res 52:1–23
Szal B, Jolivet Y, Hasenfratz-Sauderb MP, Dizengremel P, Rychter AM (2003) Oxygen concentration regulates alternative oxidase expression in barley roots during hypoxia and post-hypoxia. Physiol Plant 119:494–502
Tan Y-F, O'Toole N, Taylor NL, Millar AH (2010) Divalent metal ions in plant mitochondria and their role in interactions with proteins and oxidative stress-induced damage to respiratory function. Plant Physiol 152:747–761

Thiel J, Rolletschek H, Friedel S, Lunn JE, Nguyen TH, Feil R, Tschiersch H, Muller M, Borisjuk L (2011) Seed-specific elevation of non-symbiotic hemoglobin AtHb1: beneficial effects and underlying molecular networks in Arabidopsis thaliana. BMC Plant Biol 11:43

Vashisht D, Hesselink A, Pierik R, Ammerlaan JMH, Bailey-Serres J, Visser EJW, Pedersen O, van Zanten M, Vreugdenhil D, Jamar DCL, Voesenek LACJ, Sasidharan R (2011) Natural variation of submergence tolerance among Arabidopsis thaliana accessions. New Phytol 190: 299–310

Vestergaard CL, Flyvbjerg H, Møller IM (2012) Intracellular signalling by diffusion—Can waves of hydrogen peroxide transmit intracellular information in plant cells? Front Plant Sci 3:295. doi:10.3389/fpls.2012.00295

Vigeolas H, Huhn D, Geigenberger P (2011) Nonsymbiotic hemoglobin-2 leads to an elevated energy state and to a combined increase in polyunsaturated fatty acids and total oil content when overexpressed in developing seeds of transgenic arabidopsis plants. Plant Physiol 155:1435–1444

Wang XG, Hargrove MS (2013) Nitric oxide in plants: the roles of ascorbate and hemoglobin. PLos One 8:12

Welchen E, Garcia L, Mansilla N, Gonzales DH (2014) Coordination of plant mitochondrial biogenesis: keeping pace with cellular requirement. Front Plant Sci 4:551. doi:10.3389/fpls. 2013.00551

Winger AM, Taylor NL, Heazlewood JL, Day DA, Millar AH (2007) The cytotoxic lipid peroxidation product 4-hydroxy-2-nonenal covalently modifies a selective range of proteins linked to respiratory function in plant mitochondria. J Biol Chem 282:37436–37447

Wittenberg JB (1966) Molecular mechanism of hemoglobin-facilitated oxygen diffusion. J Biol Chem 241:104–114

Wittenberg JB (1970) Myoglobin-facilitated oxygen diffusion—role of myoglobin in oxygen entry into muscle. Physiol Rev 50:559–636

Wyman J (1966) Facilitated diffusion and possible role of myoglobin as a transport mechanism. J Biol Chem 241:115–121

Yamasaki H, Sakihama Y, Takahashi S (1999) An alternative pathway for nitric oxide production in plants: new features of an old enzyme. Trends Plant Sci 4:128–129

Zaobornyj T, Ghafourifar P (2012) Strategic localization of heart mitochondrial NOS: a review of the evidence. Am J Physiol Heart Circ Physiol 303:H1283–H1293

Feedback Loop of Non-coupled Respiration and Reactive Oxygen Species Production in Plant Mitochondria

Vasily N. Popov

1 Formation of Reactive Oxygen Species in Plant Mitochondria

The oxidation of respiratory substrates through the cytochrome pathway requires the presence of ADP for oxidative phosphorylation. When the cell's energy consumption is low, mitochondria accumulate ATP, their membrane potential is rising and, therefore, the electron transport through cytochrome pathway is suppressed. This leads to an increase in the level of ubisemiquinone, which is capable of one-electron reduction of oxygen and subsequent superoxide anion-radical ($O_2^{-\bullet}$) formation. Superoxide dismutase (SOD) quickly turns $O_2^{-\bullet}$ into hydrogen peroxide (H_2O_2). The latter can either be detoxified by catalase/peroxidase outside of mitochondria or turned into a very aggressive oxidiser, the hydroxyl-radical ($OH^{\bullet}$), which can oxidise practically all cellular components, including DNA. The protective mechanisms against this process include the reduction of intramitochondrial oxygen levels and the provision of a decreased life time of ubisemiquinone (Skulachev 1994, 1996). It has already been shown that the formation of superoxide radicals ($O_2^{-\bullet}$) is slow at low values of transmembrane potential ($\Delta\psi$) and grows exponentially with an increase of $\Delta\psi$. The rate of $O_2^{-\bullet}$-generation in Complex III correlates with the level of cytochrome b_{566} (b_L) reduction in this complex (Liu and Huang 1996).

Plants possess a branched system of noncoupled electron transport pathways, which bypass the generators of the electrochemical H^+ gradient in the electron transport chain (ETC). For mitochondria of higher plants it has been shown that the oxidation of NADH is not only carried out by Complex I but also by two or more

V.N. Popov (✉)
Department of Genetics, Cytology and Bioengineering, Voronezh State University, Voronezh 394006, Russia
e-mail: pvn@bio.vsu.ru

K.J. Gupta, A.U. Igamberdiev (eds.), *Reactive Oxygen and Nitrogen Species Signaling and Communication in Plants*, Signaling and Communication in Plants 23,
DOI 10.1007/978-3-319-10079-1_5

noncoupled NADH dehydrogenases, located on the inner (NDA) and outer (NDB) sides of the inner mitochondrial membrane. NDB can be represented by two calcium-dependent dehydrogenases, one NADH- and other NADPH dependent. In addition, there is a calcium-dependent NADPH dehydrogenase (NDC) on the inner side (Rasmusson et al. 2008). Another noncoupled pathway contains a cyanide-insensitive oxidase, also called the alternative oxidase (AOX) (Vanlerberghe and McIntosh 1997). It is one of the best studied systems of the noncoupled respiration in plants (Wagner and Moore 1997). This enzyme is located on the inner side of the inner mitochondrial membrane and catalyses the reduction of oxygen by four electrons using ubiquinol (QH_2) as the electron donor. There is no energy accumulation in the form of electrochemical proton gradient ($\Delta\mu H^+$) in this and other noncoupled processes. AOX operates at a higher level of ubiquinone (Q) reduction (the ratio of QH_2/Q) than the cytochrome oxidase (Van den Bergen et al. 1994). This ratio is usually increased during the transition of mitochondria from State 3 (ADP present) to State 4 (ADP depleted), when the risk of one-electron oxygen reductions is increased (Liu and Huang 1996). The affinity of AOX for oxygen is much lower than that of cytochrome oxidase (Solomos 1977; Millar et al. 1994) and, apparently, alternative electron transport in plant mitochondria is activated when oxygen concentrations become too high to be used by the processes coupled to ATP production. Skulachev (1996) has suggested that one of the functions of free oxidation is the reduction of the internal oxygen concentrations. This is one of the protective mechanisms preventing the superfluous generation of reactive oxygen species (ROS). In this context, the impact of ROS on the induction process of AOX is important (Wagner 1995). A protective role for AOX under oxidative stress conditions was postulated earlier (Purvis and Shewfelt 1993; Wagner 1995; Skulachev 1996; Purvis 1997) and it was shown that the AOX inhibitor salicylhydroxamate (SHAM) activates hydrogen peroxide production in plant mitochondria (Popov et al. 1997).

We report here our unpublished study of the influence of electron transport chain inhibitors on the rate of respiration, ubiquinone reduction and the rate of ROS production in cauliflower mitochondria. Mitochondria were isolated using Percoll self-generated gradients (Wagner 1995). The membrane potential was estimated by safranine method (Akerman and Wikström 1976). Superoxide production was monitored by the Aminco SPF 500 Dual Wavelength spectrophotometer as a SOD-sensitive part of epinephrine oxidation at differential optical absorbance (480 – 579 nm) in the same incubation medium as was used for oxygen consumption measurements, but with the addition of 1 mM epinephrine (Purvis et al. 1995). The extraction, HPLC separation, and determination of reduced and oxidised Q from isolated mitochondria was performed as described previously (Wagner and Wagner 1997). The respiratory control value was 3.1–3.2. ADP and the artificial uncoupler S13 effectively reduced membrane potential having the same effect on superoxide production as its reduction by inhibition of succinate oxidation by malonate (Table 1). The inhibitors of cytochrome pathway at the level of complex III myxothiazol and antimycin A inhibited respiration, reduced membrane potential but exhibited controversial effect on superoxide production (Fig. 1). It can be

Table 1 Influence of ADP, S13 (uncoupler), antimycin, and myxothiazol on superoxide production in cauliflower mitochondria

	Q reduction (%)	Rate of respiration, nmol O_2 min^{-1} mg^{-1} protein	Membrane potential (%)	ROS production, nmol $O_2^{-\bullet}$ min^{-1} mg^{-1} protein
10 mM succinate	92.0	210	100	13.00
+50 nM S13	43.2	494	85	1.88
+100 μM ADP	58.2	440	85	0.30
+1 μM antimycin	98.6	32	5	26.41
+0.6 μM myxothiazol	95.2	168	10	5.38
+4 μM myxothiazol	99.5	24	5	2.08

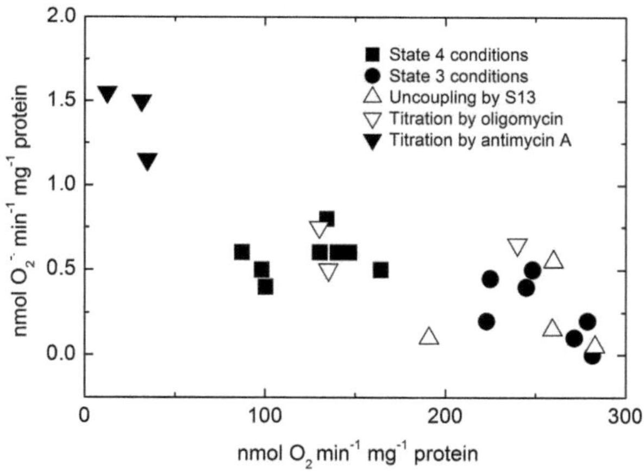

Fig. 1 Influence of antimycin (0.1–2 μM), myxothiazol (0.1–2 μM), oligomycin (0.1–1 μM), and ADP (150 μM) and the uncoupler S13 (1–10 nM) on the rate of succinate oxidation and superoxide production by cauliflower mitochondria. Incubation medium contained 20 mM HEPES (pH 7.2), 350 mM mannitol, 10 mM succinate, 1 mM potassium phosphate, 5 mM $MgCl_2$, 1 mM epinephrine, and approximately 0.5 mg of mitochondrial protein

explained by different mechanisms by which these inhibitors act. Antimycin blocks electron transfer to cytochrome *b*, causes ubisemiquinone ($CoQH^{\bullet}$) accumulation and increase in one-electron oxygen reduction. $CoQH^{\bullet}$ and other electron transport intermediates participate in the one electron O_2 reduction extending life-span when $\Delta\psi$ is sufficiently high. In the Q-cycle this is a consequence of inhibition by $\Delta\psi$ of the heme b_L–heme b_H, electron transport step. This results in complete reduction of b_H which then cannot oxidize $CoQH^{\bullet}$ formed in the Q-cycle (Skulachev, 1996). Myxothiazol in our experiments inhibited ubiquinol oxidation, preserving ubisemiquinone production and superoxide formation. This means that one of the mechanisms of superoxide formation in cauliflower mitochondria is one-electron oxygen reduction due to the presence of Q-cycle in Complex III.

The same picture was observed earlier for the mitochondria isolated from pea seedlings (Popov et al. 2003). In mitochondria from pea cotyledons the respiration is carried out mainly through AOX. 1 mM potassium cyanide (KCN) suppressed only 20 % of respiratory activity using 5 mM succinate as substrate in mitochondria from pea cotyledons. Salicylhydroxamate (AOX inhibitor) suppressed completely the residual oxygen consumption. The study of coupling of respiration and phosphorylation in pea cotyledon mitochondria has shown that respiration is stimulated at the presence of ADP or uncouplers. The uncoupler SF6847 (in concentration 69 nM) accelerated the rate of O_2 consumption only by 2.5 times. This can be explained by the high content of free fatty acids being effective uncouplers, and the high activity of noncoupled alternative electron transport.

In another study we investigated the rate of H_2O_2 production by pea mitochondria. This rate was measured with fluorescein diacetate and the basic value of H_2O_2 generation with succinate as a substrate was 0.02 nmol min^{-1} mg^{-1} protein. The addition of 2 μM antimycin A to the incubation medium increased the intensity of H_2O_2 generation only by 1.5 times, and 2 μM rotenone changed the rate of the studied process insignificantly. Inhibition of AOX by SHAM strongly activated hydrogen peroxide generation. It increased up to 2.8 nmol min^{-1} mg^{-1} protein in presence of 0.5 mM SHAM, which corresponds to 15-fold increase. The addition of 2 μM antimycin A after SHAM further accelerated H_2O_2 release (by 2 times), while 1 mM cyanide was not efficient. This can be explained by the effect of overreduction of FeS clusters in Complex III that results in inhibition of ubiquinol oxidation and limitation of ubisemiquinine formation (Popov et al. 1997).

The analysis of superoxide production by pea mitochondria using epinephrine was complicated due to a high rate of background SOD-insensitive epinephrine oxidation. Therefore we used at the same time spin traps for $O_2^{-\bullet}$ detection. The rate of superoxide generation during succinate oxidation was 0.24 nmol min^{-1} mg^{-1} protein. Addition of antimycin A increased the rate of generation up to 0.8 nmol min^{-1} mg^{-1} protein. The subsequent addition of AOX inhibitor SHAM resulted in almost 4-times increase of the rate of superoxide production: up to 3.3 nmol min^{-1} mg^{-1} protein (Popov et al. 2003). In addition to these observations it was established that the transition of mitochondria to State 3 by 150 μM ADP addition simultaneously reduced the rate of $O_2^{-\bullet}$ formation.

2 Plant Defence Against ROS Overproduction

Plant cells have developed detoxification systems to cope with free radicals. Enzymatic detoxification by superoxide dismutase, ascorbate peroxidase, monodehydroascorbate and dehydroascorbate reductases, glutathione peroxidase, glutathione reductase, catalase, and "guiacol"-peroxidase, either individually or co-operatively, removes ROS. In the normally growing plant, where there is relatively little production of ROS, enzymatic and antioxidant detoxification capacity is sufficient to keep the amount of ROS under control. Several stress conditions, however, such as wounding or other mechanical influences (radiation, drought, flooding, osmotic and temperature effects) disturb the balance between pro- and anti-oxidative potentials (Elstner and Oswald 1994). An important increase in ROS production is caused by conditions that result in the inhibition of ETC at the level of terminal oxidases (Wagner and Krab 1995; Wagner and Moore 1997; Popov 2003).

One of the many proposed mechanisms which mitochondria use to decrease ROS generation is the uncoupling between mitochondrial respiration and membrane potential, resulting in increased respiratory rates (for review see Wagner and Moore 1997; Popov 2003; Skulachev 1988). Protonophores are well known to decrease mitochondrial ROS generation (Skulachev 1994), and mitochondrial uncoupling proteins discovered in various plant tissues (Kowaltowski et al. 1998), have demonstrated the ability to produce the same effect. Based on these observations, Skulachev (1994) formulated the hypothesis that mild uncoupling of mitochondria may be an effective mechanism to reduce mitochondrial ROS generation without seriously compromising cellular energetics. The decrease in mitochondrial membrane potential by free fatty acids-induced uncoupling through plant uncoupling mitochondrial protein (PUMP) activity, inhibits mitochondrial generation of ROS at the level of the semiquinone forms of coenzyme Q in potato tuber mitochondria (Kowaltowski et al. 1998). It is important to note that accumulation of superoxide radical stimulates linoleic acid-induced proton leak in potato mitochondria that is related to the activity of uncoupling protein (Considine et al. 2003). The ATP/ADP antiporter can also provide reduction of mitochondrial membrane potential in the presence of free fatty acids in potato mitochondria (Popov et al. 2002).

Another possible mechanism of ROS regulation in plant mitochondria is AOX. In early 1990s Purvis and Shewfelt hypothesized that AOX is used in the antioxidant defence by transporting to oxygen the excess of reducing equivalents (Purvis and Shewfelt 1993). In line with this assumption, AOX was shown to be operative at much higher $CoQH_2/CoQ$ ratios than the energy coupled $CoQH_2$-cytochrome c oxidase.

It was proposed that AOX, in co-operation with the noncoupled NAD(P)H-CoQ reductases (NDA, NDB and NDC), also found in plant mitochondria, organize a respiratory chain which bypasses all three energy-coupling sites of the main respiratory chain and, hence, can perform its defensive function being no more restricted by ADP availability. It is also important that the noncoupled chain does not include the $O_2^{-\bullet}$-producing components, which obtain long life-span in State 4 such as $CoQH^\bullet$ (Skulachev 1994).

3 Reactive Oxygen Species Are Effective Secondary Messengers for Induction of the Noncoupled Respiration

In 1990s Vanlerberghe and McIntosh (1996) and Wagner and Wagner (1997) showed the induction of AOX by H_2O_2. This explains previous findings that AOX induction in plant (Minagawa et al. 1992) and yeast (Raskin 1992) cells can be achieved by adding antimycin A, which is known to strongly increase the H_2O_2 production by mitochondria. In the same way, the accumulation of H_2O_2 after the addition of catalase inhibitor salicylate can induce AOX (Purvis et al. 1995). The coexpression of the mitochondrial respiratory chain enzymes was investigated in the heat-producing plant *Arum orientale*. It was shown that mitochondria isolated from thermogenic tissues of this plant (as opposed to non-thermogenic tissues of *A. orientale* or to *Zea mays*) demonstrate significantly elevated levels of activities of two non-coupled NADH dehydrogenases (NDA and NDB) oxidizing intramitochondrial and cytoplasmic NADH pools. It was postulated that operation of a completely non-coupled respiratory chain consisting of non-coupled NADH: quinone oxidoreductases (NDA and NDB) and cyanide-resistant alternative quinol-oxidase (AOX) could be regulated by the same mechanism for heat production in thermogenic plants (Bertsova et al. 2004).

The involvement of active oxygen forms in the regulation of expression of mitochondrial respiratory chain components which are not related to energy storing has been studied in vitro and in vivo in tomato (*Solanum lycopersicum* L.). The highest level of transcription of genes encoding AOX and NADH dehydrogenase (NDB) has been observed in green tomato leaves (Eprintsev et al. 2011). It has been shown that even low H_2O_2 concentrations activate both *aox1a* and *ndb1* genes, encoding AOX and the external mitochondrial rotenone-insensitive NADH dehydrogenase (NDB), respectively. When the oxidative stress in tomato cell culture was created by treatments with H_2O_2 or antimycin A (Popov et al. 2011), it was established that the mRNA levels of AOX, of the uncoupling protein, and of the ATP/ADP transporter were controlled by ROS, and this may be the mechanism of "free" oxidation induction. The addition of 0.1 and 10 mM exogenous H_2O_2 led to 3 and 10 times lower aconitase activity, respectively, after 6 h of incubation relative to the control. Analogously, 5 and 15 μM of antimycin A resulted in 3.5- and 5.2-fold inactivation of this enzyme, respectively, with comparison to the control (Eprintsev et al. 2011; Popov et al. 2011).

The presence of 0.1 and 10 mM H_2O_2 in the callus culture medium activated lipid peroxidation with accumulation of its product malonic dialdehyde (MDA) by 144 and 208 %, respectively, with comparison to the control. MDA was also accumulated when antimycin A was added to the cultivation medium. Thus, 5 and 15 μM antimycin A elevated MDA level by 70–80 % (Eprintsev et al. 2011).

The real-time PCR showed that 0.1 and 10 mM hydrogen peroxide applied for 6 h led to a higher content of *nda1*, *ndb1*, *aox1a*, *ant*, and *pump* transcripts, encoding correspondingly the isoforms of NDA, NDB, AOX, ATP/ADP translocator, and the uncoupling protein. The study of the regulation of expression of the genes involved in alternative and uncoupled respiration in tomato cell culture in the presence of antimycin A by Real-time PCR showed that *aox1a* mRNA content grew up more than threefold relative to the control. 5 μM antimycin A did not considerably affect the *ndb1* expression and lowered the mRNA content of *nda1*. 15 μM antimycin A decreased transcript levels for *sod3f1*, *aox1a*, *nda1*, and *ndb1*.

Antimycin A in the callus cultivation medium also helped to increase mRNA levels of the genes for the respiration uncouplers. Thus, at 5 μM antimycin A, the *ant* expression was 9 times higher than in the reference. However, 15 μM antimycin A did not practically change the expression level of the gene for the ATP/ADP antiporter. For the pump gene, it was shown that 5 and 15 μM of antimycin A increased the mRNA content of this gene five- and ninefold, respectively, relative to reference (Popov et al. 2011).

Earlier it was shown that cold stress reduced the content of *nda1* and *ndb1* transcripts and corresponding proteins in potato leaves (as measured by antibody assays) and the rate of rotenone-insensitive NADH oxidation in leaf mitochondria (Svensson and Rasmusson 2001). The activity of the rotenone insensitive NADH and NADPH dehydrogenases is necessary for free oxidation of NADH and NADPH to provide normal functioning of plant mitochondria.

As we have discussed earlier free oxidation is an efficient mechanism of protection against ROS (Wagner and Krab 1995; Popov et al. 2001, 2003). In addition, it is the level of ROS that is the key factor of the AOX expression (Wagner and Wagner 1997; Minagawa et al. 1992) and possibly of the uncoupling protein regulation. The activities of the alternative respiratory pathways provide the protection of the mitochondrial electron transport chain, on the one hand, by the reduction in ROS production and, on the other, by efficient reoxidation of the cellular pools of reducing equivalents, thus preventing blocking of NADH-dependent processes.

Conclusion

In our research that we discussed in this chapter we have demonstrated that the addition of hydrogen peroxide or ROS-inducer antimycin A to cell culture causes an increase in mRNA levels of *aox1a*, *ant*, and *pump* (Fig. 2). This is similar to the increase of transcription of these genes in cold-treated green plants (Minagawa et al. 1992; Wagner 1995; Eprintsev et al. 2011). It should also be noted that in tomato cell culture the ATP/ADP antiporter was induced at low peroxide concentrations, but at higher H_2O_2 concentrations the content of the uncoupling protein mRNA grew abruptly. Both of these carriers are capable of fatty acid-induced uncoupling of respiration and oxidative phosphorylation during adaptation of plants to cold (Popov et al. 2002; Popov 2003), but it seems that the constantly expressed antiporter is activated first, and then the uncoupling protein gets involved in uncoupling process when the

(continued)

Fig. 2 Participation of reactive oxygen species (ROS) production in regulation of the non-coupled respiration pathways in plant mitochondria. AAT—ATP/ADP transporter; AOX—alternative oxidase; NDA, NDB, NDC—rotenone-insensitive NADH and NADPH dehydrogenases; UCP—uncoupling protein; bc1, aa3—Complexes III and IV of the electron transport chain

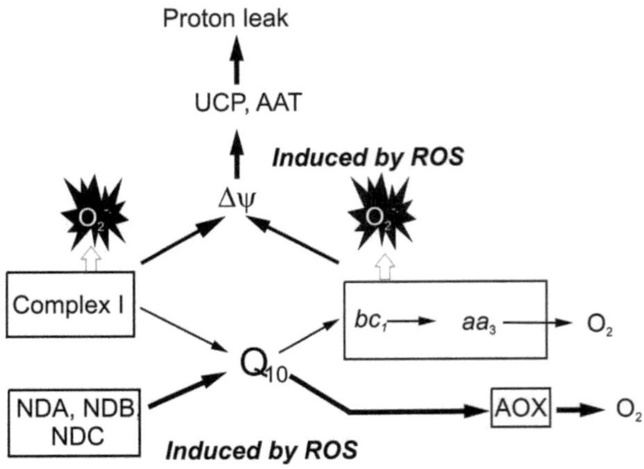

cell is considerably damaged. According to presented results, in the case of oxidative stress, AOX and NDB are coexpressed in tomato plant tissues, and the active oxygen forms serve as the secondary messengers of their coexpression.

At the same time the hypothesis that the noncoupled alternative oxidase of plant mitochondria operates as an antioxidant defence mechanism (Purvis and Shewfelt 1993; Skulachev 1994; Popov et al. 2001; Popov 2003) has been confirmed in our experiments on isolated soybean and pea cotyledon mitochondria. It was shown that inhibitors of AOX, salicylhydroxamate (SHAM) and propyl gallate strongly stimulate H_2O_2 production by these mitochondria. Effective concentrations of the inhibitors proved to be the same as those decreasing the cyanide-resistant respiration. The inhibitors proved to be ineffective in stimulating H_2O_2 formation in rat liver mitochondria lacking the alternative oxidase. Using EPR analysis it was shown that SHAM is an effective inductor of superoxide production by plant mitochondria (Popov et al. 2003).

Finally we propose that stress conditions cause oxidative burst accompanied by the dramatic increase of ROS concentration. Such oxidative burst plays a role as a secondary messenger for expression not only of the anti-oxidative enzymes, but also of the proteins involved in free (non-coupled and uncoupled) oxidation pathways. Activation of these pathways causes rapid limitation of ROS production and prevents oxidation of the ETC components and decrease of the mitochondrial membrane potential. This feedback mechanism represents an important strategy of general plant resistance to abiotic and biotic stress.

References

Akerman KE, Wikström MK (1976) Safranine as a probe of the mitochondrial membrane potential. FEBS Lett 68:191–197

Bertsova YV, Popov VN, Bogachev AV (2004) NADH oxidation by mitochondria from the thermogenic plant *Arum orientale*. Biochem Mosc 69:580–584

Considine MJ, Goodman M, Echtay KS, Laloi M, Whelan J, Brand MD, Sweetlove LJ (2003) Superoxide stimulates a proton leak in potato mitochondria that is related to the activity of uncoupling protein. J Biol Chem 278:22298–22306

Elstner EF, Oswald W (1994) Mechanisms of oxygen activation during plant stress. Proc R Soc Edinb 102B:131–154

Eprintsev AT, Mal'tseva EV, Shatskikh AS, Popov VN (2011) Involvement of hydrogen peroxide in the regulation of coexpression of alternative oxidase and rotenone-insensitive NADH dehydrogenase in tomato leaves and calluses. Izv Akad Nauk Ser Biol 1:45–51

Kowaltowski AJ, Costa ADT, Vercesi AE (1998) Activation of the potato plant uncoupling mitochondrial protein inhibits reactive oxygen species generation by the respiratory chain. FEBS Lett 425:213–216

Liu SS, Huang JP (1996) Co-existence of reactively oxygen cycle with Q-cycle in respiratory chain—a hypothesis for generation, partitioning and functioning of $O_2^{-\cdot}$ in mitochondria. In: Moores D (ed) Proceedings of the international symposium on natural antioxidants: molecular mechanisms and health effects. AOCS Press, Champaign, IL, pp 112–118

Millar AN, Bergersen FJ, Day DA (1994) Regulation of alternative oxidase activity by pyruvate in soybean mitochondria. Plant Physiol Biochem 32:847–852

Minagawa N, Koga S, Nakano M, Sakajo S, Yoshimoto A (1992) Possible involvement of superoxide anion in the induction of cyanide-resistant respiration in *Hansenula anomala*. FEBS Lett 302:217–219

Popov VN (2003) Possible role of free oxidation processes in the regulation of reactive oxygen species production in plant mitochondria. Biochem Soc Trans 31:1316–1317

Popov VN, Simonyan RA, Skulachev VP, Starkov AA (1997) Inhibition of the alternative oxidase stimulates H_2O_2 production in plant mitochondria. FEBS Lett 415:87–90

Popov VN, Purvis AC, Skulachev VP, Wagner AM (2001) Stress-induced changes in ubiquinone concentration and alternative oxidase in plant mitochondria. Biosci Rep 21:369–379

Popov VN, Markova OV, Mokhova EN, Skulachev VP (2002) Effects of cold exposure in vivo and uncouplers and recouplers in vitro on potato tuber mitochondria. Biochim Biophys Acta 1553: 232–237

Popov VN, Ruuge EK, Starkov AA (2003) Effect of electron-transport inhibitors on the generation of reactive oxygen species by pea mitochondria during succinate oxidation. Biochem Mosc 68: 747–751

Popov VN, Eprintsev AT, Maltseva EV (2011) Activation of genes encoding mitochondrial proteins involved in alternative and uncoupled respiration of tomato plants treated with low temperature and reactive oxygen species. Russ J Plant Physiol 58:914–920

Purvis AC (1997) Role of the alternative oxidase in limiting superoxide production by plant mitochondria. Physiol Plant 100:165–170

Purvis AC, Shewfelt RL (1993) Does the alternative pathway ameliorate chilling injury in sensitive plant-tissues? Physiol Plant 88:712–718

Purvis AC, Shewtelt RL, Gegogeine JW (1995) Superoxide production by mitochondria isolated from green bell pepper fruit. Physiol Plant 94:743–749

Raskin I (1992) Role of salicylic acid in plants. Annu Rev Plant Physiol Plant Mol Biol 43: 439–463

Rasmusson AG, Geisler DA, Møller IM (2008) The multiplicity of dehydrogenases in the electron transport chain of plant mitochondria. Mitochondrion 8:47–60

Skulachev VP (1988) Membrane bioenergetics. Springer, Berlin

Skulachev VP (1994) Decrease in the intracellular concentration of O_2 as a special function of the cellular respiratory system. Biochem Mosc 59:1910–1912

Skulachev VP (1996) Role of uncoupled and non-coupled oxidation in maintenance of safely low levels of oxygen and its one-electron reductants. Quart Rev Biophys 29:169–202

Solomos T (1977) Cyanide-resistant respiration in higher plants. Annu Rev Plant Physiol Plant Mol Biol 28:279–297

Svensson AS, Rasmusson AG (2001) Light dependent gene expression for proteins in the respiratory chain of potato leaves. Plant J 28:73–82

Van den Bergen CW, Wagner AM, Krab K, Moore AL (1994) The relationship between electron flux and the redox poise of the quinone pool in plant mitochondria. Interplay between quinol-oxidizing and quinone-reducing pathways. Eur J Biochem 226:1071–1078

Vanlerberghe GC, McIntosh L (1996) Mitochondrial electron transport regulation of nuclear gene expression. Studies with the alternative oxidase gene of tobacco. Plant Physiol 111:58–595

Vanlerberghe GC, McIntosh L (1997) Alternative oxidase: from gene to function. Annu Rev Plant Physiol Plant Mol Biol 48:703–734

Wagner AM (1995) A role for active oxygen species as second messengers in the induction of alternative oxidase gene expression in *Petunia hybrida* cells. FEBS Lett 368:339–342

Wagner AM, Krab K (1995) The alternative respiration pathway in plants: role and regulation. Physiol Plant 95:318–325

Wagner AM, Moore AL (1997) Structure and function of the plant alternative oxidase: its putative role in the oxygen defence mechanism. Biosci Rep 17:319–333

Wagner AM, Wagner MJ (1997) Changes in mitochondrial respiratory chain components of petunia cells during culture in the presence of antimycin A. Plant Physiol 115:617–622

Antioxidative Systems and Stress Tolerance: Insight from Wild and Cultivated Tomato Species

Valentina Mittova, Micha Volokita, and Micha Guy

1 Introduction

Oxygen-evolving microorganisms emerged about 3.2–3.5 Gyr ago (Blankenship 2010). Oxygenic photosynthetic organisms are responsible for the majority of the present (21 %) atmospheric oxygen concentration (for reviews see Hohmann-Marriott and Blankenship 2011; Nelson 2011). The buildup of the oxygenic atmosphere was a gradual process that continued along the geo-history of the planet Earth and eventually (due to oxygen toxicity) forced organisms to either adapt to oxygen or to limit their living space to anaerobic niches. Reactive oxygen species (ROS) are partially reduced forms of atmospheric oxygen (O_2) and as such are unavoidable byproducts of the oxygenic atmosphere. During the reduction of oxygen to water a portion of this oxygen is transformed into ROS such as superoxide radical (O_2^-), hydrogen peroxide (H_2O_2), and hydroxyl radical ($OH^•$) (Halliwell and Gutteridge 1999; Apel and Hirt 2004). These unstable and highly reactive ROS molecules are toxic to major cell components, lipids, proteins, and DNA (Miller et al. 2010).

To minimize the deleterious ROS effects, plant cells are capable of suppressing the buildup of harmful intracellular ROS concentrations. This is achieved by elaboration of the antioxidative defense systems consisting of enzymatic and

V. Mittova (✉)
Voronezh State Medical Academy, Studencheskaya st. 10, Voronezh 394000, Russia
e-mail: vmittova@mail.ru

M. Volokita
The National Institute for Biotechnology in the Negev, Ben-Gurion University of the Negev, Beer-Sheva, Israel

M. Guy
The Blaustein Institutes for Desert Research, Ben-Gurion University of the Negev, Sede Boqer Campus, Beer-Sheva 84990, Israel

K.J. Gupta, A.U. Igamberdiev (eds.), *Reactive Oxygen and Nitrogen Species Signaling and Communication in Plants*, Signaling and Communication in Plants 23,
DOI 10.1007/978-3-319-10079-1_6

nonenzymatic ROS scavengers. These systems are composed of: (1) Enzymes that scavenge directly toxic ROS, such as superoxide dismutase (SOD), ascorbate peroxidase (APX), glutathione peroxidase (GPX), catalase (CAT), and guaicol-peroxidase (POD); (2) Enzymes, involved in the repair of ROS-mediated oxidative damage, such as phospholipid glutathione peroxidase (PHGPX) and glutathione S-transferase (GST); (3) Enzymes that maintain the antioxidants pool in their reduced states, such as monodehydroascorbate reductase (MDHAR), dehydroascorbate reductase (DHAR), and glutathione reductase (GR); (4) Hydrophilic metabolites (ascorbate and glutathione) and lipophylic metabolites (alpha-tocopherol and carotenoids), that function either as direct antioxidants or as substrates of antioxidative enzymes. Besides their protective role, modulation of the activities of the whole or part of the antioxidative system which might change ROS cellular concentrations could be perceived by the cells as signals leading toward regulation of plethora of processes, such as development, growth, and defense (Dat et al. 2000; Vranova et al. 2002; Mittler et al. 2004, 2010; Foyer and Noctor 2005).

Environmental impact is by far, the most important factor that limits plant growth and productivity. Out of the various environmental stresses, salinity and drought are the most serious (Boyer 1982). Stomatal closure is a common response to these stressors, resulting in decreased CO_2 fixation and increased photorespiration, while at the same time the photosynthesis-driven electron transport is active. These events lead to a stress-induced increase in ROS production (Halliwell and Gutteridge 1999; Alscher et al. 1997; Noctor and Foyer 1998; Mittler 2002; Miller et al. 2010). In plants, the energy transducing organelles, chloroplasts, and by a lesser degree the mitochondria are major ROS producers (Boveris 1984; Asada 1996). In addition, leaf peroxisomes and glyoxysomes are effective ROS producers (Corpas et al. 2001). Under stress conditions, ROS production increases in all cell organelles (Noctor and Foyer 1998; Mittler 2002; Miller et al. 2010).

2 The Tomato Model System

The tomato (*Lycopersicon esculentum*) crop cultivars are the domestication products of edible wild tomato starting with the pre-Columbian inhabitants of the South American Andes. As of today, tomatoes have a global distribution as a prime crop that is important in the fresh vegetable market as well as in the food processing industry (Food and Agriculture Organization of the United Nations, FAOSTAT database). The twelve wild tomato species which constitute the *Lycopersicon* clade are native to Western South America and the Galápagos Islands. These relatives of the cultivated tomato had thus evolved in diversified and harsh habitats that extend from the high Andes to the shores of the Pacific Ocean (Peralta et al. 2005). As such, wild tomatoes have attained morphological and physiological features that are adaptive responses to their native habitats. Accordingly, resistance to specific biotic and abiotic stresses of the various wild tomatoes have been found, for example, the resistance of *L. lycopersicum* var. *cerasiforme* to various fungal infections and to water logging, the tolerance of *L. cheesmaniae* and *L. pennellii* to salinity and the drought tolerance of *L. chilense* (for a review see Moyle 2008). These stress tolerance traits are important for the enhancement of the genetic diversity of the cultivated tomatoes and for breeding purposes (Menda et al. 2013). In their review, Hajjar and Hodgkin (2007) surveyed the use of wild relatives in crop improvement. They showed that in the 20 years prior to their survey, cultivars of 13 crops of major economic importance were released with contributions of beneficial traits from their wild relative species. The total number (55) of beneficial traits contributed to the tomato crop exceeded those introduced to the rice crop (12) by far. Recently, the genome sequence of domesticated tomato and a draft sequence of *Lycopersicon pimpinellifolium*, its closest wild relative, were released (Tomato Genome Consortium 2012) and it is expected that draft genomes of other wild tomato species will be released in the near future. These research tools may facilitate the exploitation of the wild tomatoes genetic potential for complex traits such as drought and salinity tolerances. The natural habitation of the wild tomato species *Lycoperscicon pennellii* (Lpa) spans from Northern Peru (Piura) to the northern Chile (Tarapaca) deserts and thrives in dry rocky hillsides and sandy areas from sea level up to 2,300 m (Peralta et al. 2005). As such, this species has attained stress tolerance traits, including drought tolerance (Kahn et al. 1993) and salt tolerance (Tal and Shannon 1983). In comparison, the cultivated tomato cultivars (that had been selected for vigorous growth and high yields) are moderately salt sensitive (for a review see Cuartero and Fernandez-Munoz 1999) and under high NaCl concentrations significant damage is inevitable (Dasgan et al. 2002).

In this chapter the major findings of our research concerning the involvement of the antioxidative systems in the acquisition of salt-tolerance in tomato are described. The model plants used for the study were the salt-sensitive cultivated tomato, *L. esculentum* Mill. cv. M82 (Lem) and its wild salt-tolerant relative *L. pennellii* (Lpa). This work was done mostly at the organelle level where the antioxidative systems of leaf and root cell organelles were isolated from control and salt-treated plants, and then were characterized and compared. For this purpose, isolated chloroplasts/plastids, mitochondria, and peroxisomes were used. The effects of salinity on the antioxidative systems of isolated organelles (mostly leaf organelles) had been studied before however this research was probably the first to systematically characterize the effects of salinity on the major leaf and roots organelles of salt-sensitive and related salt-tolerant tomato plants grown at the same time and under the same experimental conditions.

3 Identification of Soluble and Membrane-Bound SOD and Ascorbate–Glutathione Pathway Isozymes

A comparison of the inherent activities (in the absence of stress) of SOD and the ascorbate–glutathione cycle antioxidant isozymes, in the leaf and root cell organelles of the two species (Table 1, Mittova et al. 2000) indicate that (1) The activities

Table 1 Specific activities of antioxidant enzymes in leaf and root cell organelles of the cultivated (Lem) and the wild (Lpa) tomato species

Species	Organ	Organelle	SOD	APX	MDHAR	DHAR	GR	CAT
Lem	Leaves	Chloroplast	0.59 ± 0.04	0.85 ± 0.14	0.77 ± 0.08	0.04 ± 0.01	0.02 ± 0.001	n.d.
		Mitochondria	2.80 ± 0.04	1.12 ± 0.38	3.93 ± 0.12	0.44 ± 0.13	0.05 ± 0.01	0.01 ± 0.002
		Peroxisomes	3.47 ± 0.47	1.07 ± 0.18	2.43 ± 0.02	0.16 ± 0.02	0.09 ± 0.02	0.12 ± 0.04
		Crude extract	0.13 ± 0.01	0.28 ± 0.04	0.28 ± 0.03	0.02 ± 0.001	0.02 ± 0.001	0.02 ± 0.001
	Roots	Plastids	1.41 ± 0.50	2.15 ± 0.18	2.20 ± 0.26	0.52 ± 0.11	0.26 ± 0.001	n.d.
		Mitochondria	2.05 ± 0.21	1.91 ± 0.40	9.23 ± 1.17	0.47 ± 0.08	0.11 ± 0.005	0.003 ± 0.001
		Peroxisomes	2.78 ± 0.46	2.94 ± 0.35	5.67 ± 0.65	0.77 ± 0.21	0.08 ± 0.01	0.13 ± 0.004
		Crude extract	0.30 ± 0.04	0.21 ± 0.02	0.28 ± 0.04	0.02 ± 0.001	0.02 ± 0.005	0.03 ± 0.01
Lpa	Leaves	Chloroplast	1.13 ± 0.30	0.57 ± 0.16	0.41 ± 0.03	0.14 ± 0.05	0.02 ± 0.001	0.01 ± 0.001
		Mitochondria	10.90 ± 3.10	1.80 ± 0.48	6.74 ± 0.22	0.60 ± 0.05	0.07 ± 0.02	0.004 ± 0.001
		Peroxisomes	4.90 ± 0.11	2.43 ± 0.25	4.07 ± 0.31	0.31 ± 0.03	0.05 ± 0.01	0.23 ± 0.02
		Crude extract	0.42 ± 0.08	0.14 ± 0.01	0.31 ± 0.01	0.01 ± 0.001	0.03 ± 0.01	0.02 ± 0.006
	Roots	Plastids	3.15 ± 0.70	1.49 ± 0.14	1.54 ± 0.34	0.18 ± 0.04	0.05 ± 0.01	0.01 ± 0.001
		Mitochondria	11.10 ± 0.90	2.96 ± 0.49	4.82 ± 0.18	1.46 ± 0.40	0.03 ± 0.001	0.03 ± 0.01
		Peroxisomes	10.30 ± 0.30	1.46 ± 0.15	3.43 ± 0.57	0.35 ± 0.05	0.03 ± 0.003	0.46 ± 0.08
		Crude extract	0.53 ± 0.08	0.28 ± 0.08	0.23 ± 0.01	0.02 ± 0.001	0.01 ± 0.001	0.03 ± 0.004

SOD activity is expressed as U mg^{-1} protein, all other enzyme activities are expressed as μmol min^{-1} mg^{-1} protein. Values represent the means of three independent experiments ±SD of these five isozymes are present in the chloroplasts/plastids, mitochondria, and peroxisomes of the two species. (2) The specific activity of MDHAR was similar or higher than that of APX, while the specific activity of DHAR was one order of magnitude lower than that of MDHAR (Table 1). These results suggest that the capacity to regenerate ASC from MDHA by MDHAR was similar or higher than the capacity of APX to oxidize ASC to MDHA. On the other hand, the much lower activities of DHAR in all organelles suggest that nonenzymatic oxidation of MDHA to DHA does not pose a significant problem in tomato. The lower activities of GR in all organelles (Table 1) further support this suggestion. However, the relative importance of DHAR and GR in ASC regeneration might be species dependent, as was shown for the important role of DHAR in the regeneration of ASC in pea mitochondria and peroxisomes (Jiménez et al. 1997).

The matrix (i.e., soluble) and membrane-bound activities of SOD isozymes were determined in the studied organelles (Table 2). To minimize possible artifacts due to nonspecific binding of soluble proteins released from broken organelles during the isolation procedure (inter- and intraorganellar cross-contamination) the isolated membrane fractions were resuspended in the presence of the chaotropic salt KI (Briskin and Poole 1983; Mittova et al. 2004a, b). In general, KI-washing was effective in solubilizing antioxidative isozymes (from osmotically disrupted membranes) that were either loosely attached or associated with the membranes,

Table 2 Percentage of the activity of SOD types isozymes in soluble and membrane fractions of leaf and root cell organelles isolated from the cultivated tomato (Lem) and the wild species (Lpa)

Organ/ species	Organelle	CuZnSOD Soluble (OSR)	CuZnSOD Mem bound (KIR)	FeSOD Soluble (OSR)	FeSOD Mem bound (KIR)	MnSOD Soluble (OSR)	MnSOD Mem bound (KIR)
LEAF							
Lem	Chloroplasts	63	1	10	18	n.d.	n.d.
	Mitochondria	n.d.	n.d.	n.d.	n.d.	82	5
	Peroxisomes	10	n.d	21	n.d	6	61
Lpa	Chloroplasts	32	2	17	25	n.d.	n.d.
	Mitochondria	n.d.	n.d.	n.d.	n.d.	89	10
	Peroxisomes	28	n.d.	n.d.	n.d.	5	51
ROOT							
Lem	Plastids	71	19	n.d.	n.d.	n.d.	n.d.
	Mitochondria	n.d.	n.d.	n.d.	n.d.	83	3
	Peroxisomes	92	2	n.d.	n.d.	n.d.	3
Lpa	Plastids	28	41	n.d.	13	n.d.	n.d.
	Mitochondria	n.d.	n.d.	n.d.	n.d.	93	2
	Peroxisomes	n.d.	7	n.d.	n.d.	87	2

Values represent the percentage activity of a soluble and a membrane-bound SOD type isozyme, out of the total SOD activity, of a given organelle. OSR, Osmotic Shock Residual (activity in the soluble fraction was determined after osmotic shock); KIR, KI Residual (activity in the membrane fraction was determined after KI wash treatment) [For details see Mittova et al. (2000)]. Values represent mean of three independent experiments probably as part of functional complexes. Latency values revealed that, in some cases, a given KI-solubilized isozyme was, most probably, localized to the cytosolic-faced of the organelle membranes. For instance, about 30 % of GR activity was found in the pre KI-washed Lpa peroxisomal membrane (data not shown). Following a KI-wash, this membranal activity value was diminished to a mere amount of about 0.5 %. The latency of GR in Lpa root peroxisomes was 70 % (Table 3) indicating that about 30 % of its activity was accessible to the substrate added to intact peroxisomes. These findings support the notion that the KI-solubilized GR activity originated from GR attached to the cytosolic face of the peroxisomal membrane. It is worthwhile to explore whether this adsorption reflects a functional complex or a mere cytosolic GR contaminant. Nevertheless, the data obtained by the KI-washing procedure indicated that membrane-bound, per se, GR isozyme does not exist in contrast to membrane treatment.

SOD activity was resolved into its three class members, CuZnSOD, FeSOD, and MnSOD (Table 2). Of these, MnSOD was detected as the sole SOD activity in leaf and root mitochondria of both species. This activity was mainly localized in the mitochondrial matrix as a soluble isozyme. Matrix MnSOD was detected in mitochondria of spruce needles (Sehmer and Dizengremel 1998), pea leaf (Hernández et al. 1993), Vigna and Phaseolus leaves (Corpas et al. 1991), Scots pine (Streller et al. 1994), maize (Zhu and Scandalios 1993), and bread wheat (Sheoran et al. 2011). Oddly, in Lpa root peroxisomes, but not in those of Lem, a high activity (90 %) of soluble MnSOD isozyme was detected.

In leaf peroxisomes of the two species, significant activities (more than 50 %) of the total peroxisomal SOD activity were assigned to membrane-bound MnSOD isozyme. The presence of membranal MnSOD in peroxisomes of watermelon cotyledons was confirmed by immunoelectron microscopy (Rodríguez-Serrano et al. 2007). Interestingly, low MnSOD activity (about 5–10 %) was detected in the membrane fraction of Lem and Lpa leaf mitochondria (Table 2).

CuZnSOD isozyme activity was found in chloroplasts/plastids and peroxisomes of both tomato species (Table 2). The chloroplastic CuZnSOD activities were characterized as being soluble (Table 2). However, significant activity of CuZnSOD isozyme was found at the membrane fractions of the root plastids (41 % and 19 % of total SOD activity in Lpa and Lem plastids, respectively). Interestingly, low activity of membrane-bound CuZnSOD (7 % of the total SOD activity) was determined in Lpa root peroxisomes (Table 2).

FeSOD isozyme activity comprises a significant proportion (34 % and 52 %) of the total SOD activity in Lem and Lpa chloroplasts, respectively, and 26 % of Lpa root plastids (however, it could not be detected in Lem plastids). In these organelles, FeSOD activities were equally distributed between membrane-bound and soluble isozymes (Table 2). Although CuZnSODs and FeSODs were usually characterized as soluble isozymes, they were also found as membrane-bound isozymes (Ogawa et al. 1995; Navari-Izzo et al. 1998; Gómez et al. 2004).

Soluble and membrane-bound activities of APX, MDHAR and DHAR were found in the plastids, mitochondria, and peroxisomes of leaves and roots of both tomato species (Table 3). GR activity, in contrast, was characterized solely as soluble isozymes in all the three cell organelles (Table 3). In plants, APX,

Table 3 Latency and percentage of activity of the ascorbate-glutathione cycle in soluble and membrane fractions, of leaf and root cell organelles isolated from the cultivated tomato (Lem) and the wild species (Lpa)

		Leaves						Roots					
		Lem			Lpa			Lem			Lpa		
Activity	Organelle	Laten	Solu (OSR)	Mem (KIR)	Laten	Solu (OSR)	Mem (KIR)	Laten	Solu (OSR)	Mem (KIR)	Laten	Solu (OSR)	Mem (KIR)
APX	Choropl/Plast	59	46	54	65	38	44	27	38	55	84	50	48
	Mitochondria	6	30	36	85	66	31	44	80	18	46	77	22
	Peroxisomes	32	38	58	66	65	34	59	77	17	53	45	23
MDHAR	Choropl/Plast	90	65	32	63	27	65	42	66	26	62	42	40
	Mitochondria	48	22	63	60	18	67	76	49	11	57	48	45
	Peroxisomes	71	31	48	65	70	21	75	72	18	81	78	12
DHAR	Choropl/Plast	93	42	19	95	69	18	88	12	39	80	41	21
	Mitochondria	47	31	36	78	51	34	60	45	14	69	82	16
	Peroxisomes	59	78	4	45	93	7	62	67	2	69	65	3
GR	Choropl/Plast	87	87	5	87	82	5	94	88	7	96	99	0
	Mitochondria	91	85	6	90	75	0	87	64	4	79	80	0
	Peroxisomes	92	94	2	99	91	10	88	84	0	70	70	0

Values represent the means of three independent experiments. OSR, Osmotic Shock Residual (activity in the soluble fraction was determined after osmotic shock); KIR, KI Residual (activity in the membrane fraction was determined after KI wash treatment) [For details see Mittova et al. (2000)]

MDHAR and DHAR are encoded by small gene families. In tomato, the *SlApx* gene family, for instance, consists of genes that encode for soluble (stromal) and membrane-bound APX proteins (Najami et al. 2008). Similarly, genes that encode for soluble and membrane-bound MDHAR proteins were annotated (Gest et al. 2013).

4 Assignment of Suborganellar Localization of SOD and the Ascorbate–Glutathione Isozymes

Our next aim was to localize the soluble and membrane-bound isozymes of SOD and the ascorbate–glutathione pathway proteins to their in situ suborganellar localization, in order to better understand their function under nonstressed and stressed growth conditions.

In the light of the vast scope of the study (five enzyme in three organelles, two organs and two species that were grown under two growth conditions) the relatively simple methods of latency and solubilization assays were employed, while taking into consideration possible pitfalls that might arise from the fact that other methodologies, such as immune-detection or exposure to proteases were not used.

In the case of a simple organelle that is restricted by a single membrane (like in the case of peroxisomes) two components should be considered; the intraorganellar space and its surrounding membrane. Biological membranes are asymmetrical, thus a membrane–bound protein may possess a *cis*- or *trans*-membrane orientation, relative to the intraorganellar space. This imposes a further topological complexity that arises from the orientation of the catalytic site of that membrane-bound isozyme relative to the membrane face, and hence the availability of the substrate (s) to the catalytic site (Sze 1985). In organelles with complex intraorganellar structure (chloroplasts, plastids and mitochondria) that are enclosed by a double membrane system and have also intraorganellar sub-compartments, the topologies of soluble and membrane-bound isozymes are further complicated. In the chloroplast, for instance, depending on the membrane it is bound to, the catalytic site of a given membrane bound isozyme might face four soluble phases: the lumen, stroma, inter-envelope membrane space, and the cytosol. A given soluble isozymes might be localized to the stroma, the thylakoid lumen, and/or the inter-envelope membrane space. Latency experiments expose the cryptic activity of a given isozyme, these experiments are useful in cases where the isozyme and its substrate(s) are separated by an impermeable membrane (assuming that the membrane permeability to its substrate(s) is the rate-limiting step, i.e., contribution of transporter(s) to substrate availability is negligible). Accordingly, 100 % latency indicates that all the active sites are inaccessible and hence are localized at the *trans*-membrane face. Alternatively, low latency (when membrane permeabilization does not boost the activity) indicates that most of the active site resides at the *cis*-membrane face (i.e., the side facing the external medium). Following the isolation of intact chloroplasts/

plastids, mitochondria, and peroxisomes, latent activities of the ascorbate–glutathione cycle isozymes were determined (Table 3, and see Mittova et al. 2000).

5 Production and Detoxification of ROS in Tomato Organelles

5.1 Peroxisomes

5.1.1 ROS Production

Peroxisomes are ubiquitous cell organelles surrounded by a single membrane that are found in all eukaryotic cells. A common function of the various peroxisome types is their oxidative metabolism that produces H_2O_2 as a byproduct. Plant peroxisomes participate in variety of metabolic reactions important in plant growth and development, such as, photorespiration, fatty acid β-oxidation, the glyoxylate cycle, and biosynthesis of plant phytohormones (for reviews, see Kaur et al. 2009; Hu et al. 2012a). Some of these metabolic pathways (such as photorespiration and fatty acid β-oxidation) produce high levels of H_2O_2. This is most significant during photosynthesis when high rates of H_2O_2 production occur via the photorespiration process. Thus, peroxisomes of C_3 plants were estimated to be the major producers of H_2O_2 in the light, more than the two energy-transducing organelles (for review, see Foyer and Noctor 2003). In addition to H_2O_2, superoxide radicals are produced in the peroxisomal matrix by xanthine oxidase activity (Corpas et al. 2001; Miller et al. 2010) and on the cytosolic face of the peroxisomal membrane, where a small electron transport chain appears to be involved (Lopez-Huertas et al. 1999, for review see del Río et al. 2006).

5.1.2 ROS Detoxification

The presence of SOD and ascorbate-glutathione cycle activities in the peroxisomes is well documented (del Río et al. 2002, 2006). Catalase, a matrix enzyme, which is one of most abundant peroxisomal proteins has two drawbacks; (1) a very high *K*m for H_2O_2, and thus is an inefficient detoxifier under low H_2O_2 concentrations and (2) is photo-inactivated, especially under stress conditions (Volk and Feieraben 1989; Hertwig et al. 1992; Mhamdi et al. 2010). As peroxisomal matrix APX activity was not reported so far (except for tomatoes) it was presumed that peroxisomal matrix H_2O_2 is detoxified enzymatically by catalase and none-enzymatically by ascorbate, as suggested in the models of del Río et al. (2006) and Kaur et al. (2009). It should be noted that matrix MDHAR, DHAR and GR activities were determined in pea and Arabidopsis peroxisomes (del Río et al. 1998b and Kaur et al. 2009, respectively) and thus may participate in ascorbate regeneration in this compartment. In contrast, in the matrices of tomato leaf and root peroxisomes, a significant amount of activity of soluble peroxisomal APX, together with soluble activities of MDHAR, DHAR, and GR were determined (Mittova et al. 2000, 2003a, 2004a). However, as these reports come from the same laboratory, the identity of this matrix APX activity awaits further verification.

Peroxisomal membrane-bound APX and MDHAR proteins were reported in *Arabidopsis* (Lisenbee et al. 2005), pea (del Río et al. 1998a), and castor beans (Karyotou and Donaldson 2005) where their localization had been assigned by various methodologies. In tomato peroxisomes, the activities of three membrane-bound antioxidative isozymes (APX, MDHAR, and MnSOD) had been determined (Mittova et al. 2000). Membrane-bound MnSOD was localized to the external side (i.e., cytosolic face) of glyoxysomal membranes of watermelon cotyledons and castor bean endosperm (Sandalio and del Río 1988; Sandalio et al. 1997; Rodríguez-Serrano et al. 2007). Peroxisomal membrane-bound MnSOD activity was determined also in leaf peroxisomes of Lem and Lpa plants (Mittova et al. 2000), unfortunately, in these experiments the topology of this activity was overlooked. Peroxisomal localization of APX was further confirmed by immunoblot analysis (Mittova et al. unpublished data). These results suggest that the tomato 31 kD polypeptide is orthologous to the *Arabidopsis* peroxisomal APX3 protein. The topology of the peroxisomal membrane-bound APX protein was assigned to the cytosolic face of the membrane (Lisenbee et al. 2005; del Río et al. 2006; Kaur et al. 2009). In tomato, the most plausible interpretation of latency and fractionation experiments is that the catalytic sites of the membrane-bound APX and MDHAR isozymes face the cytosol in both species. This interpretation however, is in contrast to Karyotou and Donaldson (2005) who proposed that the active site of the membrane-bound APX and MDHAR isozymes face the peroxisomal matrix. Similarly, based on protease protection experiments with intact *Arabidopsis* peroxisomes, Lisenbee et al. (2005) concluded that the bulk of the MDHAR polypeptide faces the peroxisomal matrix. It should be noted that also in the case of an opposite orientation of the membrane-bound APX and MDHAR isozymes couple, so that their active sites face the peroxisomal matrix, as suggested by Karyotou and Donaldson (2005), their function as a barrier against ROS influx and efflux is not hampered.

5.1.3 The Peroxisomal Antioxidative Systems

The peroxisomes, as all cell organelles, are threatened by self-produced superoxide radicals and H_2O_2 and by H_2O_2 influx produced in the cytosol and by other organelles. Here we suggest, based on our results (Mittova et al. 2000, 2003a, 2004a) a model for peroxisomal antioxidative system that provides protection against internal and external agents of oxidative threats (Fig. 1). Accordingly, in tomato, matrix-produced superoxide radicals are detoxified to H_2O_2 by soluble SOD isozymes (mainly CuZnSOD but also FeSOD as described below), in turn, H_2O_2, is detoxified by matrix localized isozymes of catalase and APX. The matrix APX activity can compensate for loss of catalase activity that might occur under

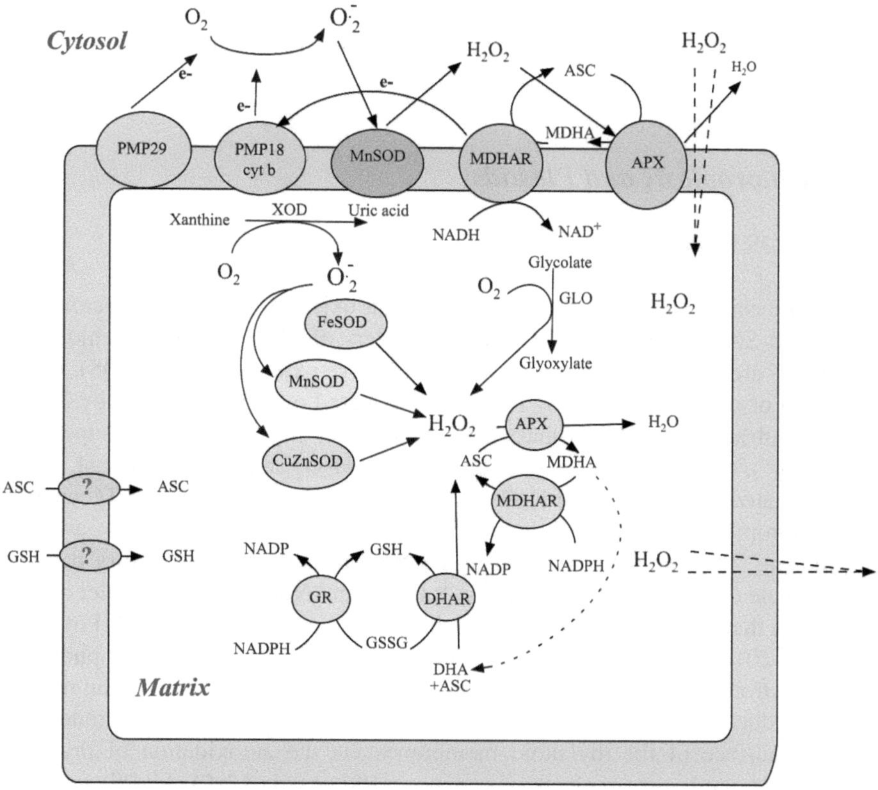

Fig. 1 A proposed model for the localization of ROS–scavenging system in peroxisomes of the cultivated (Lem) and the wild (Lpa) tomato species. *Solid line*-enzymatic reaction, *dotted line*-non-enzymatic reaction.PMP29, peroxisomal membrane polypeptide, MW 29 kDa; PMP 18, peroxisomal membrane polypeptide, MW 18 kDa. PMP29 and PMP 18 were depicted according to Del Río et al. (2006). Antioxidative isozymes are depicted as *dark circles* and *ovals*. *Ovals with a question mark* represent ASC and GSH transporters. Hydrogen peroxide fluxes thorough the membrane are depicted as *two dotted lines* connected to an *arrowhead*

adverse conditions such as salinity, cold, and high light (Volk and Feieraben 1989; Hertwig et al. 1992; Mhamdi et al. 2010). It should be noted that this model is different from those of del Río et al. (2006) and Kaur et al. (2009) for pea and *Arabidopsis* peroxisomes, respectively, where the activity of matrix APX is absent, and part of H_2O_2 detoxification is done non-enzymatically by ascorbate. Based on the data presented in Table 3, the proposed model suggests that the active sites of the membrane bound APX and MDHAR are facing toward the cytosol. Thus the function of this membrane-bound couple is to protect the cytosolic face of the peroxisomal membrane from externally produced H_2O_2 (Fang et al. 1987; Lopez-Huertas et al. 1997, 1999). In addition, this orientation can also deal with efflux of H_2O_2 from the peroxisomal matrix. Fast regeneration of ascorbate, consumed by the membrane-bound APX activity, is provided by membrane-bound MDHAR that might be in close proximity to the former isozyme (Karyotou and Donaldson 2005).

5.2 *Chloroplasts and Plastids*

5.2.1 ROS Production

Photosynthesis is an important source of cellular oxidants (Foyer and Noctor 2003; Foyer and Shigeoka 2010). The reaction centers of PSI and PSII in chloroplast thylakoids are the major generation sites of reactive oxygen species (ROS). Photoreduction of oxygen to H_2O_2 by PSI was discovered over 50 years ago by Mehler (1951). Subsequently, the primary reduced product was identified as superoxide anion (O_2^-), and its disproportionation produces H_2O_2 and O_2 (Asada et al. 1974). The estimated maximal rate of H_2O_2 formation by the photosynthetic ETC chain in the chloroplasts during photosynthesis in C_3 leaves is nearly as high (4 μmol m^{-2} s^{-1}) as that produced in the peroxisomal photorespiratory pathway by glycolate oxidation (10 μmol m^{-2} s^{-1}; Foyer and Noctor 2003). Singlet oxygen molecules that are produced at the PSII site are also considered as ROS (Foyer and Shigeoka 2010), however, they are out of the scope of this chapter. The photosynthetic electron transport process is an inherent source for superoxide anion radicals (O_2^-) production (Foyer and Shigeoka 2010). Superoxide radicals are produced at the stromal-face of the thylakoid membranes via the autoxidation of thylakoid membrane-bound primary electron acceptor in PSI (Asada 1999). All PSI acceptors have sufficiently negative redox potential to reduce O_2 to O_2^- (Foyer and Noctor 2000). The majority of O_2^- produced this way is mediated through ferredoxin (Fd) (Furbank and Badger 1983) by the Mehler reaction (Asada 1999). Superoxide radicals are produced also enzymatically by flavodehydrogenases, especially by Fd-NADP reductase and are disproportionated into H_2O_2 by either nonenzymatic reaction or enzymatically by SOD (Asada 1999).

Nonphotosynthetic plastids are derivatives of either proplastids or chloroplasts that are present in all the plant organs and tissues (Neuhaus and Emes 2000). Plastids are involved in the synthesis of starch, fatty acids, and nitrogen assimilation (Neuhaus and Emes 2000) as well as in synthesis of aromatic amino acids, isopentenyl diphosphate, and heme (Gould et al. 2008). Plastid respiratory processes, conversion of reducing power contained in carbohydrates into phosphorylating power, were documented in chloroplasts and suggested to occur also in nonphotosynthetic plastids (for a review see Peltier and Cournac 2002). Little is known whether such diverse metabolic activity is also associated with ROS production and their detoxification. However, the findings that following exposure to methyl-viologen, intact tomato root plastids accumulate H_2O_2 suggest that these organelles are capable of ROS production.

5.2.2 ROS Detoxification

The Mehler reaction is the first step in a sequence of processes that eventually lead to the formation of a *trans*-thylakoid proton gradient and to ATP synthesis. These processes, termed "the water–water cycle" (Asada 1999), include activities of APX, SOD, and the ASC–GSH cycle (Foyer and Shigeoka 2010). In chloroplasts, two types of SOD isozymes, CuZnSOD and FeSOD, exist (Asada 1999; Pilon et al. 2011). A significant portion of the apparent soluble CuZnSOD protein is attached to the stroma-facing thylakoid membranes in spinach chloroplasts (Ogawa et al. 1995). Similarly, in *Arabidopsis*, chloroplastic FeSOD proteins (that lack *trans*-membrane domains) were deduced to be tightly attached to the stromal face of the thylakoids membranes (Myouga et al. 2008). In tomato chloroplasts of both species, activities of both soluble and membrane-bound FeSOD isozymes were detected (Table 2). However, while stromal and membrane-bound isozymes of CuZnSOD were found in Lpa, in Lem only the stromal isozyme could be detected (Table 2). In Lpa and Lem root plastids, both soluble and membrane-bound isozymes of CuZnSOD were detected, however, the membrane-bound FeSOD activity was found only Lpa plastids (Table 2). The localization of FeSOD on the tomato thylakoid membrane (Table 2) probably contributes to the prompt O_2^- scavenging, just after its ejection from the PSI complex, thus providing a flux of H_2O_2 for nearby thylakoid membrane-bound tAPX. A physiological role for FeSOD, during the early chloroplast development, as a protector of its genome from ROS damage was suggested (Myouga et al. 2008).

Chloroplastic APX is classified into thylakoid-bound (tAPX) and stroma-localized forms. tAPX binds in the vicinity of PSI (Miyake and Asada 1992) and its anchor is a putative *trans*-membrane domain located at the C-terminal, hydrophobic 50 residues. In tomato genome two separate genes *SlApx6* and *SlApx7*, which are orthologous to the *Arabidopsis tApx* and *sApx* genes, respectively, had been identified and characterized (Najami et al. 2008).

In tomato chloroplasts and plastids, both soluble and membrane-bound MDHAR isozymes were detected (Table 3 and Mittova et al. 2000). To the best of our knowledge, this is the first time that a membrane-bound MDHAR is described although report for MDHAR attached to the thylakoid membranes of spinach chloroplasts are cited as unpublished results in Miyake et al. (1998). When the tAPX reduces H_2O_2, MDHA radicals are produced as the primary oxidation product. While MDA is spontaneously converted to ASC and dehydroascorbate (DHA), it is also rapidly reduced to ASC by the action of a NAD(P)H-dependent MDA reductase. In tomato chloroplasts, ASC is regenerated from MDHA via the activity of thylakoid-bound NAD(P)H-MDHAR (Hossain et al. 1984; Mittova et al. 2002a) and possibly also by its photoreduction in the thylakoids via reduced ferredoxin (Miyake and Asada 1994). In addition to MDA, MDA reductase is able to reduce phenoxyl radicals, such as quercetin radicals to their parent phenols (Sakihama et al. 2000) suggesting that this enzyme participates also in the regeneration of other antioxidants. Stromal activity of MDHAR was also demonstrated in chloroplasts of spinach (Hossain et al. 1984), cucumber (Hossain and Asada 1985), and tomato (Mittova et al. 2000).

The *Arabidopsis* DHAR gene family consists of five genes, where only the deduced polypeptide sequence for AtDHAR3 contain an N-terminal polypeptide extension with 42-residue transit peptide, which target it to the chloroplast (Dixon et al. 2002). In tomato chloroplasts and plastids, most of DHAR activity was determined as soluble stromal), however, a significant portion of its total chloroplastic and root plastidic activity (19 %) is membrane bound (Table 3).

GR is localized mainly in the chloroplast stroma (Edwards et al. 1990). In tomato chloroplasts and plastids of both species, GR activity was found mostly in the soluble fractions (Table 3).

High ASC concentrations (above 20 mM) are typical in plant chloroplasts (Foyer and Lelandais 1996; Smirnoff and Wheeler 2000). Differential distributions of ascorbate contents in the various cell organelles and compartments were determined immunocytochemically (Zechmann et al. 2011). Accordingly, in cells of *Arabidopsis* grown under high light, the highest ascorbate concentrations were found in the peroxisomes and cytoplasm (about 22 mM), while lower concentrations were measured in the chloroplasts and mitochondria. These findings suggest that a dynamic cellular ascorbate steady state exists and maintained by the fluxes of ascorbate synthesis and consumption and also by the activity of organellar ascorbate transport systems that import ascorbate from the cytoplasm (Gest et al. 2013). It is expected that under high light stress, the demand for chloroplastic ascorbate is the highest.

In addition to the above ROS detoxification pathways, an additional detoxification system that consists of glutathione peroxidase (GPX), peroxiredoxin, and thioredoxin, is operative in the chloroplast's stroma (for a review see Dietz 2011). However, the catalytic rates of plant GPXs and their affinities for H_2O_2 are rather low compared with those of APXs (Dietz et al. 2006). It is possible that this system compensates for the lability of the chloroplastic APX, which is inactivated in the absence of its electron donor, ascorbate (Asada 1999). Indeed, experiments with *Arabidopsis* double t*apx* s*apx* mutants indicated that tAPX and sAPX proteins are functionally redundant. Accumulation of chloroplast 2-Cys peroxiredoxin (PRX) was observed in the double t*apx* s*apx* mutant acclimatized to high light stress (Kangasjärvi et al. 2008).

5.2.3 Subcompartmentation of SOD and the Ascorbate–Glutatione Isozymes in Tomato Chloroplasts

In chloroplasts, a membrane-bound (thylakoidal) and stromal ROS scavenging systems operate (Asada 1999). The membranal system is composed of thylakoid-bound CuZnSOD, APX (tAPX), and loosely bound ferredoxin, which is directly involved in the regeneration of ascorbate from monodehydroascorbate. This scavenging system is responsible for efficient detoxification of superoxide radicals and H_2O_2 near the stromal face of thylakoids, where they might damage the enzymes

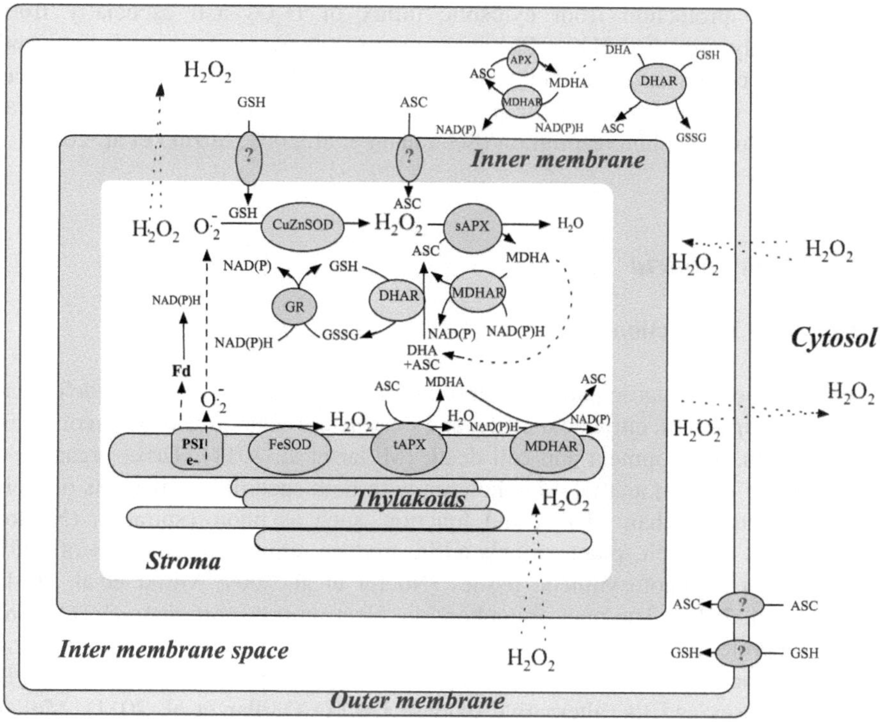

Fig. 2 A proposed model for the localization of ROS–scavenging system in chloroplasts of the cultivated (Lem) and the wild (Lpa) tomato species. *Solid line*-enzymatic reaction, *dotted line*-non-enzymatic reaction. Fd, ferredoxin; PSI, photosystem I; tAPX- thylakoid-bound APX. Antioxidative isozymes are depicted as *dark circles* and *ovals*. *Ovals with a question mark* represent ASC and GSH transporters. Hydrogen peroxide fluxes thorough membranes are depicted as *two dotted lines* connected to an *arrowhead*

participating in the thylakoidal photosynthetic system, as well as the nearby chloroplast nucleoids (Myouga et al. 2008). The second ROS scavenging system is composed of soluble CuZnSOD, APX, MDHAR, DHAR, and GR isozymes. This stromal system acts as a second line of defense under conditions when the thylakoidal system fails to scavenge O_2^- and H_2O_2. In the stroma, O_2^- is disproportionated by CuZnSOD concomitantly with the detoxification of H_2O_2 by the stromal APX (sAPX).

In the model depicted here for tomato (Fig. 2) a third scavenging system, localized to the intermembrane space of the chloroplast envelope is assigned. This suggestion is based on analyses and interpretations, as outlined above, of data obtained from latency and solubilization experiments. Ascorbate regeneration in this subcompartment may take place by the cytosolic or stromal ascorbate regeneration pathways and by the activity of ASC transport system(s) (Foyer and Lelandais 1996). The presence of APX and MDHAR isozymes in this compartment can provide protection from cytosolic influx of H_2O_2 and especially from peroxisomally produced H_2O_2. Physical proximity between the peroxisomes and chloroplasts occurs in vivo and may be important for the function of peroxisomes as intermediate organelles that participate in metabolic pathways common to several organelles, such as photorespiration (Schumann et al. 2007; Suzuki et al. 2012).

5.3 *Mitochondria*

5.3.1 ROS Production

Plant mitochondria participate in central biological processes; these include oxidative phosphorylation, citric acid cycle, and fatty acid oxidation and are involved in plant growth, development and cell death (Millar et al. 2011). These organelles, however, differ from animal mitochondria with both specific components of their electron transport chain (ETC) and function, such as photorespiration (Noctor et al. 2007). As such, plant mitochondria produce significant amounts of ROS especially in nonphotosynthetic tissues (Noctor et al. 2007; Millar et al. 2011; Lázaro et al. 2013). The plant mitochondrial electron transport chain (ETC) contains four protein complexes (I to IV) that interact through a small lipid ubiquinone and cytochrome *c* and also plant-specific non-proton-pumping NADH and NADPH dehydrogenases and the alternative oxidase (AOX) (Millar et al. 2011). Møller (2001) concluded that plant mitochondria produce both H_2O_2 and O_2^- at rates equal or greater than mammalian mitochondria, both absolutely and as a percentage of the rate of electron transport. When oxygen interacts with complex IV and alternative oxidase (AOX), two terminal oxidases in plant mitochondria, four electrons are transferred and water is produced (Moore and Siedow 1991). Several mitochondrial ROS production sites had been characterized. Oxygen might interact with the reduced forms of other electron-transport chain components, e.g., flavins and ubiquinone, however, respiratory complexes I and III are considered to be major sites of mitochondrial O_2^- production (Møller 2001; Millar et al. 2011). It is also possible that internal rotenone-insensitive NAD(P)H dehydrogenases are also a site of ROS production (Rasmusson et al. 2004).

5.3.2 ROS Detoxification

Both leaf and root mitochondria of Lpa and Lem showed high specific activity of SOD isozymes (Mittova et al. 2003a, 2004a, respectively). In these mitochondria, MnSOD was the sole SOD type, based on its tolerance to H_2O_2 and KCN inhibition (Mittova et al. 2003a). These findings are in agreement with observations indicating that mitochondria contain the MnSOD type as the sole SOD type (for a review see Møller 2001). The presence of a complete, ascorbate–glutathione cycle in plant mitochondria was shown in: pea leaf (Jiménez et al. 1997), leaf and roots tomato

(Mittova et al. 2000), bean nodules (Iturbe-Ormaetxe et al. 2001), and in dark grown cell cultures of *Arabidopsis* (Chew et al. 2003).

5.3.3 Subcompartmentation of SOD and the Ascorbate–Glutathione Isozymes in Tomato Mitochondria

The data obtained from the solubilization and latency assays of Lem leaf mitochondria were used to model the intermitochondrial localization of soluble and membrane-bound activities of MnSOD and the glutathione-ascorbate isozymes (Fig. 3). MnSOD activity was detected mostly in the soluble fraction; however, small fractions (5 % and 10 %) were characterized as membrane-bound isozymes in Lem and Lpa leaf mitochondria, respectively (Table 2). In the model, membrane-bound MnSOD is localized to the matrix-facing side ofthe inner membrane, in proximity to the production site of O_2^- that is at the matrix site of respiratory

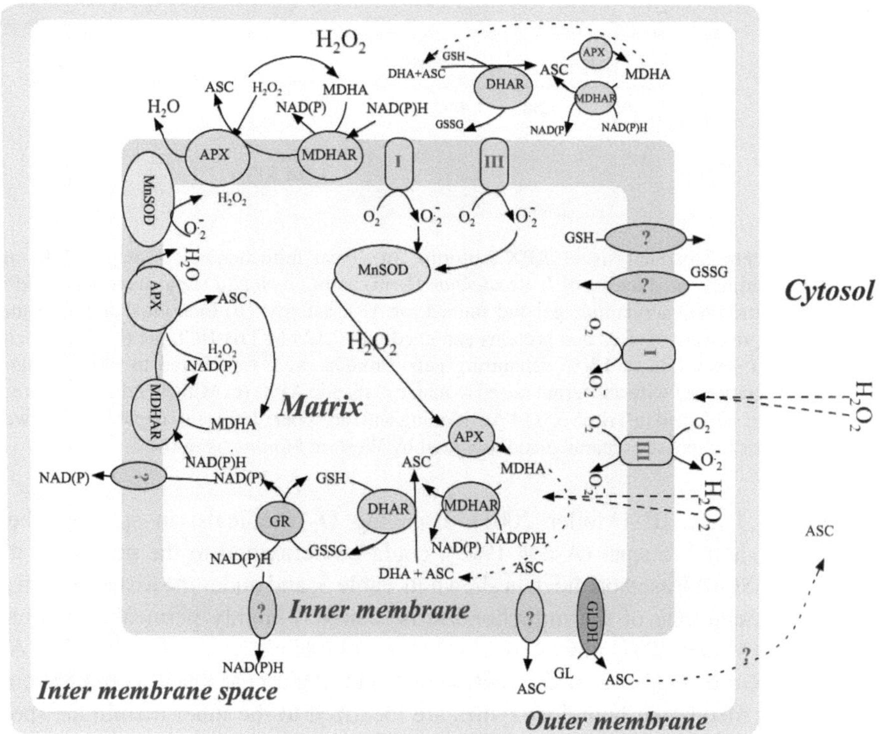

Fig. 3 A proposed model for the localization of ROS–scavenging system in mitochondria of the cultivated (Lem) and the wild (Lpa) tomato species. *Solid line*-enzymatic reaction, *dotted line*-non-enzymatic reaction. I, III, Respiratory Complexes; I, Complex I, NADH Dehydrogenase; III, Complex III, Ubiquinol-Cytochrome bc1 reductase. Antioxidative isozymes are depicted as *dark circles* and *ovals*. *Ovals with a question mark* represent ASC and GSH transporters. Hydrogen peroxide fluxes thorough membranes are depicted as *two dotted lines* connected to an *arrowhead*

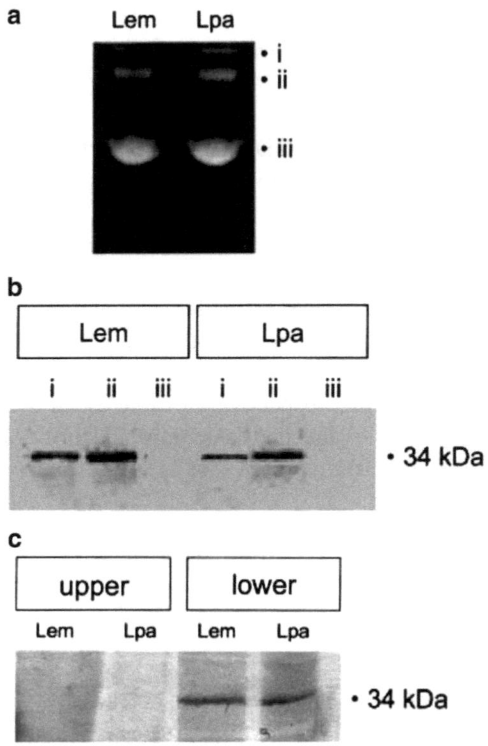

Fig. 4 Western blot analysis of APX isoforms. (**a**) Total mitochondrial proteins (100 mg) extracted from untreated leaves of *L. esculentum* (Lem) and *L. pennellii* (Lpa) were separated in a non-denaturing polyacrylamide gel and stained for APX activity. (**b**) Individual activity bands (marked i, ii, iii) were excised and proteins extracted into 0.125 M Tris-HCl, pH 6.8, 0.1% (w/v) SDS prior to separation in 12 % denaturing gels. Proteins were transferred to nitro-cellulose membrane and probed with antiserum raised to maize cytosolic APX. (**c**) Mitochondrial membrane fractions were subjected to Triton X-114 partitioning and the upper, detergent- depleted and lower, detergent-enriched phases separated and analyzed by Western blotting as in (**b**)

complexes I and III (Møller 2001). Escaping O_2^- radicals, in spite of their extremely short life span (Asada 1999), could be detoxified in the matrix by the soluble MnSOD. Based on the data shown in Table 3, and under the assumption that the outer membrane of the mitochondria is relatively highly permeable to small molecules (Møller 2001), it was concluded that soluble isozymes of APX, MDHAR and DHAR are localized in the matrix and intermembrane space compartments while their membrane-bound isozymes are localized to the inner membrane (possibly at both sides). However, localization of membrane-bound isozymes also to the outer membrane cannot be excluded. It should be emphasized that similar latency values, as reported for tomato mitochondria, were reported for the mitochondrial APX, MDHAR, DHAR and GR, in intact leaf pea mitochondria (Edwards et al. 1990; Jiménez et al. 1997) and in *Arabidopsis* mitochondria (Chew et al. 2003). In these studies, lower values were obtained for APX and MDHAR and higher values for GR, which supports the common notion of the latter as a soluble matrix isozyme. The main features of the proposed model are that the mitochondria can cope both with ROS produced inside and with those that are produced outside, i.e., in the cytosol, chloroplasts and peroxisomes. Ascorbate (ASC) is synthesized from galactono-γ-lactone (GL) by the inner membrane-bound flavoprotein, galactono-γ-lactone dehydrogenase (GLDH; black oval symbol on Fig. 3) (Siendones et al. 1999). ASC is exported to the cytosol and imported into the matrix by as yet uncharacterized mechanisms (marked as a question mark on Fig. 3). To complete the water–water cycle in the intermembrane compartment due to absence of GR activity there, separate putative transporters for GSH and its oxidized form GSSG as well as transporters for NADH and NAD^+ are depicted.

Cumulative data support the notion that mitochondria have at least two APX isozymes encoded by different nuclear genes. In tomato, the mitochondrial membrane-bound APX isozyme was characterized (using antibodies specific to the cytosolic and the peroxisomal APXs) as peroxisomal-like APX protein (Fig. 4). An APX protein was isolated from mitochondria of potato tubers is similar to that of the stromal APX (sAPX) (De Leonardis et al. 2000). In line with the latter work, Chew et al. (2003) found that, in *Arabidopsis*, one of the six APX gene products (sAPX) and one of the five MDHAR gene products are dual targeted into mitochondria and chloroplasts.

6 Species and Tissue Variations in the Intraorganellar Distribution of Antioxidative Systems

In general, the antioxidative systems of the two tomato species are highly similar; nevertheless, some differences have been determined. These include different ratios of the soluble to the membrane-bound activities of a given isozyme and different SOD types. These differences were found between "homologous organelles" of the two species ("inter") and between leaf and root cell organelles of the same species ("intra"). Such inherent differences might reflect variations in oxidative load that is exerted on the organelles when they are engaged, for instance, in photosynthetic or nonphotosynthetic metabolism in the same plant. The cultivated tomato Lem had been selected for enhanced growth and productivity, while the wild species was evolutionary selected to cope with the harsh environment of the Ands. Taken together, it is likely that the metabolic fluxes, and their resulted ROS production, in a given Lem and Lpa organelle, would differ.

Peroxisomes: The most notable difference between root and leaf peroxisomes (of both species) is that in the former most (>90 %) of SOD activity was characterized as soluble, while in leaf peroxisomes membrane-bound SOD activity comprised considerable (51–61 %) activity of total peroxisomal SOD activity (Table 2). Similarly, in root peroxisomes most of APX, MDHAR, and DHAR isozyme activities were soluble. These results may suggest that under nonstressed growth the main oxidative threats to root peroxisomes are the matrix produced ROS. Beside the organ-dependent variations discussed above, species-dependent variations were identified. Thus, while the matrix CuZnSOD isozyme comprised most of SOD activity in Lem root peroxisomes (92 %) in those of Lpa this activity could not be detected, and most of SOD activity comprised of matrix MnSOD (>90 %) (Table 2). CuZnSOD and MnSOD isozymes can be distinguished by sensitivity to H_2O_2 (Kanematsu and Asada 1994). While the activity of CuZnSOD is inhibited by H_2O_2, the MnSOD isozyme is not affected. Hence, it is possible that the inherent MnSOD activity found in Lpa root peroxisomes render these organelles to cope better with the buildup of ROS. FeSOD isozyme activity was present only in Lem leaf peroxisomes where only the activity of the soluble isozyme was measured (Table 2). The presence of CuZnSOD and FeSOD isozymes was demonstrated in peroxisomes of watermelon cotyledons (Bueno et al. 1995) and of carnation petals (Droillard and Paulin 1990). However, the function of FeSOD in peroxisomal ROS detoxification is not resolved yet.

Higher percentages of membrane-bound isozymes of APX and MDHAR (as compared to their soluble activities) were found in Lem leaf peroxisomes, these values are about double as compared to those of Lpa leaf peroxisomes (Table 3). These findings could be ascribed to the role of the peroxisomal membrane-bound antioxidative system in the protection of the organelle from influx of cytosolic H_2O_2. Cytosolic H_2O_2 is produced by various cellular sources, of which the contribution of the chloroplasts is most significant. It is highly likely that in the cultivated tomato, that was selected for high photosynthetic rates, even in the absence of stressors, the threat from photosynthetically produced H_2O_2 is higher than in Lpa.

Chloroplasts and plastids: In leaf chloroplasts of both species, the membrane-bound SOD activity is exclusively comprised of the FeSOD type, while the CuZnSOD type represents the major membrane-bound SOD activity in root plastids (Table 2). However, the processes that lead to the different membrane-bound SOD types in the leaf chloroplasts and root plastids are not clear yet. Interestingly, considerable activities of soluble and membrane-bound FeSOD isozymes were measured in Lpa root plastids but not in those of Lem.

Mitochondria: Root mitochondria of both species, are characterized by higher activity of the matrix antioxidative system as compared to the membrane bound activity, in contrast, leaf mitochondria are characterized by higher activity of the membrane-bound systems (Tables 2 and 3). These observations suggest that under nonstressed growth conditions, the main oxidative threats to root mitochondria are the matrix produced ROS, while in the leaves, where photosynthesis takes place, the influx of H_2O_2 from the chloroplasts and peroxisomes might exert a serious threat to the mitochondria.

7 Salt-Induced Oxidative Stress and Its Alleviation in the Wild Tomato Species

7.1 Opposite Responses to Salinity of the Antioxidative Systems of Lpa and Lem Cell Organelles

Various abiotic stresses, including salinity, induce oxidative-stress (Smirnoff 1993; Levine 1999). Lem and Lpa plants exposed to salinity exhibited differential salt-induced oxidative stress response. Accordingly, only in Lem a gradual increase in leaf and root lipid peroxidation was observed (Shalata and Tal 1998; Shalata et al. 2001, respectively). Thus, it was of interest to verify whether leaf and root cell organelles isolated from salt-treated Lem and Lpa plants are differentially affected by salinity. This question was addressed by: examination of changes in the activities of selected antioxidative isozymes both at the intact organelle (Fig. 6) and at the suborganellar (soluble and membrane-bound isozyme) levels (Figs. 7 and 8); determination of the contents of the soluble antioxidants ascorbate, glutathione, and their oxidized forms (Fig. 9) and the contents of H_2O_2 and MDA as indicators of oxidative damage (Fig. 5). In general, a mirror image pattern was obtained for the responses of the antioxidative systems of organelles isolated from salt-stressed Lem and Lpa plants (Figs. 5, 6, 7, 8, and 9). These findings indicate a positive correlation between stress-induced upregulation of antioxidative isozymes, improved antioxidant redox potential, and alleviation of oxidative stress in Lpa organelles. In contrast, in Lem organelles, stress-induced downregulation of antioxidative isozymes was in correlation with a moderate decrease in the ASC and GSH redox potential and increased oxidative damage (Figs. 5, 6, 7, 8, and 9).

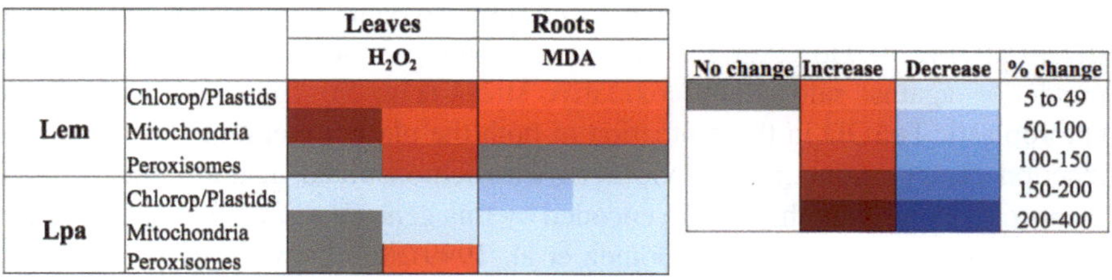

Fig. 5 Salt stress-induced changes of H_2O_2 and MDA contents in leaf and root cell organelles isolated from the cultivated tomato (Lem) and the wild species (Lpa). Changes in contents are *color coded*, relative to the values of control treatments

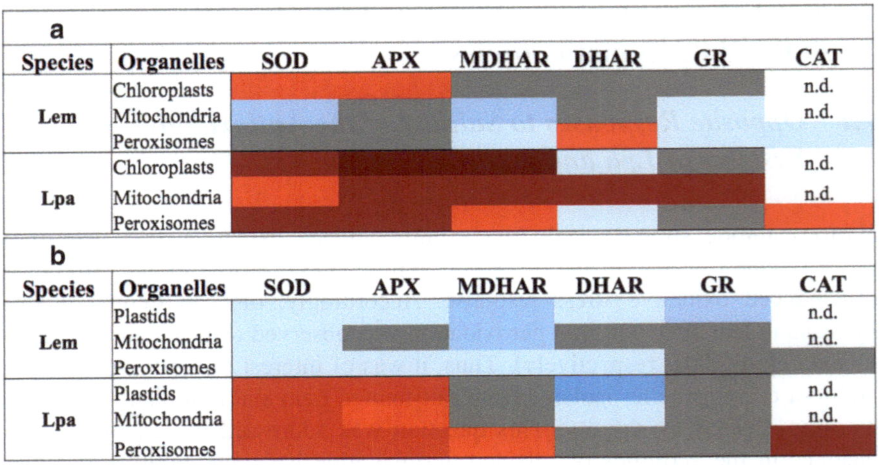

Fig. 6 Salt stress-induced changes in the activities of selected antioxidative isozymes in leaf and root cell organelles isolated from the cultivated tomato (Lem) and the wild species (Lpa). Changes in isozyme activities are *color coded*, relative to the values of control treatment

7.2 *Salt Stress-Induced Changes in the Antioxidative Systems of Lem and Lpa Organelles*

7.2.1 Chloroplasts

In salt-stressed Lpa chloroplasts, stromal and membrane-bound CuZnSOD and FeSOD isozymes were upregulated. This defensive response represent also a shift towards a higher activities of the membrane-bound isozymes (both absolutely and in percentage), especially that of the membrane-bound FeSOD isozyme (Fig. 8). In Lem chloroplasts, only the stromal CuZnSOD isozyme activity was upregulated (by 40 %) (Fig. 8). However, this response was not sufficient to provide an adequate protection against salt-induced oxidative stress (Fig. 5). The importance of the chloroplastic FeSOD in the protection of both the plasma membrane and the PSII reaction center against methyl-viologen was demonstrated in transgenic tobacco plants overexpressing the protein encoded by this gene (Van Camp et al. 1996) and in a salt-tolerant pea cultivar (Gómez et al. 1999). In chloroplasts isolated from control Lpa plants, the activity of the membrane-bound APX isozymes was much higher than that of the stromal (by 250 %). However, under salinity a shift toward equal activities of both isozymes was found, as salinity preferentially increased the stromal activity, by about 400 % (Fig. 7). The importance of the stromal APX in the detoxification of H_2O_2 molecules that escaped from the thylakoid-associated Mehler-peroxidase cycle, and in the detoxification of H_2O_2 molecules that enter the chloroplast from the cytosol was suggested by Foyer and Noctor (2000). In Lem chloroplasts, a mixed response of the APX isozymes to salinity was observed. Accordingly, a small increase (by 22 %) in the activity of the stromal APX isozyme

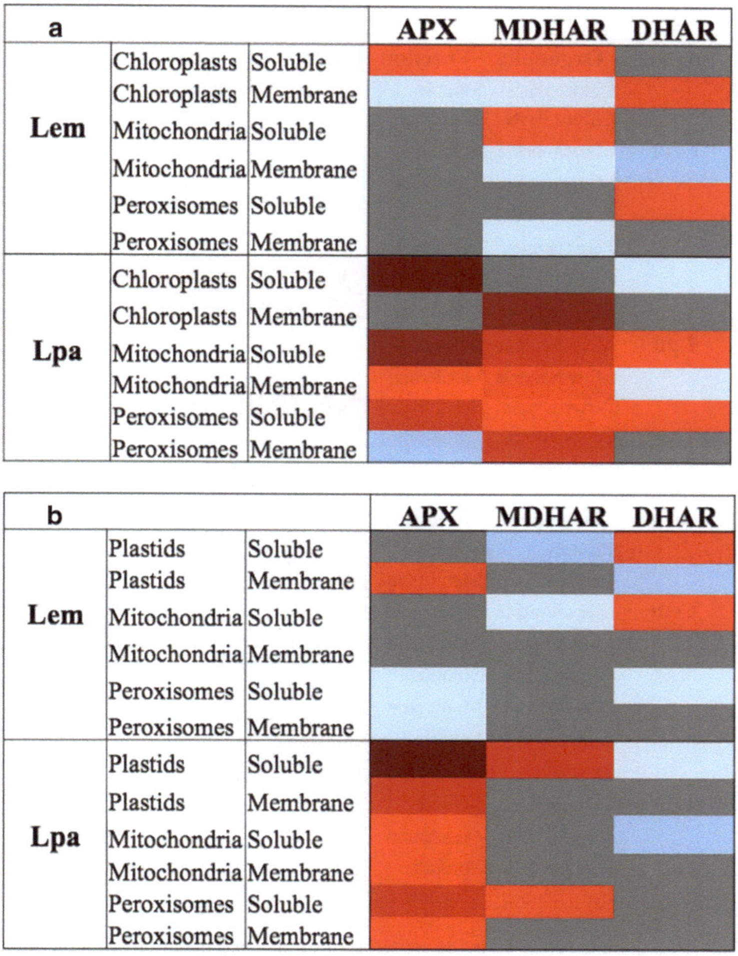

Fig. 7 Salt stress-induced changes in the activities of soluble and membrane-bound APX, MDHAR and DHARisozymes in leaf and root cell organelles isolated from the cultivated tomato (Lem) and the wild species (Lpa). Changes in isozyme activities *are color coded*, relative to the values of control treatment. (**a**) leaf organelles; (**b**) root organelles

and a double decrease (by 45 %) in that of the membrane-bound isozyme were determined; these changes were apparently not adequate to cope with the salt-induced increase in H_2O_2 and MDA contents in these organelles (Fig. 5). Taken together, these results suggest the following scenario. Under salinity, the increased rate of $O_2^{\bullet-}$ production occurs at the stromal side of thylakoids. To avoid its adverse effects, a membrane-bound FeSOD is successfully upregulated in Lpa (but not in Lem). Superoxide anion radicals that managed to escape this membrane-bound activity are then detoxified by the upregulated stromal CuZnSOD (in both, Lpa and Lem). At the same time, upregulated levels of the stromal APX isozyme (mainly that of Lpa) are recruited to cope with increased levels of stromal

a

Species	Organelle	Fraction	CuZnSOD	FeSOD	MnSOD
Lem	Chloroplasts	Soluble			n.d.
	Chloroplasts	Membrane	n.d.		n.d.
	Mitochondria	Soluble	n.d.	n.d.	
	Mitochondria	Membrane	n.d.	n.d.	
	Peroxisomes	Soluble			
	Peroxisomes	Membrane	n.d.	n.d.	
Lpa	Chloroplasts	Soluble			n.d.
	Chloroplasts	Membrane			n.d.
	Mitochondria	Soluble	n.d.	n.d.	
	Mitochondria	Membrane	n.d.	n.d.	
	Peroxisomes	Soluble		n.d.	
	Peroxisomes	Membrane	n.d.	n.d.	

b

Species	Organelle	Fraction	CuZnSOD	FeSOD	MnSOD
Lem	Plastids	Soluble		n.d.	n.d.
	Plastids	Membrane		n.d.	n.d.
	Mitochondria	Soluble	n.d.	n.d.	
	Mitochondria	Membrane	n.d.	n.d.	
	Peroxisomes	Soluble		n.d.	n.d.
	Peroxisomes	Membrane		n.d.	n.d.
Lpa	Plastids	Soluble		n.d.	n.d.
	Plastids	Membrane			n.d.
	Mitochondria	Soluble	n.d.	n.d.	
	Mitochondria	Membrane	n.d.	n.d.	
	Peroxisomes	Soluble	n.d.	n.d.	
	Peroxisomes	Membrane		n.d.	

Fig. 8 Salt stress-induced changes in the activities of soluble and membrane-bound SOD-types in leaf and root cell organelles isolated from the cultivated tomato (Lem) and the wild species (Lpa). Changes in isozyme activities are *color coded*, relative to the values of control treatment. (**a**) leaf organelles; (**b**) root organelles

H_2O_2. The increased H_2O_2 content may originate from several sources such as: residual H_2O_2 that escapes the thylakoid-bound APX isozyme; stromal production, and possibly also from an increased influx of peroxisomal H_2O_2 that is produced at a higher rate under stress.

The additional H_2O_2 detoxification activity is achieved by guaicol peroxidases (PODs). The increased activities of POD isozymes, found in chloroplasts of salt-stressed Lpa plants (Mittova et al. 2002a), may also contribute to their better ability to scavenge H_2O_2 under these conditions. Kieselbach et al. (1998) showed that peroxidase activity was mainly localized in the thylakoid lumen and Casano et al. (2000) provided evidence that the thylakoid peroxidase can scavenge H_2O_2 by oxidizing plastoquinol. However, the fact that in tomato chloroplasts, APX

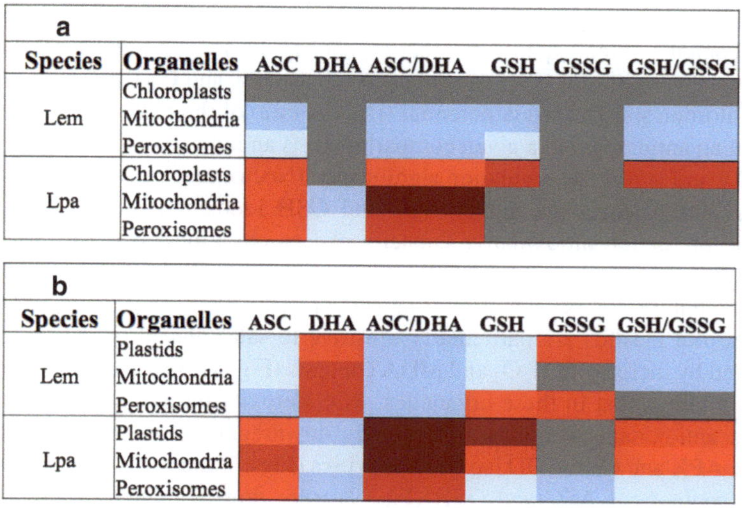

Fig. 9 Salt stress-induced changes in the contents of ASC, GSH and their oxidized forms in leaf and root cell organelles isolated from the cultivated tomato (Lem) and the wild species (Lpa). Changes in contents are *color coded*, relative to the values of control treatments. (**a**) leaf organelles; (**b**) root organelles

activity was two orders of magnitude higher than that of the nonspecific peroxidase (Mittova et al. 2002a) reaffirms the major role of APX in chloroplastic H_2O_2 scavenging as suggested by Asada (1996). It is possible, though, that POD isozymes have a role in H_2O_2 scavenging in the thylakoid lumen as suggested by Kieselbach et al. (1998). Increased POD activity was demonstrated also in chloroplasts of *Cucumis sativus* L. under salt stress (Shu et al. 2013).

The differential effect of salinity on Lem and Lpa chloroplastic contents of the major soluble antioxidants ASC and GSH and their oxidized forms were found (Fig. 9). In Lpa, ASC and GSH contents increased in response to salinity, while those of their oxidized forms were not affected (Fig. 9). As a result, the increased ASC and GSH redox potentials are established in these organelles (Fig. 9). These findings can be ascribed to salt-induced increase in the rates of ASC and GSH synthesis and possibly also to their increased influx into the chloroplasts via specific transporters, located in the chloroplast envelope (Foyer and Lelandais 1996). The need for enhanced regeneration of ASC and GSH under stress is also supported by the findings that the activities of chloroplastic GST, PHGPX, and γ-ECS increased under salt stress (Mittova et al. 2002a). In contrast, the contents of ASC, GSH, and their respective oxidized forms in salt-stressed Lem chloroplasts remained as those in control chloroplasts. Considering the observation that Lem chloroplasticγ-ECS activity decreased under salinity (Mittova et al. 2003b) it is conceivable that GSH synthesis rate decreased too. Indeed, in Lem leaves and more pronounced in Lem root homogenates, a decrease in the total level of GSH was detected (Mittova et al. 2003b). Such a decrease was also observed in Lem leaf mitochondria and peroxisomes (Fig. 9). It is possible that under salt-stress conditions priority is given to maintain the chloroplastic GSH redox potential by replenishing their GSH (and ASC) contents at the expense of other organelles. In line with this scenario, under salinity the activities of GSH consumers, such as GPX and GST did not changed in Lem chloroplasts. The redox potential (i.e., the ratio of the reduced to the oxidized form of an antioxidant in a given compartment) is an indicative for its antioxidative capacity and is of gene regulation significance (Noctor and Foyer 1998; Foyer and Noctor 2005) hence, the higher ASC and GSH redox potentials under salinity indicate the better antioxidative capacity of Lpa chloroplasts. It was also shown that the level and redox state of GSH might serve as markers of plant responses to environmental stresses such as chilling (Foyer et al. 1997; Tausz et al. 2004).

Lem root plastids are particularly sensitive to salt-induced oxidative stress, as indicated by increase in H_2O_2 and MDA contents (Fig. 5). This situation is reflected by the findings that in these organelles, ROS detoxifying isozymes activities and soluble antioxidants contents were mostly downregulated by salinity. Thus, the soluble APX and CuZnSOD isozyme activities (Figs. 7 and 8, respectively), as well as the contents of ASC and GSH (Fig. 9) were downregulated. These findings, together with the observations that the contents of the oxidized forms, DHA and GSSG, were not affected by salinity resulted in the decreased redox potentials of ascorbate and glutathione in Lem root plastids. It should be noted however, that the inherent APX isozyme activities(i.e., that of none-stressed plants) of Lpa chloroplasts are lower (by 33 %) than those of Lem's (Table 1), while its H_2O_2 and MDA contents were about twofold higher than that of Lem's, suggesting that in the absence of external stress Lpa cells exhibit moderate oxidative stress.

7.2.2 Mitochondria

Similarly to chloroplasts, a salt-dependent oxidative stress was evident in Lem leaf and root mitochondria, as indicated by their increased levels of H_2O_2 and lipid peroxidation (Fig. 5 and Mittova et al. 2003a, 2004a). As was shown in pea leaf mitochondria (Hernández et al. 1993), the increase of H_2O_2 in tomato may result from a salt-induced increase of production of O_2^- which is considered as the main precursor of mitochondrial H_2O_2 production (Møller 2001; Vanlerberghe 2013). For root mitochondria similar mechanism was demonstrated by Dixit et al. (2002) who showed that treatment of pea plants with hexavalent chromium increased the generation of O_2^- radicals by root mitochondria. In both leaf and root Lem mitochondria, increased H_2O_2 content was accompanied by decreased activity of soluble MnSOD (H_2O_2 producer), while the activities of APX and GPX (H_2O_2 detoxifiers) remained as in the control (Mittova et al. 2003a, 2004a). The fact that in contrast to Lem, mitochondria of salt-treated Lpa did not exhibit salt-induced oxidative stress at the time of leaf harvest, as indicated by the unchanged H_2O_2 content and the decrease of lipid peroxidation, can be explained by the salt-induced increased activity of soluble MnSOD (Fig. 8) responsible for O_2^- detoxification in the mitochondrial matrix (Mittova et al. 2003a). The salt-induced increase in leaf and root mitochondrial APX observed in Lpa is reminiscent of that reported in salt-stressed pea plants (Gómez et al. 1999). Our results suggest that the presence of an inducible APX, either in the matrix or facing it (Mittova et al. 2003a), is important in defense against the build-up of H_2O_2 produced by the action of MnSOD. This type of APX is clearly present in the salt-tolerant Lpa but is minimal in the salt-sensitive Lem. The salt-induced increase in activity of APX in Lpa leaf mitochondria correlated with an increase of soluble MDHAR and DHAR activities (Fig. 6 and Mittova et al. 2003a), indicating that ASC regeneration takes place in Lpa mitochondrial matrix. Hamilton and Heckathorn (2001) demonstrated that during salt stress Complex I of maize mitochondria is protected against ROS by ascorbate–glutathione cycle and that the tight control of ascorbate and glutathione levels are necessary for ascorbate-glutathione cycle to be effective in this protection. Interestingly, in Lpa root mitochondria, the increased ASC and GSH levels (Fig. 9) were not accompanied with increased activities of enzymes involved in their regeneration (MDHAR, DHAR and GR) (Fig. 6; Mittova et al. 2004a). This can be explained by the import of ASC and GSH into the root mitochondria.

In both Lpa leaf and root mitochondria, salt stress increased activity of POD isozymes (Mittova et al. 2003a, 2004a), which indicates that POD(s) is also involved in controlling mitochondrial H_2O_2 level. Prasad et al. (1995) demonstrated increased POD activity in leaf mitochondria of chilling-acclimated maize seedlings and suggested that these enzymes have a role in the control of mitochondrial H_2O_2 level. Increased APX and POD activities were demonstrated in mitochondria isolated from lead-treated pea roots (Malecka et al. 2001). In contrast to Lpa leaf mitochondria (Mittova et al. 2003a), GPX activity was unaffected by salinity in root mitochondria of Lpa (Mittova et al. 2004a), indicating that APX and POD are the main H_2O_2 scavengers in root organelles.

7.2.3 Peroxisomes

Leaf Peroxisomes

As in mitochondria, salt-induced oxidative stress was indicated in peroxisomes of Lem by increased lipid peroxidation (Fig. 5) and decreased levels of reduced ascorbate and glutathione (Fig. 9). This salt-induced peroxisomal oxidative stress probably resulted from an increased rate of ROS generation by the photorespiratory peroxisomal glycolate oxidase under conditions in which salt-induced stomata closure occurs (Smirnoff 1993). It is conceivable that salinity also increased the rate of ROS production in Lpa peroxisomes. The latter organelles, however, could cope better with this stress as indicated by the minimal decrease in their lipid peroxidation (Fig. 5), which correlated with increased activities of SOD, CAT, APX, and MDHAR (Fig. 9). Peroxisomes of higher plants contain multiple forms of SOD (reviewed in Corpas et al. 2001). MnSOD activity is localized mainly as associated with the membrane of tomato leaf peroxisomes (Table 2) and peroxisomes of pea (del Río et al. 1998a, b). CuZnSOD activity is localized in the peroxisomal matrix in watermelon (Bueno et al. 1995) and tomato (Table 2). It is possible that the matrix CuZnSOD functions as the main detoxifier of superoxide radicals produced by xanthine oxidase activity (Sandalio and del Río 1988), whereas MnSOD scavenges O_2^- generated by membranal NAD(P)H-dependent O_2^- production (Lopez-Huertas et al. 1999).

FeSOD activity is fourfold higher in Lpa chloroplasts in comparison with those of Lem (Mittova et al. 2002b). The detection of Fe-SOD activity might conceivably be the result of non-specific adsorption of chloroplastic FeSOD to the peroxisomal membrane. This is unlikely however, since FeSOD activity could be detected only in Lem peroxisomes (Fig. 8) where it was exclusively confined to the soluble (matrix) fraction (Table 2). These results suggest that FeSOD is a genuine Lem peroxisomal protein. The presence of FeSOD activity in peroxisomes of carnation petals was reported by Droillard and Paulin (1990) however, its intraorganellar localization was not determined. FeSOD has been shown to be less sensitive than CuZnSOD to inhibition by H_2O_2 (Kanematsu and Asada 1994). It is possible that the presence of FeSOD in the matrix of Lem peroxisomes enables these organelles to dismutate superoxide radicals under conditions in which the increased H_2O_2 level suppresses the activity of the H_2O_2-sensitive CuZnSOD. This may explain, in part, why 3-amino-1,2,4-triazole(3-AT)-induced chlorosis was less pronounced in leaf discs excised from control Lem plants as compared with those of Lpa (Fig. 10). Salinity did not alter H_2O_2 level in the peroxisomes of either species (Fig. 6), although the activities of APX and CAT, which are both involved in H_2O_2 metabolism, were affected in the two species to different extent (Fig. 6). This can be explained by the presence of porins in the peroxisomal membranes (Reumann et al. 1997) which facilitate the efflux of H_2O_2 from these organelles.

The salt-induced increase in peroxisomal Lpa APX activity (Fig. 6) seems to be responsible for the higher tolerance toward 3-AT-induced oxidative stress in leaf discs. Salt- stressed Lem plants, which failed to upregulate their peroxisomal APX (Fig. 6), were rendered more sensitive to 3-AT-induced oxidative stress (Fig. 10), which is in agreement with the findings of Wang et al. (1999) in transgenic tobacco. Overexpression of the gene encoding peroxisomal APX3 in these transgenic tobacco plants increased resistance to 3-AT-imposed oxidative stress. These transgenic plants, however, did not exhibit cross-tolerance toward paraquat-imposed oxidative stress, which indicates that APX3 overexpression confers tolerance mainly against peroxisomal-originated oxidative stress.

Root Peroxisomes

In both tomato species, SOD (Table 2) and the ascorbate–glutathione cycle isozymes (Mittova et al. 2004a) were localized to the matrix of root peroxisomes. These results are in contrast to the situation with other leaf and root organelles where also membrane-bound isozyme activities have been reported (Jiménez et al. 1997; Mittova et al. 2002a, b, 2003a). It is therefore likely that metabolic processes responsible for ROS production in root peroxisomes are localized mostly in their matrix. In root peroxisomes of salt-stressed Lem plants, the activities of SOD and APX decreased and that of CAT remained unchanged (Fig. 6). In agreement, in these organelles H_2O_2 content did not change (Fig. 5). Similar responses to salinity were exhibited by Lem leaf peroxisomes (Fig. 5), although in these organelles lipid peroxidation increased, whereas in the root peroxisomes it

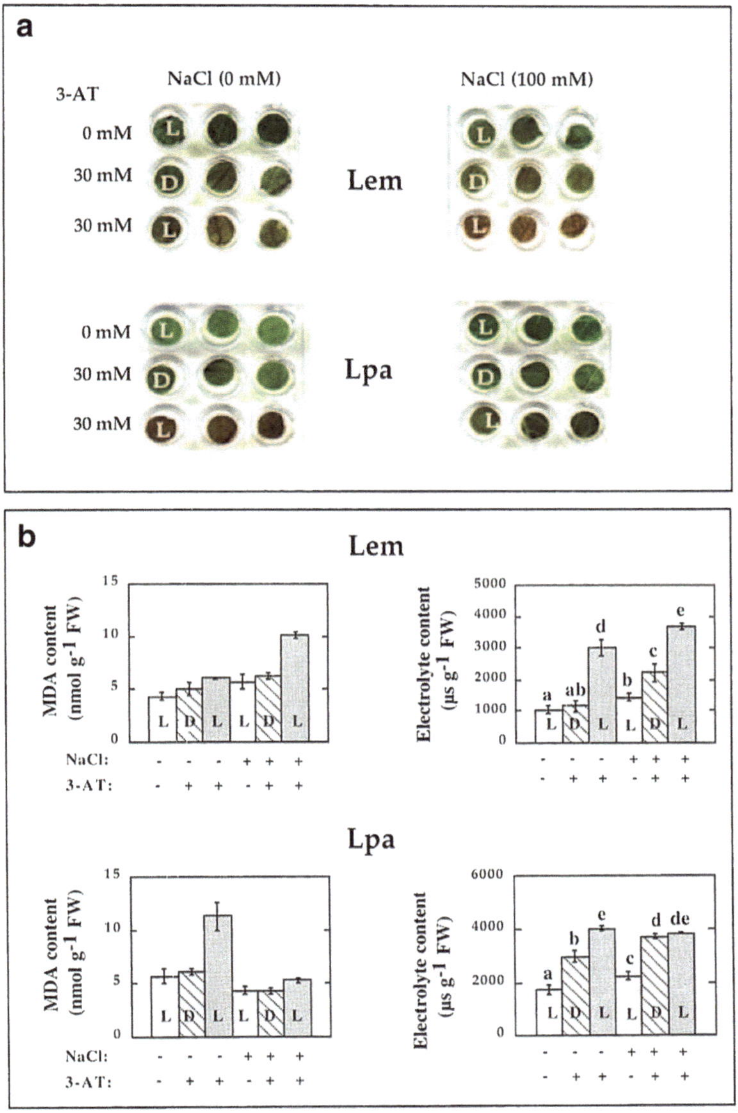

Fig. 10 3-AT-induced oxidative stress in leaf discs excised from plants of the cultivated tomato (Lem) and the wild species (Lpa) grown for 14 d in either control or salinized (100 mM NaCl) media. The oxidative stress effects were expressed as (**a**) chlorophyll loss (leaf discs images) and (**b**) lipid peroxidation (MDA content) and electrolyte leakage. Leaf discs were incubated for 4 h in the light (L) or in darkness (D) in the absence or presence of 30 mM 3-AT. Additional details as in Fig. 11.

remained unchanged. By contrast to Lem, decreased H_2O_2 and lipid peroxidation levels were found in root peroxisomes of salt-treated Lpa plants (Fig. 5). These responses to salinity resulted, at least partially, from differential increased activities of APX and CAT over that of SOD, similarly to leaf peroxisomes of salt-treated Lpa

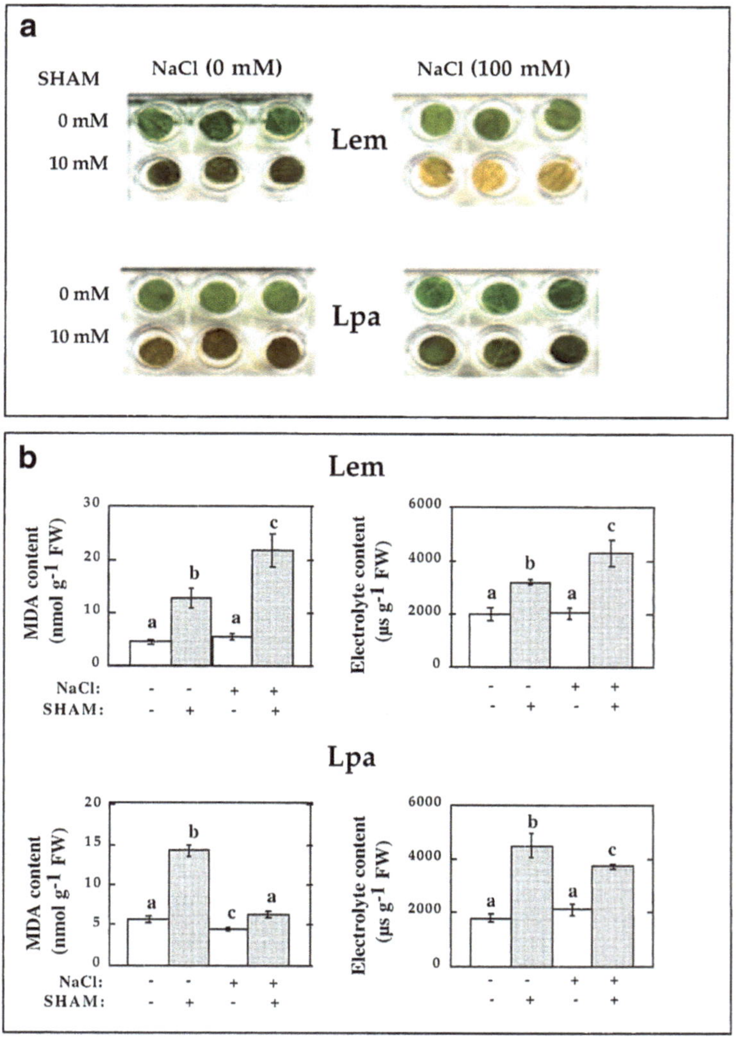

Fig. 11 SHAM-induced oxidative stress in leaf discs excised from plants of the cultivated tomato (Lem) and the wild species (Lpa) grown for 14 d in either control or salinized (100 mM NaCl) media. The oxidative stress effects were expressed as (**a**) chlorophyll loss (leaf discs images) and (**b**) lipid peroxidation (MDA content) and electrolyte leakage. Leaf discs were incubated for 4 h in the light in the absence or presence of 10 mM SHAM. The experiment was repeated twice, values represent the means ± SE of three replicates. Data was analyzed by four-way ANOVA. *Bars* with different letters are significantly different according to LSD test ($P = 0.05$)

plants (Fig. 6). It is possible that the decreased MDHAR activity in peroxisomes of salt-stressed Lem plants (Fig. 6) is responsible for the lower capacity of these organelles to regenerate ASC under these conditions, as reflected by their increased

DHA content (Fig. 9). By contrast, MDHAR activity increased in peroxisomes of salt-treated Lpa plants (Fig. 6) concomitant with their decreased DHA content (Fig. 9). Similarly, the increased MDHAR activity was found in leaf peroxisomes of salt-treated Lpa plants while in those of Lem it slightly decreased (Fig. 6). Interestingly, the activities of matrix DHAR and membrane-bound MDHAR in Lem root peroxisomes were not affected by salinity, while in Lpa that of the matrix MDHAR increased (Fig. 7). Taken together, these results suggest that regeneration of ASC in peroxisomes of salt-treated Lpa plants occurs in the matrix via MDHAR activity. Different SOD types were detected in the root peroxisomes of the two species: in Lem, SOD activity was comprised of matrix CuZnSOD (mainly the matrix isozyme) while that of Lpa was comprised of both MnSOD (mainly the matrix isozyme) and CuZnSOD (exclusively the membrane-bound isozyme) (Table 2). It is suggested that the matrix SOD functions as the main detoxifier of O_2^- produced by xanthine oxidase activity (Sandalio and del Río 1988), while the membrane-bound SOD scavenges O_2^- generated by the membranal NAD(P)H-dependent O_2^- production site as described by Lopez-Huertas et al. (1999). The finding that matrix MnSOD activity increased in root peroxisomes of salt- treated Lpa plants (Fig. 8), suggests that under salinity the matrix production of O_2^- exceeds that of the membrane. Like in the case of leaf peroxisomes, it is possible that the inherent MnSOD activity observed only in Lpa peroxisomes enables these organelles to cope better with a high H_2O_2 content (Fig. 5). However, under salinity conditions excessive H_2O_2 produced by the matrix MnSOD is efficiently scavenged by the matrix APX and CAT, as these activities were differentially increased in these organelles (Fig. 6).

7.3 *Salt-Induced Cross-tolerance Response Is Different in Lpa an Lem Plants*

Growth of Lem plants in the presence of NaCl resulted, during the first stage, in a significant growth decline that is consequently manifested as overt salt-stress damages, such as yellowing of leaves, necrotic spots, and eventually death of the plants. When grown under the same conditions, Lpa plants do not exhibit theses long-term salt-induced damages. It should be noted that Lpa plants are not inherently stress tolerant but rather acquire tolerance during a gradual exposure to stress. For instance, leaf disks that were excised from Lpa plants grown in non-salinized medium and then were exposed to high dose of paraquat and subsequently exhibited severe paraquat-induced oxidative stress (as indicated by chlorophyll bleaching) similarly to the response of leaf disks excised from control and salt-stressed Lem plants Mittova et al. 2002a).

Cross-tolerance is a phenomenon in which the exposure of plants to a particular moderate stress induces resistance to other stresses. This phenomenon was shown in numerous cases like: salt-stress that induces cold hardiness in potato and spinach seedlings, mechanical-stress that increases chilling tolerance in tomato, and heat-stress that protects against heavy metal toxicity (for a review see Sabehat et al. 1998). Two experiments have brought evidence for the contribution of the upregulated antioxidative system(s) in the acquisition of cross-stress-tolerance by Lpa. In these experiments, leaf disks that were excised from salt-stressed and control Lem plants and from control Lpa plants and then were exposed to the mitochondrial AOX inhibitor salicylhydroxamic acid(SHAM) or to the peroxisomal catalase inhibitor (3-AT), exhibited oxidative damage, as indicated by enhanced chlorophyll loss, lipid peroxidation, and electrolyte leakage (Figs. 11 and 10, respectively). In contrast leaf disks that were excised from salt-treated Lpa plants were tolerant to the oxidative stress imposed by these stressors. It should be noted that in these experiments, exposure of the leaf disks to SHAM and 3-AT was done 14 days after the exposure of the plants to NaCl, a time period in which the antioxidative system of Lpa had been already fully upregulated.

7.4 *Coordinate Induction of Antioxidant Enzymes and Glutathione Biosynthesis Is Correlated with Salt-Tolerance in Tomato*

Evidence for the important role of GSH in the tolerance toward slat-induced oxidative stress in Lpa was found by the use of buthionine sulfoximine (BSO), a specific inhibitor of γ-glutamylcysteine synthetase (γ-ECS) activity, a key enzyme in GSH biosynthesis (Noctor et al. 2002). In these experiments, salinity differentially increased GSH content and the GSH/GSSG ratios in Lpa leaves and roots, but not in those of Lem (Fig. 9). In agreement, salt-induced upregulation of γ-ECS protein (Fig. 12) and of the activities of GST and GPX (enzymes that modulate GSH content) but not of GR, were determined in Lpa leaves and roots but not in those of Lem (Mittova et al. 2003b). In both tomato species, exposure of the plants to BSO (5 days before salinization) significantly decreased the leaf and root GSH contents, however, under these conditions, BSO also abolished Lpa's capacity to tolerate salt-induced oxidative stress as indicated by increased hydrogen peroxide and lipid peroxidation levels in BSO treated plants (Fig. 13). Similar experiments showed that decrease in GSH content, by exposure to BSO, resulted with decreased Cd-tolerance of *Arabidopsis* (Wójcik and Tukiendorf 2011) and decreased chilling tolerance of mung bean seedlings (Yu et al. 2002).

These data demonstrate that the ability of Lpa plants to withstand a potentially damaging salt-induced oxidative stress is dependent, at least in part; upon the capacity for de novo GSH synthesis and that the observed salt-induced increase in the level of GSH can play a crucial role in this protection. Moreover, the coordinated salt-induced upregulation of γ-ECS, GST and GPX, but not of GR, found only in Lpa suggests that the two species differ in the mechanism employed for salt-induced oxidative stress tolerance.

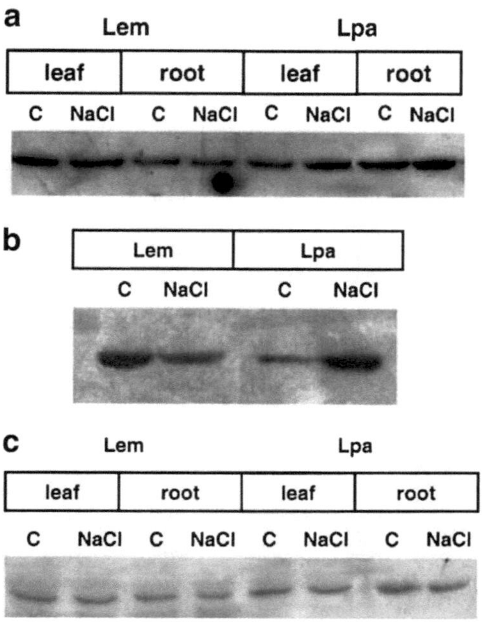

Fig. 12 Western blot analysis of γ-ECS and GR in tomato leaf and root extracts. Crude extracts and isolated chloroplasts were prepared from the cultivated tomato Lem and wild salt-tolerant species Lpa grown for 14 days in control (−) or salinized (+) media. Crude extracts (20 μg protein) were loaded onto 10 % (w/v) denaturing acrylamide gels, transferred to PVDF membranes and probed with antibodies raised to either maize γ-ECS (**a**) or maize GR (**c**). (**b**) Chloroplastic proteins (30μg) were loaded onto 10 % (w/v) denaturing acrylamide gels, transferred to PVDF membranes and probed with antibodies raised against maize γ-ECS

Concluding Remarks

In general, our findings demonstrate that inherent activities of the isozymes of the ascorbate–glutathione cycle enzymes are present in tomato leaf and root cell organelles (Table 1) and that these activities are segregated between the soluble and membrane-bound fractions of these organelles (Table 3). However, varied inherent antioxidative activities were determined in the same organelle of the two species or in homologous leaf and root organelles of the same species (Table 1). Similarly, different activity ratios were calculated for the antioxidative activity of a given isozyme in the organellar soluble and membrane-bound fractions (Table 3). It is suggested that these variations reflect differential tissue and species-dependent expression of these isozymes. The absence of FeSOD activity in Lpa peroxisomes and Lem root plastids (Fig. 8) can be explained as the result of tissue-specific silencing or downregulation of the *FeSOD* gene products that are targeted to these organelles.

(continued)

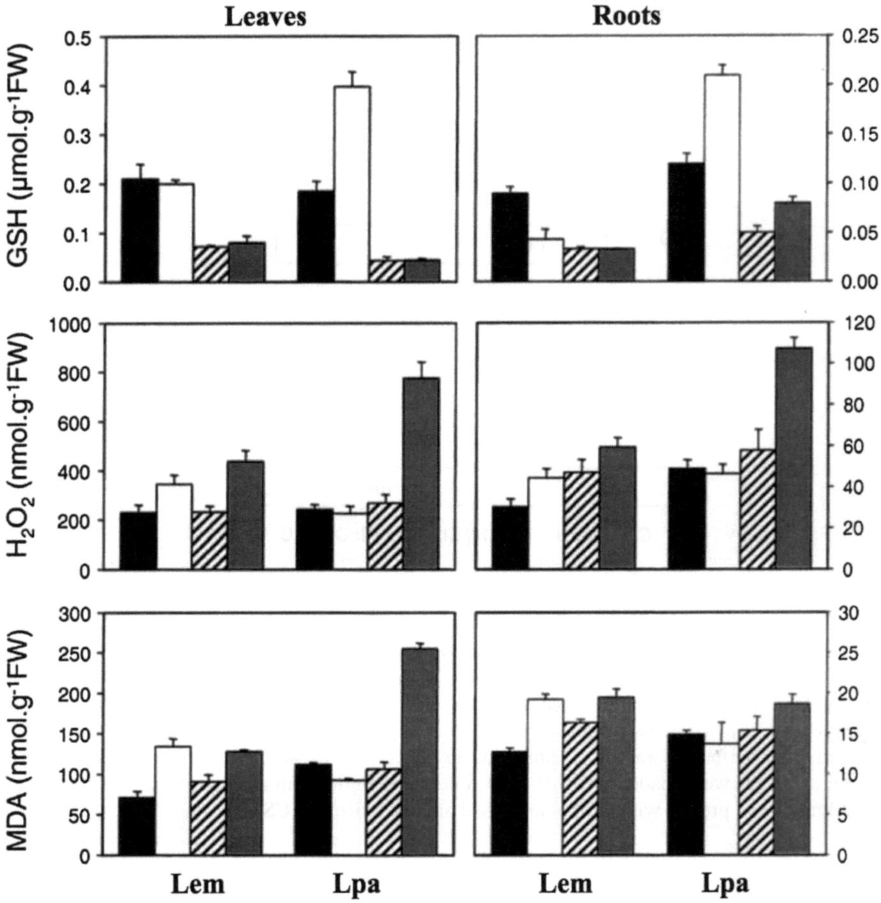

Fig. 13 The effect of BSO on GSH, MDA and H_2O_2 contents in leaves and roots of the cultivated tomato (Lem) and the wild species (Lpa) in the absence and presence of salt. *Black bars*: control; *open bars*: 150 mM NaCl; *hatched bars*: 1 mM BSO; *gray bars*: 1 mM BSO, 150 mM NaCl. Plants were pretreated with 1 mM BSO from the two leafs stage. Salt treatment started at the stage of four true leaves, 5 days after addition of BSO by successively increasing the NaCl concentration by 50 mM per day over a period of 3 days, to a final concentration of 150 mM with samples harvested after 5 days of treatment. Note the different scales on the *y* axes. Values represent the means of three independent experiments ±S.D.

Opposite responses to salinity by the two species were not restricted to the SOD and the ascorbate–glutathione cycle isozymes. Such a pattern was observed also for ascorbate and glutathione contents (Fig. 14). Accordingly, the increased GSH contents in Lpa organelles was associated with salt-induced upregulation of γ-ECS protein and induced GPX and GST activities

(continued)

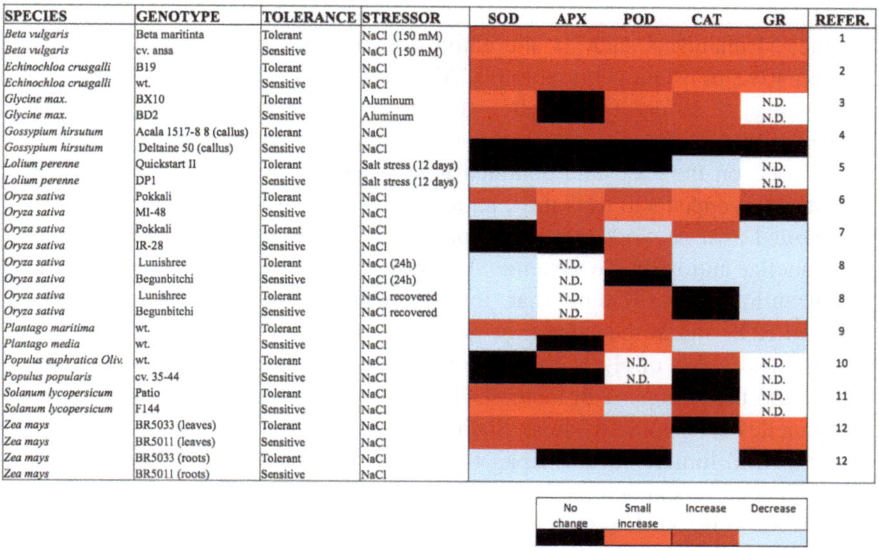

Fig. 14 Literature survey of salt stress induced changes in activities of selected antioxidant isozymes. Changes in activities are *color coded*, relative to values of control treatment. References cited: 1. Bor et al (2003); 2. Abogadallah et al (2010); 3. Zhen et al (2009); 4. Gossett et al (1994); 5. Hu et al (2012b); 6. Chawla et al (2013); 7. Demiral and Turkan (2005); 8. Khan and Panda (2008); 9. Sekmen et al (2007); 10. Wang et al (2008); 11. Chen and Heuer (2013); 12. Neto et al (2006). ND not determined

(Mittova et al. 2003b). Similar upregulation of enzymes and transporters involved in the biosynthesis and intracellular transport of ascorbate, respectively, is expected to characterize the response of Lpa to salinity, but not that of Lem.

In all leaf and root cell organelles of both species, the activity of MDHAR was similar to, or higher than, that of APX, while the activity of DHAR was one order of magnitude lower than that of MDHAR (Table 1 and Mittova et al. 2000). Based on these results, it had been suggested that in the organelles of both tomato species, ascorbate is regenerated mainly by MDHAR (Mittova et al. 2000). This suggestion was substantiated later when evidence was found that in organelles isolated from salt-stressed Lem and Lpa plants, the activities of DHAR and GR were not upregulated. Moreover, in Lpa's organelles, the activity ratios of APX to DHAR or to GR and of MDHAR to those of GR or of MDHAR even increased in response to salinity (Fig. 6).

The number of the various isozymes of a given antioxidant enzyme, for instance APX, exceeds the number of genes that comprise the small *Apx* gene family. This fact suggests that some isozymes are the products of the same

(continued)

gene. Mechanisms such as alternative splicing (Staiger and Brown 2013), alternative translation initiation(s) (Morgane and Duchene 2012) and dual targeting to different organelles (Chew et al. 2003) may underlie these phenomena.

Based on the vast data obtained for the tomato organellar antioxidative systems, it appears to us that the antioxidative system of the plant cell can be regarded as a network that is comprised of its overall multicompartmental organellar antioxidative systems. Accordingly, each organelle contains several antioxidative systems that, together, detoxify the intraorganellar produced ROS and at the same time act also as gatekeepers that diminish both efflux and influx of H_2O_2. Such a cellular antioxidative network can coordinate and better cope with increased ROS production in one or several of its organelles in response to stress. Such a scenario is the stress-induced response of the wild tomato species Lpa, where in all organelles of this species the coordinated upregulation of most of the ascorbate-glutathione isozymes was observed (Fig. 7). It seems that in the cultivated species Lem, the inherent cellular antioxidative network functions adequately when the plants are not confronted by salinity (or other stresses). However, under prolonged stress conditions, when the reinforcement of the antioxidative systems fails, increasing salt-induced oxidative damages are inevitable till the progressive death of the cell.

The failure of Lem to upregulate its antioxidative systems can be explained by yet unknown fault(s) in its signal-transduction pathway and/or in the *cis*-regulatory elements of the antioxidative genes of this species. With the completion of the tomato genome-sequencing project, a powerful tool has been added to the tomato research community. Thus, cloning of sequences, such as entire genes or of sequences that comprise putative promoters, is becoming a simple task. Following this approach, functional analyses and comparisons of orthologous promoters of selected antioxidant genes of the two species can be carried out in order to distinguish between the two possibilities outlined above.

Modern cultivars are the outcome of repeated crosses among closely related high-yield varieties. High yield is influenced by many genes, including those that originated from low-yield wild relatives and as a result, some beneficial alleles have been excluded from the modern cultivar genome (Tanksley and McCouch 1997). This situation is even more complex, as it has been shown recently for *Arabidopsis* ecotypes from northern habitats. These ecotypes employ different strategies to adjust their chloroplastic antioxidative defense system, thus, "domestication" of one of these ecotypes would fix only its own strategy, in the progenies of this domesticated cultivar (Juszczak et al. 2012). It is also possible, as speculated, that trade-offs, during

(continued)

domestication, for high yield might be detrimental to the regulation of stress responses in the cultivated tomato species (Gest et al. 2013).

A comparison of a literature survey of salt stress-induced changes in the activities of selected antioxidant isozymes (Fig. 14) with those of Lem and Lpa (Fig. 6) reveals that as compared with the mirror image pattern, obtained for the tomato isozymes, much milder responses, mainly quantitative but in the same direction, are depicted in Fig. 14. This can be explained by the fact that the latter works were performed with domesticated cultivars that exhibit differential salt sensitivity or tolerance. It is possible that these observed mild responses are the result of the domestication process, where beneficial alleles were excluded from their genome. Our study, where a cultivated tomato species was compared with its wild-related tomato species, emphasizes the importance of the use of wild species in stress tolerance studies.

References

Abogadallah GM, Serag MM, Quick WP (2010) Fine and coarse regulation of reactive oxygen species in the salt tolerant mutants of barnyard grass and their wild-type parents under salt stress. Physiol Plant 138:60–73

Alscher RG, Donahue JL, Cramer CL (1997) Reactive oxygen species and antioxidants: relationships in green cells. Physiol Plant 100:224–233

Apel K, Hirt H (2004) Reactive oxygen species: metabolism, oxidative stress, and signal transduction. Annu Rev Plant Biol 55:373–399

Asada K (1996) Radical production and scavenging in chloroplasts. In: Baker NR (ed) Photosynthesis and environment. Kluwer, Dordrecht, pp 123–150

Asada K (1999) The water-water cycle in chloroplasts: scavenging of active oxygens and dissipation of excess photons. Annu Rev Plant Physiol Plant Mol Biol 50:601–639

Asada K, Kiso K, Yoshikawa K (1974) Univalent reduction of molecular oxygen by spinach chloroplasts on illumination. J Biol Chem 249:2175–2181

Blankenship RE (2010) Early evolution of photosynthesis. Plant Physiol 154:434–438

Bor M, Ozdemir F, Turkan I (2003) The effect of salt stress on lipid peroxidation and antioxidants in leaves of sugar beet *Beta vulgaris* L. and wild beet Beta *maritima* L. Plant Sci 164:77–84

Boyer JS (1982) Plant productivity and the environment. Science 218:443–448

Boveris A (1984) Determination of the production of superoxide radicals and hydrogen peroxide in mitochondria. Methods Enzymol 105:429–435

Briskin DP, Poole RJ (1983) Plasma membrane ATPase of red beet forms a phosphorylated intermediate. Plant Physiol 71:507–512

Bueno P, Varela J, Gimenezgallego G et al (1995) Peroxisomal copper, zinc superoxide-dismutase—characterization of the isoenzyme from watermelon cotyledons. Plant Physiol 108:1151–1160

Casano LM, Zapata JM, Martin M, Sabater B (2000) Chlororespiration and poising of cyclic electron transport—Plastoquinone as electron transporter between thylakoid NADH dehydrogenase and peroxidase. J Biol Chem 275:942–948

Chawla S, Jain S, Jain V (2013) Salinity induced oxidative stress and antioxidant system in salt-tolerant and salt-sensitive cultivars of rice (*Oryza sativa* L.). J Plant Biochem Biotechnol 22:27–34

Chen SB, Heuer B (2013) Effect of genotype and exogenous application of glycinebetaine on antioxidant enzyme activity in native gels of 7-day-old salt-stressed tomato (Solanum lycopersicum) seedlings. Sci Hortic 162:106–116

Chew O, Whelan J, Millar H (2003) Molecular definition of the ascorbate-glutathione cycle in *Arabidopsis* mitochondria reveals dual targeting of antioxidant defenses in plants. J Biol Chem 278:46869–46877

Corpas FJ, Sandalio LM, Palma JM, Leidi EO, Hernandez JA, Sevilla F, del Rio LA (1991) Subcellular-distribution of superoxide dismutase in leaves of ureide-producing leguminous plants. Physiol Plant 82:285–291

Corpas FJ, Barroso JB, del Río LA (2001) Peroxisomes as a source of reactive oxygen species and nitric oxide signal molecules in plant cells. Trends Plant Sci 6:145–150

Cuartero J, Fernandez-Munoz R (1999) Tomato and salinity. Sci Hortic 78:83–125

Dasgan HY, Aktas H, Abak K, Cakmak I (2002) Determination of screening techniques to salinity tolerance in tomatoes and investigation of genotype responses. Plant Sci 163:695–703

Dat J, Vandenabeele S, Vranova E, Van Montagu M, Inze D, Van Breusegem F (2000) Dual action of the active oxygen species during plant stress responses. CMLS Cell Mol Life Sci 57:779–795

De Leonardis S, Dipierro N, Dipierro S (2000) Purification and characterization of an ascorbate peroxidase from potato tuber mitochondria. Plant Physiol Biochem 38:773–779

del Río LA, Sandalio LM, Corpas FJ, Lopez-Huertas E, Palma JM, Pastori GM (1998a) Activated oxygen-mediated metabolic functions of leaf peroxisomes. Physiol Plant 104:673–680

del Río LA, Pastori GM, Palma JM, Sandalio LM, Sevilla F, Corpas FJ, Jiménez A, López-Huertas E, Hernández JA (1998b) The activated oxygen role of peroxisomes in senescence. Plant Physiol 116:1195–1200

del Río LA, Corpas FJ, Sandalio LM, Palma JM, Gomez M, Barroso JB (2002) Reactive oxygen species, antioxidant systems and nitric oxide in peroxisomes. J Exp Bot 53:1255–1272

del Río LA, Sandalio LM, Corpas FJ, Palma JM, Barroso JB (2006) Reactive oxygen species and reactive nitrogen species in peroxisomes. Production, scavenging, and role in cell signaling. Plant Physiol 141:330–335

Demiral T, Turkan I (2005) Comparative lipid peroxidation, antioxidant defense systems and proline content in roots of two rice cultivars differing in salt tolerance. Environ Exp Bot 5:247–257

Dietz KJ (2011) Peroxiredoxins in plants and cyanobacteria. Antioxid Redox Signal 15:1129–1159

Dietz KJ, Jacob S, Oelze ML, Laxa M, Tognetti V, de Miranda SMN, Baier M, Finkemeier I (2006) The function of peroxiredoxins in plant organelle redox metabolism. J Exp Bot 57:1697–1709

Dixit V, Pandey V, Shyam R (2002) Chromium ions inactivate electron transport and enhance superoxide generation in vivo in pea (*Pisum sativum* L. cv. Azad) root mitochondria. Plant Cell Environ 25:687–693

Dixon DP, Davis BG, Edwards R (2002) Functional divergence in the glutathione transferase superfamily in plants—identification of two classes with putative functions in redox homeostasis in *Arabidopsis thaliana*. J Biol Chem 277:30859–30869

Droillard M, Paulin A (1990) Isozymes of superoxide dismutase in mitochondria and peroxisomes isolated from petals of carnation (*Dianthus caryophyllus*) during senescence. Plant Physiol 94:1187–1192

Edwards EA, Rawsthorne S, Mullineaux PM (1990) Subcellular-distribution of multiple forms of glutathione-reductase in leaves of pea (*Pisum sativum* L.). Planta 180:278–284

Fang TK, Donaldson RP, Vigil EL (1987) Electron transport in purified glyoxysomal membranes from castor bean endosperm. Planta 172:1–13

Foyer CH, Lelandais M (1996) A comparison of the relative rates of transport of ascorbate and glucose across the thylakoid, chloroplast and plasmalemma membranes of pea leaf mesophyll cells. J Plant Physiol 148:391–398

Foyer CH, Noctor G (2000) Oxygen processing in photosynthesis: regulation and signalling. New Phytol 146:359–388
Foyer CH, Noctor G (2003) Redox sensing and signalling associated with reactive oxygen in chloroplasts, peroxisomes and mitochondria. Physiol Plant 119:355–364
Foyer CH, Noctor G (2005) Oxidant and antioxidant signalling in plants: a re-evaluation of the concept of oxidative stress in a physiological context. Plant Cell Environ 28:1056–1071
Foyer CH, Shigeoka S (2010) Understanding oxidative stress and antioxidant functions to enhance photosynthesis. Plant Physiol 155:93–100
Foyer CH, Lopez-Delgado H, Dat JF, Scott IM (1997) Hydrogen peroxide- and glutathione-associated mechanisms of acclimatory stress tolerance and signaling. Physiol Plant 100:241–254
Furbank RT, Badger MR (1983) Oxygen exchange associated with electron transport and photophosphorylation in spinach thylakoids. Biochim Biophys Acta 723:400–409
Gest N, Gautier H, Stevens R (2013) Ascorbate as seen through plant evolution: the rise of a successful molecule? J Exp Bot 64:33–53
Gómez JM, Hernández JA, Jiménez A, del Río LA, Sevilla F (1999) Differential response of antioxidative system of chloroplasts and mitochondria to long-term NaCl stress of pea plants. Free Radic Res 31:11–18
Gómez JM, Jiménez A, Olmos E, Sevilla F (2004) Location and effects of long-term NaCl stress on superoxide dismutase and ascorbate peroxidase isoenzymes of pea (*Pisum sativum* cv. Puget) chloroplasts. J Exp Bot 55:119–130
Gossett DR, Millhollon EP, Lueas MC (1994) The effects of NaCI on antioxidant enzyme activities in callus tissue of salt-tolerant and salt-sensitive cotton cultivars (*Gossypium hirsutum* L.). Plant Cell Rep 13:498–503
Gould SB, Waller RF, McFadden GI (2008) Plastid evolution. Annu Rev Plant Biol 59:491–517
Hajjar R, Hodgkin T (2007) The use of wild relatives in crop improvement: a survey of developments over the last 20 years. Euphytica 156:1–13
Halliwell B, Gutteridge JMC (1999) Free radicals in biology and medicine. Oxford University Press, Oxford
Hamilton EW, Heckathorn SA (2001) Mitochondrial adaptations to NaCl. Complex I is protected by anti-oxidants and small heat shockproteins, whereas complex II is protected by proline and betaine. Plant Physiol 126:1266–1274
Hernández JA, Corpas FJ, Gómez M, del Rio LA, Sevilla F (1993) Salt-induced oxidative stress mediated by activated oxygen species in pea leaf mitochondria. Physiol Plant 89:103–110
Hertwig B, Streb P, Feierabend J (1992) Light dependence of catalase synthesis and degradation in leaves and the influence of interfering stress conditions. Plant Physiol 100:1547–1553
Hohmann-Marriott MF, Blankenship RE (2011) Evolution of photosynthesis. Annu Rev Plant Biol 62:515–548
Hossain MA, Asada K (1985) Monodehydroascorbate reductase from cucumber is a flavin adenine dinucleotide enzyme. J Biol Chem 260:12920–12926
Hossain MA, Nakano Y, Asada K (1984) Monodehydroascorbate reductase in spinach chloroplasts and its participation in regeneration of ascorbate for scavenging hydrogen peroxide. Plant Cell Physiol 25:385–395
Hu JP, Baker A, Bartel B, Linka N, Mullen RT, Reumann S, Zolman BK (2012a) Plant peroxisomes: biogenesis and function. Plant Cell 24:2279–2303
Hu LX, Li HY, Pang HC, Fu JM (2012b) Responses of antioxidant gene, protein and enzymes to salinity stress in two genotypes of perennial ryegrass (*Lolium perenne*) differing in salt tolerance. J Plant Physiol 169:146–156
Iturbe-Ormaetxe I, Matamoros MA, Rubio MC, Dalton DA, Becana M (2001) The antioxidants of legume nodule mitochondria. Mol Plant Microbe Interact 14:1189–1196
Jiménez A, Hernández JA, del Río LA, Sevilla F (1997) Evidence for the presence of the ascorbate-glutathione cycle in mitochondria and peroxisomes of pea leaves. Plant Physiol 114:275–284

Juszczak I, Rudnik R, Pietzenuk B, Baier M (2012) Natural genetic variation in the expression regulation of the chloroplast antioxidant system among *Arabidopsis thaliana* accessions. Physiol Plant 146:53–70

Kahn TL, Fender SE, Bray EA, O'Connell MA (1993) Characterization of expression of drought- and abscisic acid-regulated tomato genes in the drought-resistant species *Lycopersicon pennellii*. Plant Physiol 103:597–605

Kanematsu S, Asada K (1994) Superoxide dismutase. In: Fukui T, Soda K (eds) Molecular aspects of enzyme catalysis. Kondansha Ltd, Tokyo, pp 191–210

Kangasjärvi S, Lepistö A, Hännikäinen K, Piippo M, Luomala EM, Aro EM, Rintamäki E (2008) Diverse roles for chloroplast stromal and thylakoid-bound ascorbate peroxidases in plant stress responses. Biochem J 412:275–285

Karyotou K, Donaldson RP (2005) Ascorbate peroxidase, a scavenger of hydrogen peroxide in glyoxysomal membranes. Arch Biochem Biophys 434:248–257

Kaur N, Reumann S, Hu J (2009) Peroxisome biogenesis and function. Arabidopsis Book 7:e0123. doi:10.1199/tab.0123

Khan MH, Panda SK (2008) Alterations in root lipid peroxidation and antioxidative responses in two rice cultivars under NaCl-salinity stress. Acta Physiol Plant 30:81–89

Kieselbach T, Hagman A, Andersson B, Schroder W (1998) The thylakoid lumen of chloroplasts—isolation and characterization. J Biol Chem 273:6710–6716

Lázaro JJ, Jiménez A, Camejo D, Iglesias-Baena I, Martí Mdel C, Lázaro-Payo A, Barranco-Medina S, Sevilla F (2013) Dissecting the integrative antioxidant and redox systems in plant mitochondria. Effect of stress and S-nitrosylation. Front Plant Sci 4:460

Levine A (1999) Oxidative stress as a regulator of environmental responses in plants. In: Lerner HR (ed) Plant responses to environmental stresses from phytohormones to genome reorganization. Marcel Dekker, New York, pp 247–264

Lisenbee CS, Lingard MJ, Trelease RN (2005) *Arabidopsis* peroxisomes possess functionally redundant membrane and matrix isoforms of monodehydroascorbate reductase. Plant J 43:900–914

Lopez-Huertas E, Sandalio LM, Gómez M, del Rio LA (1997) Superoxide radical generation in peroxisomal membranes: evidence for the participation of the 18-kDa integral membrane polypeptide. Free Radic Res 26:497–506

Lopez-Huertas E, Corpas FJ, Sandalio LM, del Rio LA (1999) Characterization of membrane polypeptides from pea leaf peroxisomes involved in superoxide radical generation. Biochem J 337:531–536

Malecka A, Jarmuszkiewicz W, Tomaszewska B (2001) Antioxidative defense to lead stress in subcellular compartments of pea root cells. Acta Biochim Pol 48:687–698

Mehler AH (1951) Studies on reactivities of illuminated chloroplasts. I. Mechanism of the reduction of oxygen and other Hill reagents. Arch Biochem Biophys 33:65–77

Menda N, Strickler SR, Mueller LA (2013) Advances in tomato research in the post-genome era. Plant Biotechnol 30:243–256

Mhamdi A, Guillaume Q, Sejir C, Vanderauwera S, Breusegem FV, Noctor G (2010) Catalase function in plants: a focus on *Arabidopsis* mutants as stress-mimic models. J Exp Bot 6:4197–4220

Millar AH, Whelan J, Soole KL (2011) Organization and regulation of mitochondrial respiration in plants. Annu Rev Plant Biol 62:79–104

Miller G, Nobuhiro Suzukis N, Ciftci-Yilmaz S, Mittler R (2010) Reactive oxygen species homeostasis and signalling during drought and salinity stresses. Plant Cell Environ 33:453–467

Mittler R (2002) Oxidative stress, antioxidants and stress tolerance. Trends Plant Sci 7:405–410

Mittler R, Vanderauwera S, Gollery M, Van Breusegem F (2004) Reactive oxygen gene network of plants. Trends Plant Sci 9:490–498

Mittova V, Volokita M, Guy M, Tal M (2000) Activities of SOD and the ascorbate-glutathione cycle enzymes in subcellular compartments in leaves and roots of the cultivated tomato and its wild salt tolerant relative Lycopersicon pennellii. Physiol Plant 110:42–51

Mittova V, Tal M, Volokita M, Guy M (2002a) Salt stress induces upregulation of an efficient chloroplast antioxidant system in the salt-tolerant wild tomato species *Lycopersicon pennellii* but not in the cultivated species. Physiol Plant 115:393–400

Mittova V, Guy M, Tal M, Volokita M (2002b) Response of the cultivated tomato and its wild salt-tolerant relative *Lycopersicon pennellii* to salt-dependent oxidative stress: increased activities of antioxidant enzymes in root plastids. Free Radic Res 36:195–202

Mittova V, Tal M, Volokita M, Guy M (2003a) Upregulation of the leaf mitochondrial and peroxisomal antioxidative systems in response to salt-induced oxidative stress in the wild salt-tolerant tomato species *Lycopersicon pennellii*. Plant Cell Environ 26:845–856

Mittova V, Theodoulou FL, Kiddle G, Gomez L, Volokita M, Tal M, Foyer CH, Guy M (2003b) Co-ordinate induction of glutathione biosynthesis and glutathione-metabolizing enzymes is correlated with salt tolerance in tomato. FEBS Lett 554:417–421

Mittova V, Guy M, Tal M, Volokita M (2004a) Salinity up-regulates the antioxidative system in root mitochondria and peroxisomes of the wild salt-tolerant tomato species Lycopersicon pennellii. J Exp Bot 55:1105–1113

Mittova V, Theodoulou F, Kiddle G, Volokita M, Tal M, Foyer CH, Guy M (2004b) Comparison of mitochondrial ascorbate peroxidase in the cultivated tomato, *Lycopersicon esculentum*, and its wild, salt-tolerant relative, *L. pennellii*—a role for matrix isoforms in protection against oxidative damage. Plant Cell Environ 27:237–250

Miyake C, Asada K (1992) Thylakoid-bound ascorbate peroxidase in spinach-chloroplasts and photoreduction of its primary oxidation product monodehydroascorbate radicals in thylakoids. Plant Cell Physiol 33:541–553

Miyake C, Asada K (1994) Ferredoxin-dependent photoreduction of the monodehydroascorbate radical in spinach thylakoids. Plant Cell Physiol 35:539–549

Miyake C, Schreiber U, Hormann H, Sano S, Asada K (1998) The FAD-enzyme monodehydroascorbate radical reductase mediates photoproduction of superoxide radicals in spinach thylakoid membranes. Plant Cell Physiol 39:821–829

Møller IM (2001) Plant mitochondria and oxidative stress: electron transport, NADPH turnover, and metabolism of reactive oxygen species. Annu Rev Plant Physiol Plant Mol Biol 52:561–591

Moore AL, Siedow JN (1991) The regulation and nature of the cyanide-resistant alternative oxidase of plant mitochondria. Biochim Biophys Acta 1059:121–140

Morgane M, Duchene AM (2012) Macromolecules trafficking to plant mitochondria. Adv Bot Res 63:347–421 (Mitochondrial genome evolution book series)

Moyle CL (2008) Ecological and evolutionary genomics in the wild tomatoes (*Solanumsect. lycopersicon*). Evolution 62:2995–3013

Myouga F, Hosoda C, Umezawa T, Iizumi H, Kuromori T, Motohashi R (2008) A heterocomplex of iron superoxide dismutases defends chloroplast nucleoids against oxidative stress and is essential for chloroplast development in Arabidopsis. Plant Cell 20:3148–3162

Najami N, Janda T, Barriah W, Kayam G, Tal M, Guy M, Volokita M (2008) Ascorbate peroxidase gene family in tomato: its identification and characterization. Mol Genet Genomics 279:171–182

Navari-Izzo F, Quartacci MF, Pinzino C, Dalla Vecchia F, Sgherri CLM (1998) Thylakoid-bound and stromal antioxidative enzymes in wheat treated with excess copper. Physiol Plant 104:630–638

Nelson N (2011) Photosystems and global effects of oxygenic photosynthesis. Biochim Biophys Acta 1807:856–863

Neto AD, Prisco JT, Eneas J et al (2006) Effect of salt stress on antioxidative enzymes and lipid peroxidation in leaves and roots of salt-tolerant and salt-sensitive maize genotypes. Environ Exp Bot 56:87–94

Neuhaus HE, Emes MJ (2000) Nonphotosynthetic metabolism in plastids. Annu Rev Plant Physiol Plant Mol Biol 51:111–140

Noctor G, Foyer CH (1998) Ascorbate and glutathione: keeping active oxygen under control. Annu Rev Plant Physiol Plant Mol Biol 49:249–279

Noctor G, Gómez L, Vanacker H, Foyer CH (2002) Interactions between biosynthesis, compartmentation and transport in the control of glutathione homeostasis and signaling. J Exp Bot 53:1283–1304

Noctor G, De Paepe R, Foyer CH (2007) Mitochondrial redox biology and homeostasis in plants. Trends Plant Sci 12:125–134

Ogawa K, Kanematsu S, Takabe K, Asada K (1995) Attachment of CuZn-superoxide dismutase to thylakoid membranes at the site of superoxide generation (PSI) in spinach chloroplasts: detection by immuno-gold labeling after rapid freezing and substitution method. Plant Cell Physiol 36:565–573

Peltier G, Cournac L (2002) Chlororespiration. Annu Rev Plant Biol 53:523–550

Peralta IE, Sandra K, Spooner DM (2005) New species of wild tomatoes (Solanum Section *Lycopersicon*: *Solanaceae*) from Northern Peru. Syst Bot 30:424–434

Pilon M, Ravet K, Tapken W (2011) The biogenesis and physiological function of chloroplast superoxide dismutases. Biochim Biophys Acta 1807:989–998

Prasad TK, Anderson MD, Stewart CR (1995) Localization and characterization of peroxidases in the mitochondria of chilling-acclimated maize seedlings. Plant Physiol 108:1597–1605

Rasmusson AG, Soole KL, Elthon TE (2004) Alternative NAD(P)H dehydrogenases of plantmitochondria. Annu Rev Plant Biol 55:23–39

Reumann S, Bettermann M, Bent R, Heldt HW (1997) Evidence for the presence of a porin in the membrane of glyoxysomes of castor bean. Plant Physiol 115:891–899

Rodríguez-Serrano M, Romero-Puertas M, Pastori G, Corpas FJ, Sandalio LM, del Río LA, Palma JM (2007) Peroxisomal membrane manganese superoxide dismutase: characterization of the isozyme from watermelon (*Citrullus lanatus Schrad.*) cotyledons. J Exp Bot 58:2417–2427

Sabehat A, Weiss D, Lurie S (1998) Heat-shock proteins and cross-tolerance in plants. Physiol Plant 103:437–441

Sakihama Y, Mano J, Sano S, Asada K, Yamasaki H (2000) Reduction of phenoxyl radicals mediated by monodehydroascorbate reductase. Biochem Biophys Res Commun 279:949–954

Sandalio LM, del Río LA (1988) Intraorganellar distribution of superoxide dismutase in plant peroxisomes (glyoxisomes and leaf peroxisomes). Plant Physiol 88:1215–1218

Sandalio LM, López-Huertas E, Bueno P, del Río LA (1997) Immunocytochemical localization of copper, zinc superoxide dismutase in peroxisomes from watermelon (Citrullus vulgaris Schrad.). Free Radic Res 26:187–194

Schumann U, Prestele J, O'Geen H, Brueggeman R, Wanner G, Gietl C (2007) Requirement of the C3HC4 zinc RING finger of the *Arabidopsis* PEX10 for photorespiration and leaf peroxisome contact with chloroplasts. Proc Natl Acad Sci USA 104:1069–1074

Sehmer L, Dizengremel P (1998) Contribution to subcellular localization of superoxide dismutase isoforms of spruce needles and oak leaves. J Plant Physiol 153:545–551

Sekmen AH, Turkan I, Takio S (2007) Differential responses of antioxidative enzymes and lipid peroxidation to salt stress in salt-tolerant *Plantago maritima* and salt-sensitive *Plantago media*. Physiol Plant 131:399–411

Shalata A, Tal M (1998) The effect of salt stress on lipid peroxidation and antioxdants in the leaf of the cultivated tomato and its wild salt-tolerant relative *Lycopersicon pennelii*. Physiol Plant 104:169–174

Shalata A, Mittova V, Volokita M, Guy M, Tal M (2001) Response of the cultivated tomato and its wild salt-tolerant relative *Lycopersicon pennellii* to salt-dependent oxidative stress: the root antioxidative system. Physiol Plant 112:487–494

Sheoran S, Pandey B, Singh R, Sharma P, Chatrath R (2011) Modeling and phylogeny analysis of bread wheat MnSOD. Bioinformation 6:209–211

Shu S, Yuan LY, Guo SR, Sun J, Yuan YH (2013) Effects of exogenous spermine on chlorophyll fluorescence, antioxidant system and ultrastructure of chloroplasts in *Cucumis sativus* L. under salt stress. Plant Physiol Biochem 63:209–216

Siendones E, Gonzáz-Reyes JA, Santos-Ocaña C, Navas P, Córdoba F (1999) Biosynthesis of ascorbic acid in kidney bean. L-Galactono-γ-lactone dehydrogenase is an intrinsic protein located at the inner mitochondrial membrane. Plant Physiol 120:907–912
Smirnoff N (1993) The role of active oxygen in response of plants to water deficit and desiccation. New Phytol 125:27–58
Smirnoff N, Wheeler GL (2000) Ascorbic acid in plants: biosynthesis and function. Crit Rev Biochem Mol Biol 35:291–314
Staiger D, Brown JWS (2013) Alternative splicing at the intersection of biological timing, development, and stress responses. Plant Cell 10:3640–3656
Streller S, Kromer S, Wingsle G (1994) Isolation and purification of mitochondrial mn-superoxide dismutase from the gymnosperm *Pinus sylvestris* L. Plant Cell Physiol 35:859–867
Suzuki N, Koussevitzky S, Ron Mittler R (2012) ROS and redox signalling in the response of plants to abiotic stress. Plant Cell Environ 35:259–270
Sze H (1985) H^+-Translocating atpases: advances using membrane vesicles. Annu Rev Plant Physiol 36:175–208
Tal M, Shannon MC (1983) Salt tolerance in the wild relatives of the cultivated tomato: responses of *Lycopersicon esculentum, L. peruvianum and Solanum pennellii* to high salinity. Z Pflanzenphysiol 86:231–240
Tanksley SD, McCouch SR (1997) Seed banks and molecular maps: unlocking genetic potential from the wild. Science 277:1063–1066
Tausz M, Sircelj H, Grill D (2004) The glutathione system as a stress marker in plant ecophysiology: is a stress–response concept valid? J Exp Bot 55:1955–1962
Tomato Genome Consortium (2012) The tomato genome sequence provides insights into fleshy fruit evolution. Nature 485:635–641
Van Camp W, Capiau K, Van Montagu M, Inze D, Slooten L (1996) Enhancement of oxidative stress tolerance in transgenic tobacco plants overproducing Fe-superoxide dismutase in chloroplasts. Plant Physiol 112:1703–1714
Vanlerberghe GC (2013) Alternative oxidase: a mitochondrial respiratory pathway to maintain metabolic and signaling homeostasis during abiotic and biotic stress in plants. Int J Mol Sci 14:6805–6847
Volk S, Feieraben J (1989) Photoinactivation of catalase at low temperature and its relevance to photosynthetic and peroxide metabolism in leaves. Plant Cell Environ 12:701–712
Vranova E, Inze D, Van Breusegen F (2002) Signal transduction during oxidative stress. J Exp Bot 53:227–1236
Wang J, Zhang H, Allen RD (1999) Overexpression of an *Arabidopsis* peroxisomal ascorbate peroxidase gene in tobacco increases protection against oxidative stress. Plant Cell Physiol 40:725–732
Wang RG, Chen SL, Zhou XY et al (2008) Ionic homeostasis and reactive oxygen species control in leaves and xylem sap of two poplars subjected to NaCl stress. Tree Physiol 28:947–957
Wójcik M, Tukiendorf A (2011) Glutathione in adaptation of *Arabidopsis thaliana* to cadmium stress. Biol Plant 55:125–132
Yu CW, Murphy TM, Sung WW, Lin CH (2002) H_2O_2 treatment induces glutathione accumulation and chilling tolerance in mung bean. Funct Plant Biol 29:1081–1087
Zechmann B, Stumpe M, Mauch F (2011) Immunocytochemical determination of the subcellular distribution of ascorbate in plants. Planta 233:1–12
Zhen Y, Miao L, Su J et al (2009) Differential responses of anti-oxidative enzymes to aluminum stress in tolerant and sensitive soybean genotypes. J Plant Nutr 32:1255–1270
Zhu D, Scandalios JG (1993) Maize mitochondrial manganese superoxide dismutases are encoded by a differentially expressed multigene family. Proc Natl Acad Sci USA 90:9310–9314

The Role of Reactive Oxygen Species Under Ammonium Nutrition

Anna Podgórska and Bożena Szal

1 Nitrogen Assimilation in Plants

Nitrogen (N) is one of the most important macroelements for living organisms, serving as a central precursor for nucleic acids, amino acids, and other organic molecules. Because N is a ubiquitous intermediate of plant metabolic pathways, the element is undoubtedly crucial to optimal plant growth and development. Plants are capable of utilizing different inorganic N sources, mainly nitrate (NO_3^-) and ammonium (NH_4^+) (Barker and Mills 1980; Lewis 1992). NO_3^- and NH_4^+ are taken up from the soil by roots; both N forms are actively absorbed via several low- and high-affinity transporters (Glass et al. 2002). Nitrate is cotransported with H^+ and is then directly reduced or stored in vacuoles after importation; however, in many plants, such as *Arabidopsis*, most NO_3^- is translocated to the shoots (Masclaux-Daubresse et al. 2010). NO_3^- is reduced to NO_2^- by the cytosolic nitrate reductase (NR), which uses NAD(P)H. NO_2^- is then further reduced by the plastidic nitrite reductase (NiR), which consumes 6 electrons from reduced ferredoxin (Fd). The resulting product (NH_4^+) is incorporated into amino acid structures in the glutamine synthetase-glutamine:2-oxoglutarate (2-OG) amino-transferase (GS-GOGAT) cycle (Noctor and Foyer 1998; Masclaux-Daubresse et al. 2010).

Ammonium ions can be imported into the plant cell through plasmalemma transporters, but free diffusion of NH_3 is also possible (Loqué et al. 2007; Ludewig et al. 2007). Primary NH_4^+ assimilation occurs mainly in root plastids, and the major responsible GS has a very high affinity for NH_4^+ (Miflin and Habash 2002). The resulting amide product, glutamine (Gln), is transported via xylem sap to the

A. Podgórska • B. Szal (✉)
Faculty of Biology, Institute of Experimental Plant Biology and Biotechnology, University of Warsaw, I. Miecznikowa 1, 02-096 Warsaw, Poland
e-mail: szal@biol.uw.edu.pl

K.J. Gupta, A.U. Igamberdiev (eds.), *Reactive Oxygen and Nitrogen Species Signaling and Communication in Plants*, Signaling and Communication in Plants 23,
DOI 10.1007/978-3-319-10079-1_7

shoots. In *Arabidopsis*, only a very small amount of NH_4^+ is translocated to the leaves (Schjoerring et al. 2002). In leaf cells, ammonium ions are substrates for chloroplastic GS. The Gln transported from the roots simultaneously serves as a substrate for chloroplastic Fd-GOGAT (Podgórska et al. 2013), as glutamate (Glu) is a major donor for the biosynthesis of other amino acids and metabolites (Masclaux-Daubresse et al. 2006). When NH_4^+ is metabolized, the highly reductant-consuming NO_3^- and NO_2^- reduction reactions performed by cytosolic NR and plastidal NiR do not occur. It should be also noted that the aminating reaction catalysed by mitochondrial glutamate dehydrogenase (GDH) may consume NH_4^+ in an NADH-dependent manner; this enzyme is therefore speculated to be engaged in ammonium assimilation (Frechilla et al. 2002; Cruz et al. 2006). However, recent data indicate that in vivo the principal role of GDH is the deamination of Glu (Skopelitis et al. 2007; Labboun et al. 2009), and its functioning has an anaplerotic role (Ariz et al. 2013). The mechanism for incorporating NH_4^+ into nontoxic intermediates must be very effective because high concentrations of free NH_4^+ may be toxic at the cellular level (Cruz et al. 2006; Hoffmann et al. 2007).

2 Plant Sensitivity to Ammonium Ions

NH_4^+ might be expected to be the most favourable N form for plants, as its assimilation requires less energy than does nitrate nutrition (Noctor and Foyer 1998). Nevertheless, the use of NH_4^+ as the sole N source leads to serious growth and developmental disorders in many crop plants, which are together commonly referred as the ammonium syndrome (Gerendás et al. 1997; Britto and Kronzucker 2013). The observed symptoms depend on the plant species and vary from NH_4^+ sensitive to tolerant plants (Britto and Kronzucker 2002). Rice is regarded as unique in its high degree of ammonium tolerance. But it was shown that despite that rice prefers ammonium than nitrate nutrition, the partial nitrate nutrition (PNN) may also improve its growth (Duan et al. 2007). The effect of NH_4^+ supply in ammonium sensitive plants is already visible after short-term treatments and increasing the NH_4^+ concentration severely enhances the symptoms of toxicity (Domínguez-Valdivia et al. 2008; Cruz et al. 2011).

In extensive previous studies, researchers have attempted to determine the mechanism underlying the toxicity of NH_4^+ to plants at the physiological level (see the following reviews: Huppe and Turpin 1994; Wiesler 1997; Britto and Kronzucker 2002), and multiple factors have been proposed to account for the ammonium syndrome. Among the difficulties that may be faced by NH_4^+-fed plants is carbohydrate limitation, because the considerable carbon skeleton requirements for NH_4^+ assimilation in the roots can cause carbon deprivation in the shoots (Schortemeyer et al. 1997; Walch-Liu et al. 2000). The diminution of essential cations such as K^+, Mg^{2+}, and Ca^{2+} may also occur in plant tissues because of the direct competition between NH_4^+ and these cations for entry through specific transporters, thereby leading to nutrient imbalances (Szczerba et al. 2008). The toxicity of NH_4^+ has also been attributed to its impairment of photosynthetic processes: because NH_4^+ is a photophosphorylation uncoupler, its accumulation may lead to the disruption of photosynthesis and chloroplast damage (Raab and Terry 1994; Gerendás et al. 1997; Bendixen et al. 2001). Additionally, futile NH_4^+ cycling over the plasma membrane for NH_4^+ removal in roots is very energy demanding and may be an energetic burden for whole plants (Britto et al. 2001; Kronzucker et al. 2001; Coskun et al. 2013). Parallel proton export under NH_4^+ supply could result in the transient alkalisation of the cytosol (Britto and Kronzucker 2005) and the overacidification of the apoplast (Husted and Schjoerring 1995). The passive influx of undissociated NH_3 (Ariz et al. 2011), which is a weak base, would also increase the cytosolic pH (Gerendás and Ratcliffe 2000). The disruption of the hormonal balance under NH_4^+ supply has been proposed as well (Walch-Liu et al. 2001). However, new experimental evidence has indicated that these factors may not underlie the toxicity symptoms of NH_4^+ (Britto and Kronzucker 2002 and references therein). It has recently been proposed that the negative effects of sole NH_4^+ nutrition result from a redox imbalance in plant metabolism (Zhu et al. 2000; Guo et al. 2005). A consequence of NH_4^+ nutrition as compared to NO_3^- nutrition is a surplus of reductants (Escobar et al. 2006). Recently, Podgórska et al. (2013) have shown that both $NADH/NAD^+$ and $NADPH/NADP^+$ ratios are significantly higher in the leaf tissue of ammonium-grown *Arabidopsis*.

This redox imbalance (reductive stress, Podgórska et al. 2013) may result in oxidative stress under sole NH_4^+ nutrition. The stress symptoms observed in NH_4^+-grown plants may therefore be related to the generation of toxic oxygen radicals (Noctor and Foyer 1998; Zhu et al. 2000). Strong indications of increased ROS

Table 1 The effect of ammonium nutrition on hydrogen peroxide content

Method	Plant	The period of ammonium nutrition	Effect	References
DAB staining (cyto-chemical localisation)	*Vallisneria natans*	4 days	Increased	Wang et al. (2008)
Spectrophotometric assay (concentration)	*Vallisneria natans*	4 days	Increased	Wang et al. (2008)
Fluorymetric assay (concentration)	*Arabidopis thaliana*	0.5 h	Increased	Patterson et al. (2010)
Fluorymetric assay (concentration)	*Arabidopis thaliana*	1–8 h	Not changed	Patterson et al. (2010)
Spectrophotometric assay (concentration)	*Arabidopis thaliana*	7 weeks	Increased	Podgórska et al. (2013)
$CeCl_3$ staining (cyto-chemical localisation)	*Arabidopis thaliana*	7 weeks	Increased (mito-chondrial membranes)	Podgórska et al. (2013)
DAB staining (cyto-chemical localisation)	*Arabidopis thaliana*	2 days	Increased	Li et al. (2013b)

levels in NH_4^+-fed plants have been observed by multiple approaches (Table 1). An increase in the ROS concentration of *Arabidopsis* leaves has been observed shortly after NH_4^+ supply (Patterson et al. 2010), as well as after long-term growth on ammonium (Hoffmann et al. 2007; Podgórska et al. 2013). Patterson et al. (2010) have found an NH_4^+-specific plant response associated with the biotic stress and defence systems. Increased ROS contents were also revealed by cytochemical staining in *Arabidopsis* seedlings (Li et al. 2013b) and the macrophyte *Vallisneria natans* (Wang et al. 2008).

3 Proposed ROS Production Sites in Ammonium-Fed Plants

Apoplastic ROS production is an important response to many biotic and abiotic stress signals. Apoplastic ROS can be generated via the enhanced enzymatic activities of plasma membrane-bound NADPH oxidases (RBOH), cell wall-bound peroxidases, and amine oxidases (Wojtaszek 1997). Unfortunately, data regarding the direct impacts of NH_4^+ on these apoplastic reactions are very rare. Patterson et al. (2010) have defined an NH_4^+-specific plant response, as connected with the cell wall that was also associated with a defence stress mechanism similar to the hypersensitivity response (HR). Increased apoplastic ROS content has been detected by cytochemical H_2O_2 staining in NH_4^+-grown *Arabidopsis* (Fig. 1). NH_4^+-mediated changes in apoplastic pH (Husted and Schjoerring 1995) may induce ROS generation, possibly by the modulation of the activities of cell wall peroxidases (POX) (Lager et al. 2010). Increased ROS production may also result from RBOH on the apoplastic side because of the accumulation of cytosolic NADPH (Podgórska et al. 2013), which is the substrate for RBOH, under NH_4^+ supply. The upregulation of RBOH has previously been observed in cucumber plants in response to increased cytosolic NAD(P)H pools (Łukawska-Kuźma et al. 2012).

The cytosol is not thought to be a major site of ROS production under NH_4^+ nutrition. NR could generate nitric oxide (NO) or produce superoxide anions ($O_2^{\bullet-}$) as byproducts (Kaiser et al. 2002; Yamasaki and Sakihama 2000); however, the action of NR is diminished under ammonium nutrition.

The electron-consuming reaction catalysed by NiR does not occur under ammonium supply; it has therefore been proposed that in the overreduced chloroplast ETC (chlETC), an increased amount of $O_2^{\bullet-}$ could be generated by photosystem I (PSI) (Zhu et al. 2000; Guo et al. 2005; Hoffmann et al. 2007). However, Podgórska et al. (2013) have suggested that Fd-GOGAT is an efficient electron sink under NH_4^+ supply and plays an important role in balancing the reduction state of the chlETC. In confirmation of this hypothesis, the redox state of the chloroplast was found to be unchanged after long-term ammonium nutrition (Podgórska et al. 2013), and no negative effects on the operation of the chlETC have been

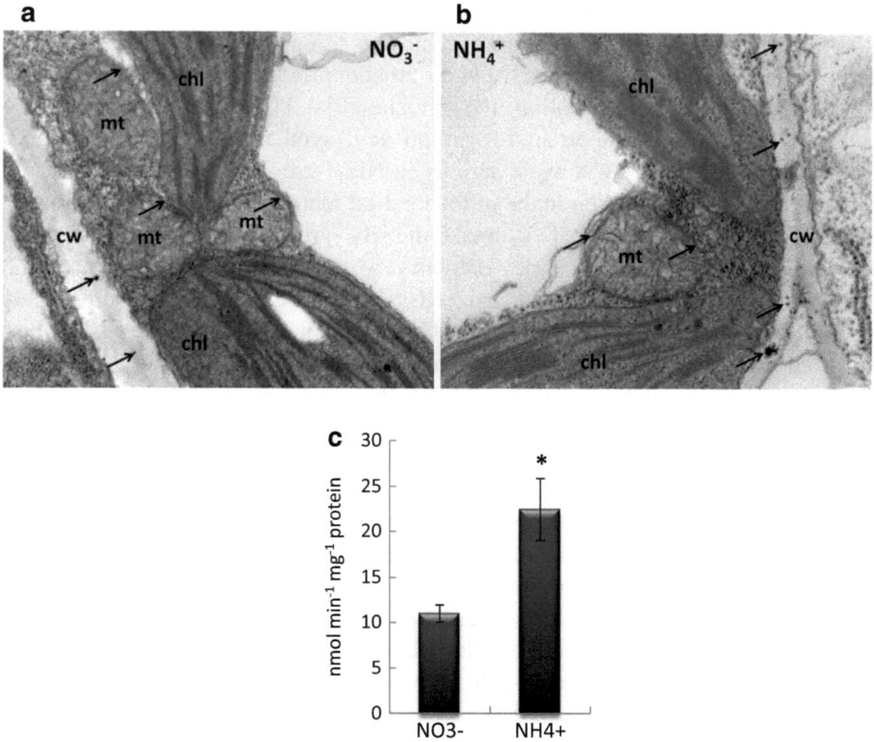

Fig. 1 Cytochemical localisation of hydrogen peroxide in nitrate- (**a**) and ammonium-grown (**b**) *Arabidopsis* (Col-0 ecotype) mesophyll cells. Chl, chloroplast; cw, cell wall; mt, mitochondrion. *Arrows* indicate precipitates of cerium peroxides in mitochondrial membranes and in the cell walls. (**c**) Superoxide ion production by isolated mitochondria towards the intermembrane space. Mitochondria were isolated from leaves of 8-week-old *Arabidopsis* (Col-0 ecotype) plants grown on nitrate or ammonium as the sole nitrogen source. Superoxide generation was determined as described in Szal et al. (2009)

reported in NH_4^+-grown *Arabidopsis* (Podgórska et al. 2013) and French bean (Zhu et al. 2000; Brück and Guo 2006). Conversely, some plants have shown a depression of photosynthesis when grown on NH_4^+, e.g., spinach (Lasa et al. 2002), sugar beet (Raab and Terry 1994), and maize (Foyer et al. 1994). Therefore, it remains unclear whether a net increase in photosynthetic ROS formation occurs under NH_4^+ nutrition. Injury to chloroplasts would be easily detectable in PSII, which is very sensitive to oxidative damage, but no harmful effects to this photosystem have been found to date in NH_4^+-grown plants; additionally, no increases in nonphotochemical quenching have been observed (Zhu et al. 2000; Bendixen et al. 2001; Hoffmann et al. 2007; Hachiya et al. 2010; Podgórska et al. 2013).

Mitochondria play a key role in regulating cellular redox homeostasis in plants (Noctor et al. 2007; Noguchi and Yoshida 2007). Guo et al. (2005) proposed that, during NH_4^+ nutrition, excess redox equivalents may be oxidised in the mitochondrial electron transport chain (mtETC). The link between NH_4^+ toxicity and mitochondrial ROS production may not be straightforward but has garnered increased attention in recent years. Plant mitochondria are a source of metabolites needed during NH_4^+ assimilation; the mitochondrial TCA cycle, as the origin of 2-OG, is particularly important for amino acid synthesis (Szal and Podgórska 2012). The increase in TCA cycle activity in NH_4^+-grown plants may also lead to increased NADH production in the mitochondrial matrix (Guo et al. 2005; Escobar et al. 2006). Increased rates of O_2 uptake in NH_4^+-grown roots have been reported from gas flux experiments in barley (Bloom et al. 1992), pea (Frechilla et al. 2002), wheat (Barneix et al. 1984), and maize (Cramer and Lewis 1993). Additionally, respiration measured as O_2 uptake in shoots or isolated mitochondria has been reported as elevated in many experiments using plants such as *Arabidopsis* (Escobar et al. 2006; Hachiya et al. 2010; Podgórska et al. 2013), French bean (Guo et al. 2005), and spinach and pea (Lasa et al. 2002). The increase in mtETC activity concomitantly triggers elevated electron leakage (Foyer and Noctor 2003) and leads to excessive mitochondrial ROS (mtROS) production (Zhu et al. 2000; Podgórska et al. 2013). ROS are generated during mtETC functioning at Complex I and III toward the mitochondrial matrix, and Complex III can simultaneously release ROS into the intermembrane space (Fig. 2).

The photorespiratory cycle can generate large amounts of H_2O_2. Glycolate oxidase (GOX) is a flavin mononucleotide-dependent peroxisomal enzyme that consumes glycolate and O_2 to produce glyoxylate and H_2O_2 as a byproduct. Higher rates of photorespiration under NH_4^+ supply have been documented in bean plants (Blackwell et al. 1988; Zhu et al. 2000; Guo et al. 2007). However, enzyme, microarray, and northern-blot analyses have shown that the genes encoding GOX and alanine glyoxylate aminotransferase (AGT) are slightly downregulated in NH_4^+-grown seedlings compared to those grown under NO_3^- nutrition (Hoffmann et al. 2007).

4 Enhancement of Antioxidant Defence in Ammonium-Grown Plants

To counteract the toxic effects of ROS, plants have developed several defence systems in all cellular compartments. This response consists of enzymatic and low molecular weight antioxidants that reduce or buffer the excess ROS. This network of enzymatic antioxidants, including the superoxide dismutases (SODs), which is important for reducing $O_2^{\bullet -}$ to H_2O_2 and, in combination with catalases (CAT), peroxidases (glutathione peroxidase, GPX; guaiacol peroxidase, GP), and ascorbate–glutathione cycle enzymes (ascorbate peroxidase, APX; glutathione reductase, GR; dehydroascorbate reductase, DHAR; monodehydroascorbate reductase, MDHAR), which decompose H_2O_2, is responsible for ROS scavenging. The upregulation of many antioxidant enzymes has been observed during NH_4^+

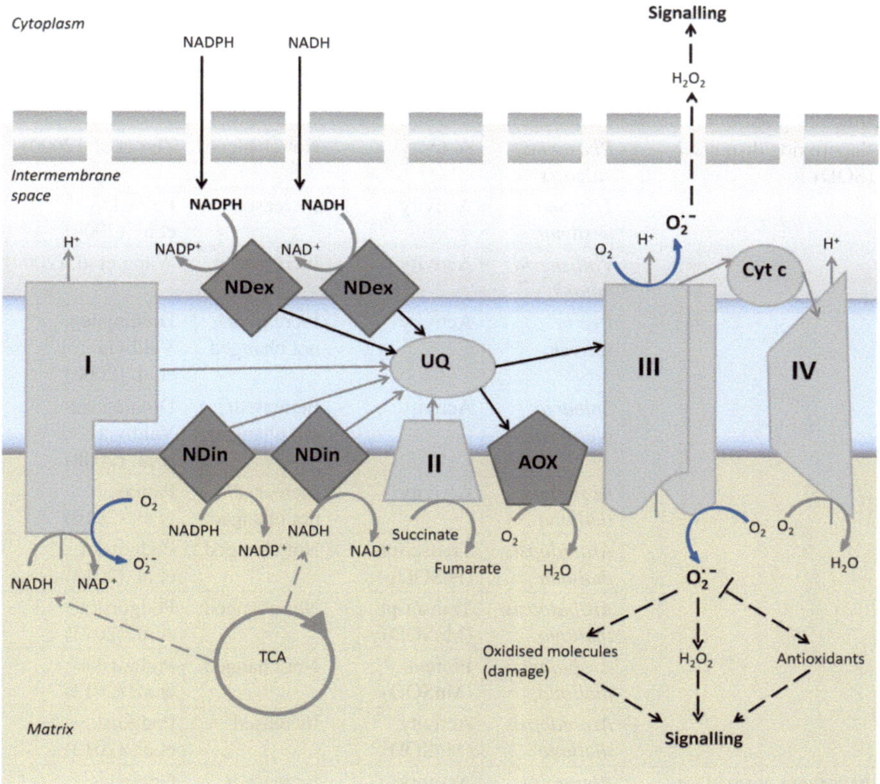

Fig. 2 Plant mitochondrial electron transport chain (mtETC) and proposed scheme for the engagement of mitochondrial reactive oxygen species in retrograde signalling. Plant specific components: external (NDex) and internal (NDin) type-II dehydrogenases and alternative oxidase (AOX) are indicated by a *dark grey colour*. The sites of superoxide ion production in the mtETC are indicated by *blue arrows*. Superoxide ion generation towards the intermembrane space may serve as a direct signal from the mitochondria. Superoxide generation towards the matrix leads mainly to the oxidative damage of different molecules, which in turn may also be involved in signal transduction from mitochondria

nutrition (Table 2). Podgórska et al. (2013) have also analysed several antioxidant enzymes in organelles isolated from NH_4^+-fed plants, observing that the mitochondrial manganese-dependent SOD was particularly upregulated.

The nonenzymatic scavenger pool, consisting primarily of ascorbate and glutathione, is also elevated in NH_4^+-fed plants (Zhu et al. 2000; Podgórska et al. 2013). Additionally, glutathione levels have been shown to increase in both chloroplast and mitochondrial fractions of these plants (Podgórska et al. 2013). Because ROS have an oxidising action, measuring the oxidation states of glutathione and ascorbate provides important information. Indeed, most plants exhibit more oxidised states for both antioxidants under ammonium supply (see Table 2). The interaction of carotenoids, another group of antioxidants, with chlorophyll molecules excited

Table 2 The changes in antioxidative systems in response to ammonium nutrition

Enzyme/low mass antioxidant	Plant	Parameter	Effect	References
Enzymatic system				
Superoxide dismutase (SOD)	*Phaseolus vulgaris*	Activity	Not changed	Zhu et al. (2000)
	Triticum aestivum	Activity	Increased	Polesskaya et al. (2004)
	Vallisneria natans	Activity	Increased	Wang et al. (2004)
	Pisum sativum	Activity	Increased/ not changed	Domínguez-Valdivia et al. (2008)
	Spinacia oleracea	Activity	Increased/ not changed	Domínguez-Valdivia et al. (2008)
	Arabidopsis thaliana	Activity	Increased/ not changed	Patterson et al. (2010)
	Arabidopsis thaliana	Transcript (FeSOD)	Not changed	Podgórska et al. (2013)
	Arabidopsis thaliana	Transcript (MnSOD)	Not changed	Podgórska et al. (2013)
	Arabidopsis thaliana	Protein (MnSOD)	Not changed	Podgórska et al. (2013)
	Arabidopsis thaliana	Activity (MnSOD)	Increased	Podgórska et al. (2013)
Ascorbate peroxidase (APX)	*Phaseolus vulgaris*	Activity	Increased/ not changed	Gerendás et al. (1997)
	Phaseolus vulgaris	Activity	Increased	Zhu et al. (2000)
	Triticum aestivum	Activity	Increased	Polesskaya et al. (2004)
	Vallisneria natans	Activity	Increased	Wang et al. (2004)
	Pisum sativum	Activity	Increased/ not changed	Domínguez-Valdivia et al. (2008)
	Spinacia oleracea	Activity	Increased	Domínguez-Valdivia et al. (2008)
	Arabidopsis thaliana	Activity	Decreased	Podgórska et al. (2013)
	Arabidopsis thaliana	Transcript (APX1)	Not changed	Podgórska et al. (2013)
	Arabidopsis thaliana	Transcript (sAPX)	Not changed	Podgórska et al. (2013)
	Arabidopsis thaliana	Transcript (tAPX)	Not changed	Podgórska et al. (2013)

(continued)

Table 2 (continued)

Enzyme/low mass antioxidant	Plant	Parameter	Effect	References
Dehydroascorbate reductase (DHAR)	*Phaseolus vulgaris*	Activity	Increased/ not changed	Gerendás et al. (1997)
	Phaseolus vulgaris	Activity	Increased/ not changed	Zhu et al. (2000)
	Arabidopsis thaliana	Activity	Not changed	Podgórska et al. (2013)
	Arabidopsis thaliana	Activity (chl)	Not changed	Podgórska et al. (2013)
	Arabidopsis thaliana	Activity (mt)	Not changed	Podgórska et al. (2013)
Monodehydroascorbate reductase (MDHAR)	*Phaseolus vulgaris*	Activity	Increased/ not changed	Gerendás et al. (1997)
	Phaseolus vulgaris	Activity	Increased/ not changed	Zhu et al. (2000)
	Arabidopsis thaliana	Activity	Not changed	Podgórska et al. (2013)
	Arabidopsis thaliana	Activity (chl)	Not changed	Podgórska et al. (2013)
	Arabidopsis thaliana	Activity (mt)	Increased	Podgórska et al. (2013)
Glutatione reductase (GR)	*Phaseolus vulgaris*	Activity	Not changed/ decreased	Gerendás et al. (1997)
	Phaseolus vulgaris	Activity	Not changed/ increased	Zhu et al. (2000)
	Triticum aestivum	Activity	Increased	Polesskaya et al. (2004)
	Vallisneria natans	Activity	Increased	Wang et al. (2004)
	Pisum sativum	Activity	Increased/ not changed	Domínguez-Valdivia et al. (2008)
	Spinacia oleracea	Activity	Decreased	Domínguez-Valdivia et al. (2008)
	Arabidopsis thaliana	Activity	Increased/ not changed	Patterson et al. (2010)
	Arabidopsis thaliana	Activity	Increased	Podgórska et al. (2013)
	Arabidopsis thaliana	Activity (mt)	Decreased	Podgórska et al. (2013)

(continued)

Table 2 (continued)

Enzyme/low mass antioxidant	Plant	Parameter	Effect	References
Catalase (CAT)	*Triticum aestivum*	Activity	Increased	Polesskaya et al. (2004)
	Vallisneria natans	Activity	Increased	Wang et al. (2004)
	Pisum sativum	Activity	Not changed	Domínguez-Valdivia et al. (2008)
	Spinacia oleracea	Activity	Increased/ not changed	Domínguez-Valdivia et al. (2008)
	Arabidopsis thaliana	Activity	Not changed	Patterson et al. (2010)
	Arabidopsis thaliana	Transcript (CAT1)	Not changed	Hachiya et al. (2010)
Glutathione peroxidase (GPX)	*Pisum sativum*	Activity	Not changed	Domínguez-Valdivia et al. (2008)
	Spinacia oleracea	Activity	Not changed/ decreased	Domínguez-Valdivia et al. (2008)
	Arabidopsis thaliana	Transcript (GPX6)	Increased	Hachiya et al. (2010)
	Arabidopsis thaliana	Transcript (GPX6)	Not changed	Podgórska et al. (2013)
	Arabidopsis thaliana	Transcript (GPX1)	Increased	Podgórska et al. (2013)
Guaiacol peroxidase (GP)	*Triticum aestivum*	Activity	Increased	Polesskaya et al. (2004)
	Vallisneria natans	Activity	Increased	Wang et al. (2004)
	Arabidopsis thaliana	Activity	Increased/ not changed	Patterson et al. (2010)
Low-mass antioxidants				
Ascorbate	*Phaseolus vulgaris*	Oxidation level	Not changed/ increased	Zhu et al. (2000)
	Pisum sativum	Oxidation level	Increased/ decreased	Domínguez-Valdivia et al. (2008)
	Spinacia oleracea	Oxidation level	Not changed	Domínguez-Valdivia et al. (2008)
	Arabidopsis thaliana	Oxidation level	Increased	Podgórska et al. (2013)

(continued)

Table 2 (continued)

Enzyme/low mass antioxidant	Plant	Parameter	Effect	References
Glutathione	*Pisum sativum*	Oxidation level	Not changed/ decreased	Domínguez-Valdivia et al. (2008)
	Spinacia oleracea	Oxidation level	Not changed	Domínguez-Valdivia et al. (2008)
	Arabidopsis thaliana	Oxidation level	Increased	Podgórska et al. (2013)
	Arabidopsis thaliana	Oxidation level (mt)	Increased	Podgórska et al. (2013)
Carotenoids	*Triticum aestivum*	Concentration	Increased	Polesskaya et al. (2004)

by light prevents the formation of ROS in PSII. An increase in carotenoids under NH_4^+ supply has been observed in wheat plants (Polesskaya et al. 2004). Conversely, a decrease in zeaxanthin level has been reported in NH_4^+-grown bean plants (Bendixen et al. 2001).

5 The Engagement of Alternative Pathways in mtETC Under Ammonium Nutrition

Plant mitochondria possess unique alternative nonphosphorylating electron routes that can bypass the linear pathway from Complex I to Complex IV. These alternative electron pathways include additional external and internal NAD(P)H type II dehydrogenases (NDex and NDin, respectively) and the alternative oxidase (AOX) (Møller 2001; Rasmusson et al. 2008). Plant mitochondria have the ability to use NAD(P)H from the matrix via Complex I or alternatively via NDin, and they can also use cytosolic NAD(P)H via NDex activity (Fig. 2). Type II dehydrogenases have low substrate affinity (Møller et al. 1993) but may play an important role under NH_4^+ supply in oxidation the excess reducing power from the cytosol or from the mitochondrial matrix (Guo et al. 2005; Escobar et al. 2006). An increase in type II dehydrogenase transcripts has also been observed under these conditions; the most abundant transcripts are those of the external NADH isoforms, namely *NDB2* and *NDB4* (Escobar et al. 2006; Hachiya et al. 2010, 2012; Podgórska et al. 2013). Escobar et al. (2006) have also found increased NDex activity in *Arabidopsis* mitochondria isolated from ammonium-treated plants, confirming that superfluous cytosolic NAD(P)H may be used by these organelles during NH_4^+ nutrition.

The increase in electron transport from Complex I to Complex IV may lead to an overreduction of the ubiquinone pool or hyperpolarization of the mitochondrial membrane potential, thereby causing the higher generation of ROS. The AOX pathway is known to be ATP-uncoupled, in contrast to the COX pathway, the activity of which is under adenylate control. Cruz et al. (2011) suggested that the rate of AOX contribution to total respiration can to some extent alleviate NH_4^+ toxicity. Therefore, AOX allows for the dissipation of excess reductants without proton motive force generation, particularly when linked to NDex or NDin. The AOX pathway is upregulated when superfluous reductants must be oxidized, thereby preventing ROS production (Noguchi and Yoshida 2007; Raghavendra and Padmasree 2003). Hence, the combined action of NDex, NDin, and AOX prevents excess ROS formation in the mtETC when enhanced respiration rates are required during NH_4^+ growth (Zhu et al. 2000; Guo et al. 2005; Escobar et al. 2006; Podgórska et al. 2013). In line with this response, increased AOX capacity and protein level have been observed in many plant species after transfer to NH_4^+ (Barneix et al. 1984; Frechilla et al. 2002; Lasa et al. 2002; Escobar et al. 2006; Podgórska et al. 2013). Additionally, the transcript levels of several isoforms of AOX have been reported as elevated in response to ammonium stress, the most abundant of which are those of the major isoform *AOX1a* (Escobar et al. 2006; Hachiya et al. 2010, 2012; Patterson et al. 2010). In a limited number of experiments, the contribution of the COX pathway was increased during NH_4^+ nutrition, as has been observed in spinach (Lasa et al. 2002) and *Arabidopsis* (Hachiya et al. 2010).

6 Oxidative Stress and Cellular Damage During Ammonium Supply

A common response to biotic and abiotic stresses is the perturbation of ROS and antioxidant homeostasis (Foyer and Noctor 2005a; De Gara et al. 2010). When ROS production is not effectively balanced by scavenging mechanisms, it may lead to oxidative stress and cellular damage. ROS have the capability of injuring membranes, proteins, and nucleic acids in plants (Gill and Tuteja 2010). As previously described, NH_4^+ supply can lead to increased ROS production (Table 1), and consequent cellular damage has been reported in several plant species (Table 3), despite the upregulation of antioxidant defence (Table 2). This damage can be considered as a marker of oxidative stress. The oxidation of proteins is a very common symptom of increased ROS. Elevated protein carbonylation has been observed in NH_4^+-fed *Arabidopsis* and spinach leaves (Domínguez-Valdivia et al. 2008; Podgórska et al. 2013). The extent of lipid peroxidation, measured as the decomposition product malondialdehyde (MDA), has also been reported as elevated in NH_4^+-grown *Arabidopsis* (Hachiya et al. 2010; Podgórska et al. 2013) and French bean (Gerendás et al. 1997; Zhu et al. 2000).

Table 3 The effect of ammonium nutrition on lipid peroxidation or protein oxidation in leaf tissue

Plant	The period of ammonium nutrition/ conditions	Effect	References
Lipid peroxidation (TBARS content)			
Phaseolus vulgaris	14 days/low light	Not changed	Gerendás et al. (1997)
Phaseolus vulgaris	14 days/high light	Increased	Gerendás et al. (1997)
Phaseolus vulgaris	7–21 days	Increased	Zhu et al. (2000)
Arabidopsis thaliana	3–5 days	Increased	Hachiya et al. (2010)
Arabidopsis thaliana	7 weeks	Increased	Podgórska et al. (2013)
Protein oxidation			
Spinacia oleracea	28 days/low light	Not changed	Domínguez-Valdivia et al. (2008)
Spinacia oleracea	28 days/high light	Increased	Domínguez-Valdivia et al. (2008)
Pisum sativum	28 days/low light	Not changed	Domínguez-Valdivia et al. (2008)
Pisum sativum	28 days/high light	Not changed	Domínguez-Valdivia et al. (2008)
Arabidopsis thaliana	7 weeks	Increased	Podgórska et al. (2013)

7 NO Generation Under Ammonium Nutrition

Most abiotic stresses induce NO production, which is considered as a part of the stress response. Sources of NO in plant cells may include cytosolic NR, the mitochondrial nitrite:NO reductase reaction, mtETC Complexes III and IV, the plasma membrane nitrite:NO reductase (PM-NiNOR) reaction, xanthine oxidoreductase, polyamine (PA)- and hydroxylamine (HA)-mediated pathways, NO synthase-like enzyme (putative), and nonenzymatic chemical reduction (Gupta et al. 2010; Igamberdiev et al. 2014). Although NH_4^+-grown plants do not synthesize NO due to the major NO-producing reaction of NR, significant evidence of NO production in NH_4^+-supplied plants has been observed (Hachiya and Noguchi 2011). Gupta et al. (2013) found lower NO concentrations in the cytosol of NH_4^+-grown tobacco plants compared to that of NO_3^--grown plants, which indirectly determined higher resistance to HR and elicited cell death. Moreover, Planchet et al. (2004) found significant amounts of NO production in NH_4^+-grown tobacco cultures, probably due to mitochondrial emission.

8 ROS Signalling in NH_4^+-Supplied Plants

To cope with fluctuations in the availability of nutrients, including N, higher plants have developed a variety of adaptations that enable them to respond to their internal nutritional status as well as to the external availability of nutrients (Stitt 1999; Coruzzi and Bush 2001; Suzuki et al. 2012). The detection of ammonium stress follows pathways distinct from those of other N sources; nevertheless, the possibility that NH_4^+ itself could act as a signalling molecule is questionable (Coruzzi and Bush 2001; Li et al. 2014). NO_3^- and NH_4^+ growth have been shown to affect distinct signalling networks; specific NH_4^+ responses include increased expression of biotic stress response and plant defence genes (Patterson et al. 2010). ROS are known to play a very important role in signalling and response under stressful conditions (Møller and Sweetlove 2010; Mittler et al. 2011). While $O_2^{\bullet -}$ is very reactive and can easily generate damage, the reduction product H_2O_2 has a longer lifetime and may be considered as an important part of the signalling cascade. ROS can influence the expression of a number of genes and thereby control processes such as growth, cell cycle, programmed cell death, abiotic stress response, pathogen defence, and development (Neill et al. 2002; Gill and Tuteja 2010). The increased ROS level in the NH_4^+-specific pathway may therefore also play a role in signalling, but potential signal transduction intermediates are unknown. The importance of chloroplast ROS (chlROS) in retrograde signalling under NH_4^+ supply has been discussed by Li et al. (2013a, 2014) based on data obtained using the *Arabidopsis ammonium-overly-sensitive 1* (*amos1*) mutant, which possesses a defect in the plastid metalloprotease EGY1. The *Amos1/egy1* mutant was found to reduce H_2O_2 accumulation in chloroplasts (Li et al. 2012). Upon exposure to ammonium, the expression of H_2O_2-responsive genes such as small heat shock proteins (HSPs) was induced in wild-type *Arabidopsis* but not in the *amos1/egy1* mutant (Li et al. 2012). Li et al. (2014) have proposed that, under ammonium stress, chloroplasts receive a signal from the plasmalemma, triggering the AMOS1/EGY1-dependent signalling pathway in which chlROS are involved and regulating the expression of NH_4^+-responsive genes.

Data regarding the contribution of mitochondria to signalling during NH_4^+ growth are rare but confirm that mitochondrial ROS could participate in this signalling network (Hachiya and Noguchi 2011; Podgórska et al. 2013). The rate of mitochondrial ROS production in photosynthetic cells represents only a small portion of overall ROS generation (Foyer and Noctor 2003); however, this site of ROS production may play a notable role. Most interesting in the context of retrograde signalling is the ROS production by mtETC Complex III towards the intermembrane space (Fig. 2), as redox buffering is low in this compartment. The results of cytochemical staining performed in our laboratory indicate that the ROS concentration in the mitochondrial membranes of ammonium-grown plants is higher than in the membranes of nitrate-grown plants (Fig. 1b versus a). Moreover, mitochondrial superoxide production towards the intermembrane space is higher in mitochondria isolated from ammonium-grown plants than in those isolated from nitrate-grown plants (Fig. 1c). Increased mtROS may be necessary as a signal under ammonium nutrition to induce responses such as AOX expression, as has been observed under other stresses (Clifton et al. 2006).

Reactive nitrogen species have emerged as key molecules in plant signal transduction pathways (Wojtaszek 2000; Neill et al. 2002). NO features as a regulatory factor in the algae *Chlamydiomonas reinhardii*, in which the compound impedes N uptake under NH_4^+ nutrition (Sanz-Luque et al. 2013). NO has been proposed as a direct signalling trigger under different N supplies, thereby regulating mtETC activity, inducing AOX, and inhibiting COX activity (Hachiya and Noguchi 2011) .

The major antioxidants regulate ROS concentrations and thereby determine signal transduction. However, oxidised molecules of ascorbate and glutathione may be considered as independent signals in this network (Foyer and Noctor 2005a, b). Additionally, the cellular damage generated by ROS production can be seen as a signal that induces a plant response. The most prominent signalling molecules are oxidised proteins (Møller and Sweetlove 2010) and oxidised lipids (Farmer and Mueller 2013) generated in the chloroplasts or the mitochondria. During NH_4^+-nutrition, these mitochondrial oxidised proteins and lipids are elevated in plant tissues (Podgórska et al. 2013). It should be noted that several other molecules unconnected with ROS metabolism can act in the metabolic interface as signals derived downstream from NH_4^+ assimilation and from the environment; these molecules include Glu and 2-OG (Stitt et al. 2002).

9 Mutants with Modified Sensitivity to Ammonium Related to ROS Metabolism

Mutants are powerful tools for connecting physiological processes and plant development. Recently, Li et al. (2014) have described a possible connection between specific genetic loci related to NH_4^+ sensitivity in *Arabidopsis* and chlROS formation based on results obtained using the *amos1/egy1* mutant (discussed above). However, it should be noted that these *amos1/egy1* seedlings exhibited ammonium syndrome symptoms only when their shoots were in direct contact with ammonium ions, growing normally when only their roots were in contact with NH_4^+. Therefore, although the results reported by Li et al. (2012) are interesting, the proposed signalling pathway may not be universal.

Furthermore, the connection of NH_4^+ toxicity and ascorbic acid deficiency has been analysed using *Arabidopsis vtc1* and *hsn* mutants with defects in the gene encoding GDP mannose pyrophosphorylase (GMPase), which is essential for synthesizing the GDP-mannose crucial to ascorbate and N-glycoprotein biosynthesis (Barth et al. 2010; Qin et al. 2008; Li et al. 2010). The *vtc1* mutant possesses 70 % less ascorbate and increased peroxidase activity in the leaves compared with the wild type (Veljovic-Jovanovic et al. 2001). However, the symptoms of ammonium toxicity in the *vtc1* mutant were most likely independent of ascorbate deficiency. Instead, the hypersensitivity to NH_4^+ in these plants has been suggested to be related to alterations in NH_4^+ metabolism, disturbed N-glycosylation, defects in cell wall synthesis, or increased NO levels (Barth et al. 2010; Kempinski et al. 2011; Li et al. 2011).

Additional interesting data have been obtained by Cao et al. (1993), who observed that *aux1*, *axr1*, and *axr2* mutants were resistant to ammonium-induced inhibition of growth. *Arabidopsis* aux1, *axr1*, and *axr2* mutants have been reported as auxin resistant (Cao et al. 1993 and references therein). The link between auxin action and ammonium-induced symptoms is interesting in the context of apoplast ROS metabolism in response to both stimuli (Schopfer et al. 2002; Patterson et al 2010); however, these phenomena require further study.

Molecular, genetic, and biochemical investigations using different mutant plants support the correlation between NH_4^+ nutrition and ROS signalling. However, considering the complexity of the N assimilation network and the reactions involved in ROS production, large numbers of mutants, mutant combinations, or overexpressor lines must to be analysed in the future.

Concluding Remarks

The metabolic pathways engaged in the interaction of chloroplastic, cytosolic, and mitochondrial processes are different in NH_4^+- and NO_3^--grown plants. As a consequence of NH_4^+ nutrition, enhanced rates of ROS production lead to oxidative stress in plants. In ammonium-grown plants, sufficient reductant sinks are crucial for alleviating NH_4^+ toxicity, compensating for the lack of NAD(P)H consumption by NR and NiR. However, the unbalanced increase in ROS can lead to oxidative stress. In this review, we highlight the role of mitochondria in maintaining redox homeostasis under ammonium supply. Although AOX activity may reduce ROS production in the mtETC, the generation of ROS at Complex III towards the intermembrane space may play an important role in retrograde signalling in NH_4^+-grown plants (Figs. 1 and 2).

Acknowledgments We are grateful to Prof. A.M. Rychter (University of Warsaw) for a critical reading of the manuscript.

This work was supported by Grant 2011/01/N/NZ3/02953 from the National Science Centre (NCN, Poland) given A.P.

References

Ariz I, Cruz C, Moran JF et al (2011) Depletion of the heaviest stable N isotope is associated with NH_4^+/NH_3 toxicity in NH_4^+-fed plants. BMC Plant Biol 11:83–96

Ariz I, Asensio AC, Zamarreño AM et al (2013) Changes in the C/N balance caused by increasing external ammonium concentrations are driven by carbon and energy availabilities during ammonium nutrition in pea plants: the key roles of asparagine synthetase and anaplerotic enzymes. Physiol Plant 148:522–537

Barker AV, Mills HA (1980) Ammonium and nitrate nutrition of horticultural crops. In: Janick J (ed) Horticultural reviews, vol 2. Wiley, New York, pp 395–423

Barneix AJ, Breteler H, van de Geijn SC (1984) Gas and ion exchanges in wheat roots after nitrogen supply. Physiol Plant 61:357–362

Barth C, Gouzd ZA, Steele HP et al (2010) A mutation in GDP-mannose pyrophosphorylase causes conditional hypersensitivity to ammonium, resulting in Arabidopsis root growth inhibition, altered ammonium metabolism, and hormone homeostasis. J Exp Bot 61:379–394

Bendixen R, Gerendás J, Schinnera K et al (2001) Difference in zeaxanthin formation in nitrate- and ammonium -grown *Phaseolus vulgaris*. Physiol Plant 111:255–261

Blackwell RD, Murray AJS, Lea PJ et al (1988) Photorespiratory amino donors, sucrose synthesis and the induction of CO2 fixation in barley deficient in glutamine synthetase and/or glutamate synthase. J Exp Bot 39:845–858

Bloom AJ, Sukrapanna SS, Warner RL (1992) Root respiration associated with ammonium and nitrate absorption and assimilation by barley. Plant Physiol 99:1294–1301

Britto DT, Kronzucker HJ (2002) NH_4^+ toxicity in higher plants: a critical review. J Plant Physiol 159:567–584

Britto DT, Kronzucker HJ (2005) Nitrogen acquisition, PEP carboxylase, and cellular pH homeostasis: new views on old paradigms. Plant Cell Environ 28:1396–1409

Britto DT, Kronzucker HJ (2013) Ecological significance and complexity of N-source preference in plants. Ann Bot 112:957–963

Britto DT, Siddiqi MY, Glass ADM et al (2001) Futile transmembrane NH_4 cycling: a cellular hypothesis to explain ammonium toxicity in plants. PNAS 98:4255–4258

Brück H, Guo S (2006) Influence of N form on growth and photosynthesis of *Phaseolus vulgaris* L. plants. J Plant Nutr Soil Sci 169:849–856

Cao Y, Glass AD, Crawford NM (1993) Ammonium inhibition of Arabidopsis root growth can be reversed by potassium and by auxin resistance mutations aux1, axr1, and axr2. Plant Physiol 102:983–989

Clifton R, Millar AH, Whelan J (2006) Alternative oxidases in Arabidopsis: a comparative analysis of differential expression in the gene family provides new insights into function of non-phosphorylating bypasses. Biochim Biophys Acta 1757:730–741

Coruzzi G, Bush DR (2001) Nitrogen and carbon nutrient and metabolite signaling in plants. Plant Physiol 125:61–64

Coskun D, Britto DT, Li M et al (2013) Rapid ammonia gas transport accounts for futile transmembrane cycling under NH_3/NH_4^+ toxicity in plant roots. Plant Physiol 163:1859–1867

Cramer MD, Lewis OAM (1993) The influence of NO_3^- and NH_4^+ nutrition on the gas exchange characteristics of the roots of wheat (*Triticum aestivum*) and maize (*Zea mays*) plants. Ann Bot 72:37–46

Cruz C, Bio AFM, Domínguez-Valdivia MD et al (2006) How does glutamine synthetase activity determine plant tolerance to ammonium? Planta 223:1068–1080

Cruz C, Domínguez-Valdivia MD, Aparicio-Tejo PM et al (2011) Intra-specific variation in pea responses to ammonium nutrition leads to different degrees of tolerance. Environ Exp Bot 70:233–243

De Gara L, Locato V, Dipierro S et al (2010) Redox homeostasis in plants. The challenge of living with endogenous oxygen production. Respir Physiol Neurobiol 173S:S13–S19

Domínguez-Valdivia MD, Aparicio-Tejo PM, Lamsfus C et al (2008) Nitrogen nutrition and antioxidant metabolism in ammonium-tolerant and -sensitive plants. Physiol Plant 132: 359–369

Duan YH, Zgang YL, Ye LT et al (2007) Responses of rice cultivars with different nitrogen use efficiency to partial nitrate nutrition. Ann Bot 99:1153–1160

Escobar MA, Geisler DA, Rasmusson AG (2006) Reorganization of the alternative pathways of the Arabidopsis respiratory chain by nitrogen supply: opposing effects of ammonium and nitrate. Plant J 45:775–788

Farmer EE, Mueller MJ (2013) ROS-mediated lipid peroxidation and RES-activated signaling. Annu Rev Plant Biol 64:429–450

Foyer CH, Noctor G (2003) Redox sensing and signalling associated with reactive oxygen in chloroplasts, peroxisomes and mitochondria. Physiol Plant 119:355–364

Foyer CH, Noctor G (2005a) Oxidant and antioxidant signalling in plants: a re-evaluation of the concept of oxidative stress in a physiological context. Plant Cell Environ 28:1056–1071

Foyer CH, Noctor G (2005b) Redox homeostasis and antioxidant signaling: a metabolic interface between stress perception and physiological responses. Plant Cell 17:1866–1875

Foyer CH, Noctor G, Lelandais M et al (1994) Short-term effects of nitrate, nitrite and ammonium assimilation on photosynthesis, carbon partitioning and protein phosphorylation in maize. Planta 192:211–220

Frechilla S, Lasa B, Aleu M et al (2002) Short-term ammonium supply stimulates glutamate dehydrogenase activity and alternative pathway respiration in roots of pea plants. J Plant Physiol 159:811–818

Gerendás J, Ratcliffe RG (2000) Intracellular pH regulation in maize root tips exposed to ammonium at high external pH. J Exp Bot 51:207–219

Gerendás J, Zhu Z, Bendixen R et al (1997) Physiological and biochemical processes related to ammonium toxicity in higher plants. Z Pflanzenernähr Bodenk 160:239–251

Gill SS, Tuteja N (2010) Reactive oxygen species and antioxidant machinery in abiotic stress tolerance in crop plants. Plant Physiol Biochem 48:909–930

Glass AD, Britto DT, Kaiser BN et al (2002) The regulation of nitrate and ammonium transport systems in plants. J Exp Bot 53:855–864

Guo S, Schinner K, Sattelmacher B et al (2005) Different apparent CO_2 compensation points in nitrate- and ammonium-grown *Phaseolus vulgaris* and the relationship to non-photorespiratory CO_2 evolution. Physiol Plant 123:288–301

Guo S, Zhou Y, Gao Y et al (2007) New insights into the nitrogen form effect on photosynthesis and photorespiration. Pedosphere 17:601–610

Gupta KJ, Fernie AR, Kaiser WM et al (2010) On the origins of nitric oxide. Trends Plant Sci 16: 160–168

Gupta KJ, Brotman Y, Segu S et al (2013) The form of nitrogen nutrition affects resistance against *Pseudomonas syringae* pv. *phaseolicola* in tobacco. J Exp Bot 64:553–568

Hachiya T, Noguchi K (2011) Integrative response of plant mitochondrial electron transport chain to nitrogen source. Plant Cell Rep 30:195–204

Hachiya T, Watanabe CK, Boom C et al (2010) Ammonium-dependent respiratory increase is dependent on the cytochrome pathway in *Arabidopsis thaliana* shoots. Plant Cell Environ 33:1888–1897

Hachiya T, Watanabe CK, Fujimoto M et al (2012) Nitrate addition alleviates ammonium toxicity without lessening ammonium accumulation, organic acid depletion and inorganic cation depletion in Arabidopsis thaliana shoots. Plant Cell Physiol 53:577–591

Hoffmann A, Milde S, Desel C et al (2007) N form–dependent growth retardation of *Arabidopsis thaliana* seedlings as revealed from physiological and microarray studies. J Plant Nutr Soil Sci 170:87–97

Huppe HC, Turpin DH (1994) Integration of carbon and nitrogen metabolism in plant and algal cells. Annu Rev Plant Physiol Plant Mol Biol 45:577–607

Husted S, Schjoerring JK (1995) Apoplastic pH and ammonium concentration in leaves of *Brassica napus* L. Plant Physiol 109:1453–1460

Igamberdiev AU, Ratcliffe RG, Gupta KJ (2014) Plant mitochondria: source and target for nitric oxide. Mitochondrion. http://dx.doi.org/10.1016/j.mito.2014.02.003

Kaiser WM, Weiner H, Kandlbinder A et al (2002) Modulation of nitrate reductase: some new insights, an unusual case and a potentially important side reaction. J Exp Bot 53:875–882

Kempinski CF, Haffar R, Barth C (2011) Toward the mechanism of NH_4^+ sensitivity mediated by *Arabidopsis* GDP-mannose pyrophosphorylase. Plant Cell Environ 34:847–858

Kronzucker HJ, Britto DT, Davenport RJ et al (2001) Ammonium toxicity and the real cost of transport. Trends Plant Sci 6:335–337

Labboun S, Tercé-Laforgue T, Roscher A et al (2009) Resolving the role of plant glutamate dehydrogenase. I. In vivo real time nuclear magnetic resonance spectroscopy experiments. Plant Cell Physiol 50:1761–1773

Lager I, Andréasson O, Dunbar TL et al (2010) Changes in external pH rapidly alter plant gene expression and modulate auxin and elicitor responses. Plant Cell Environ 33:1513–1528

Lasa B, Frechilla S, Apricio-Tejo PM et al (2002) Alternative pathway respiration is associated with ammonium ion sensitivity in spinach and pea plants. Plant Growth Regul 37:49–55

Lewis OAM (1992) Plants and nitrogen. In: Lewis OAM (ed) Studies in biology series. Cambridge University Press, London, pp 1–112

Li Q, Li B, Kronzucker HJ et al (2010) Root growth inhibition by NH_4^+ in Arabidopsis is mediatedby the root tip and is linked to NH_4^+ efflux and GMPase activity. Plant Cell Environ 33:1529–1542

Li G, Dong G, Li B et al (2011) Isolation and characterization of a novel ammonium overly sensitive mutant, amos2, in Arabidopsis thaliana. Planta 235:239–252

Li B, Li Q, Xiong L et al (2012) Arabidopsis plastid AMOS1/EGY1 integrates abscisic acid signaling to regulate global gene expression response to ammonium stress. Plant Physiol 160: 2040–2051

Li B, Kronzucker HJ, Shi W (2013a) Molecular components of stress-responsive plastid retrograde signaling networks and their involvement in ammonium stress. Plant Signal Behav 8:e23107

Li G, Li B, Dong G et al (2013b) Ammonium-induced shoot ethylene production is associated with the inhibition of lateral root formation in *Arabidopsis*. J Exp Bot 64:1413–1425

Li B, Li G, Kronzucker HJ et al (2014) Ammonium stress in Arabidopsis: signaling, genetic loci, and physiological targets. Trends Plant Sci 19:107–114

Loqué D, Lalonde S, Looger LL et al (2007) A cytosolic trans-activation domain essential for ammonium uptake. Nature 446:195–198

Ludewig U, Neuhäuser B, Dynowski M (2007) Molecular mechanisms of ammonium transport and accumulation in plants. FEBS Lett 581:2301–2308

Łukawska-Kuźma K, Podgórska A, Rychter AM (2012) Plasma membrane-generated ROS and their role in cell growth of cucumber mitochondrial mutant. Acta Physiol Plant 34:721–730

Masclaux-Daubresse C, Reisdorf-Cren M, Pageau K et al (2006) Glutamine synthetase-glutamate synthase pathway and glutamate dehydrogenase play distinct roles in the sink-source nitrogen cycle in tobacco. Plant Physiol 140:444–456

Masclaux-Daubresse C, Daniel-Vedele F, Dechorgnat J et al (2010) Nitrogen uptake, assimilation and remobilization in plants: challenges for sustainable and productive agriculture. Ann Bot 105:1141–1157

Miflin BJ, Habash DZ (2002) The role of glutamine synthetase and glutamate dehydrogenase in nitrogen assimilation and possibilities for improvement in the nitrogen utilization of crops. J Exp Bot 53:979–987

Mittler R, Vanderauwera S, Suzuki N et al (2011) ROS signaling: the new wave? Trends Plant Sci 6:300–309

Møller IM (2001) Plant mitochondria and oxidative stress: electron transport, NADPH turnover, and metabolism of reactive oxygen species. Annu Rev Plant Physiol Plant Mol Biol 52: 561–591

Møller IM, Sweetlove LJ (2010) ROS signalling—specificity is required. Trends Plant Sci 15: 370–374

Møller IM, Rasmusson AG, Fredlund KM (1993) NAD(P)H-ubiquinone oxidoreductases in plant mitochondria. J Bioenerg Biomembr 25:377–384

Neill SJ, Desikan R, Clarke A et al (2002) Hydrogen peroxide and nitric oxide as signalling molecules in plants. J Exp Bot 53:1237–1247

Noctor G, Foyer CH (1998) A re-evaluation of the ATP: NADPH budget during C3 photosynthesis: a contribution from nitrate assimilation and its associated respiratory activity? J Exp Bot 49:1895–1908
Noctor G, De Paepe R, Foyer CH (2007) Mitochondrial redox biology and homeostasis in plants. Trends Plant Sci 12:125–134
Noguchi K, Yoshida K (2007) Interaction between photosynthesis and respiration in illuminated leaves. Mitochondrion 8:87–99
Patterson K, Cakmak T, Cooper A et al (2010) Distinct signalling pathways and transcriptome response signatures differentiate ammonium and nitrate-supplied plants. Plant Cell Environ 33:1486–1501
Planchet E, Gupta KJ, Sonoda M et al (2004) Nitric oxide emission from tobacco leaves and cell suspensions: rate limiting factors and evidence for the involvement of mitochondrial electron transport. Plant J 41:732–743
Podgórska A, Gieczewska K, Łukawska-Kuźma K et al (2013) Long-term ammonium nutrition of Arabidopsis increases the extrachloroplastic NAD(P)H/NAD(P)$^+$ ratio and mitochondrial reactive oxygen species level in leaves but does not impair photosynthetic capacity. Plant Cell Environ 36:2034–2045
Polesskaya OG, Kashirina EI, Alekhina ND (2004) Changes in the activity of antioxidant enzymes in wheat leaves and roots as a function of nitrogen source and supply. Russ J Plant Physiol 51:615–620
Qin C, Qian W, Wang W et al (2008) GDP-mannose pyrophosphorylase is a genetic determinant of ammonium sensitivity in Arabidopsis thaliana. PNAS 105:18308–18313
Raab TK, Terry N (1994) Nitrogen source regulation of growth and photosynthesis in *Beta vulgaris* L. Plant Physiol 105:1159–1166
Raghavendra AS, Padmasree K (2003) Beneficial interactions of mitochondrial metabolism with photosynthetic carbon assimilation. Trends Plant Sci 8:546–553
Rasmusson AG, Geisler DA, Møller IM (2008) The multiplicity of dehydrogenases in the electrontransport chain of plant mitochondria. Mitochondrion 8:47–60
Sanz-Luque E, Ocaña-Calahorro F, Llamas A et al (2013) Nitric oxide controls nitrate and ammonium assimilation in *Chlamydomonas reinwardtii.* J Exp Bot 64:3373–3383
Schjoerring JK, Husted S, Mäck G et al (2002) The regulation of ammonium translocation in plants. J Exp Bot 53:883–890
Schopfer P, Liszkay A, Bachtold M et al (2002) Evidence that hydroxyl radicals mediate auxin-induced extension growth. Planta 214:821–828
Schortemeyer M, Stamp P, Feil B (1997) Ammonium tolerance and carbohydrate status in maize cultivars. Ann Bot 79:25–30
Skopelitis DS, Paranychianakis NV, Kouvarakis A et al (2007) The isoenzyme 7 of tobacco NAD (H)-dependent glutamate dehydrogenase exhibits high deaminating and low aminating activities in vivo. Plant Physiol 145:1726–1734
Suzuki N, Koussevitzky S, Mittler R et al (2012) ROS and redox signalling in the response of plants to abiotic stress. Plant Cell Environ 35:259–270
Stitt M (1999) Nitrate regulation of metabolism and growth. Curr Opin Plant Biol 2:178–186
Stitt M, Müller C, Matt P et al (2002) Steps towards an integrated view of nitrogen metabolism. J Exp Bot 53:959–970
Szal B, Podgórska A (2012) The role of mitochondria in leaf nitrogen metabolism. Plant Cell Environ 35:1756–1768
Szal B, Łukawska K, Zdolińska I et al (2009) Chilling stress and mitochondrial genome rearrangement in the MSC16 cucumber mutant affect the alternative oxidase and antioxidant defense system to a similar extent. Physiol Plant 137:435–445
Szczerba MW, Britto DT, Balkos KD et al (2008) Alleviation of rapid, futile ammonium cycling at the plasma membrane by potassium reveals K^+-sensitive and -insensitive components of NH_4^+ transport. J Exp Bot 59:303–313

Veljovic-Jovanovic SD, Pignocchi C, Noctor G et al (2001) Low ascorbic acid in the *vtc-1* mutant of Arabidopsis is associated with decreased growth and intracellular redistribution of the antioxidant system. Plant Physiol 127:426–435

Walch-Liu P, Neumann G, Bangerth F et al (2000) Rapid effects of nitrogen form on leaf morphogenesis in tobacco. J Exp Bot 51:227–237

Walch-Liu P, Neumann G, Engels C (2001) Response of shoot and root growth to supply of different nitrogen forms is not related to carbohydrate and nitrogen status of tobacco plants. J Plant Nutr Soil Sci 164:97–103

Wang R, Tischner R, Gutiérrez RA et al (2004) Genomic analysis of the nitrate response using a nitrate reductase-null mutant of Arabidopsis. Plant Physiol 136:2512–2522

Wang C, Zhang SH, Wang PF et al (2008) Metabolic adaptations to ammonia-induced oxidative stress in leaves of the submerged macrophyte *Vallisneria natans* (Lour.) Hara. Aquat Toxicol 87:88–98

Wiesler F (1997) Agronomical and physiological aspects of ammonium and nitrate nutrition of plants. Z Pflanzenernähr Bodenk 160:227–238

Wojtaszek P (1997) Oxidative burst: an early plant response to pathogen infection. Biochem J 322: 681–692

Wojtaszek P (2000) Nitric oxide in plants to NO or not to NO. Phytochemistry 54:1–4

Yamasaki H, Sakihama Y (2000) Simultaneous production of nitric oxide and peroxynitrite by plant nitrate reductase: in vitro evidence for the NR-dependent formation of active nitrogen species. FEBS Lett 468:89–92

Zhu Z, Gerendás J, Bendixen R et al (2000) Different tolerance to light stress in NO_3^-- and NH_4^+-grown *Phaseolus vulgaris* L. Plant Biol 2:558–570

Allelopathic Compounds as Oxidative Stress Agents: Yes or NO

Agnieszka Gniazdowska, Urszula Krasuska, Olga Andrzejczak, and Dorota Soltys

1 Introduction

In the middle of twentieth century reactive oxygen species (ROS) have been recognized as toxic molecules, responsible for oxidative damage to cell macromolecules, mainly lipids, proteins, and nucleic acids. In early 1990s ROS generation was reported in plants and associated initially with photosynthetic and photochemical reactions, then the burst of ROS production was accepted as a stimulator of the plant defense mechanisms induced by attack of pathogenic organisms (for review, see Nanda et al. 2010; Torres 2010). Production of ROS and induction of oxidative stress was pointed also as a common plant reaction to environmental stresses such as temperature, salinity, drought, heavy metals, or pollutants (Kacperska-Lewak 2004; Miller et al. 2008; Kreslavski et al. 2012 and references therein, Sharma et al. 2012 and references therein). Nowadays, there is a lot of evidence that ROS play crucial role in plant physiology, acting as signaling molecules in regulation of all plant's ontogenetic phases starting from seed germination, while terminating at seed formation and fruit ripening, plant senescence or death (Gapper and Dolan 2006).

Allelopathy, according to resolution of the International Allelopathy Society from 1998 is defined as biotic interactions between plants, microorganisms and

A. Gniazdowska (✉) • U. Krasuska • O. Andrzejczak
Department of Plant Physiology, Warsaw University of Life Sciences-SGGW, Nowoursynowska 159, 02-776 Warsaw, Poland
e-mail: agnieszka_gniazdowska@sggw.pl; urszula_krasuska@sggw.pl; olga_andrzejczak@sggw.pl

D. Soltys
Plant Breeding and Acclimatization Institute, Młochów Research Center, Platanowa 19, 05-831 Młochów, Poland
e-mail: d.soltys@ihar.edu.pl

K.J. Gupta, A.U. Igamberdiev (eds.), *Reactive Oxygen and Nitrogen Species Signaling and Communication in Plants*, Signaling and Communication in Plants 23,
DOI 10.1007/978-3-319-10079-1_8

algae induced by chemicals (allelochemicals) released into the environment. In general, most allelopathic interactions are negative due to phytotoxicity of allelochemicals. Numerous allelochemicals belonging to the phenolic, flavonoid, terpenoid classes have been suggested as the compounds implicated in plant growth suppression, including model (*Arabidopsis thaliana*) or weed and crop plant species. When these compounds suppress plant growth, the phenomenon is considered to be a biotic stress termed "allelochemical stress" (Cruz-Ortega et al. 2002). Although, in some circumstances low concentrations of allelochemicals may stimulate growth of acceptor plants, and this phenomenon is recognized as hormesis.

As suggested for other biotic stresses, in allelopathic interaction an essential function of ROS was indicated several years ago by some authors (Weir et al. 2004; Gniazdowska and Bogatek 2005; Cruz-Ortega et al. 2007). Similarly to other stresses, in allelopathy, a shift from a regulatory role of ROS in cell signaling to their toxicity is probably related to changes in homeostasis of ROS maintained by imbalance of ROS production and ROS scavenging. Plants contain cellular system responsible for control of concentration of ROS and consisting of soluble antioxidant compounds, such as ascorbic acid, vitamin E, and glutathione, and a battery of enzymes that can efficiently scavenge ROS: superoxide dismutase (SOD), catalase (CAT), peroxidase (POX), thioredoxin (Trx), and the enzymes of Asada–Halliwell–Foyer pathway (Foyer and Noctor 2005).

Based on many reports published in last 10 years there is no doubt that induction of oxidative stress is one of the putative mode of actions of many allelochemicals. Despite an increasing amount of data on ROS in allelopathy, it should be emphasized that relatively little is known about the primary mechanism by which allelopathic compounds induce oxidative stress in plants. There are still very rare reports on molecular mechanisms; it is difficult to distinguish whether they are causes or consequences of ROS production. In this particular review, we try to present a comprehensive view on ROS generation in plants subjected to allelopathy stress, and mechanisms of ROS scavenging in comparison with physiological effects of ROS imbalance induced by allelochemicals. We have focused also on ROS acting as signaling agents that participate in formation of plant response to allelopathic compounds. Most of presented data are based on experiments conducted in laboratory conditions, using solutions of isolated chemicals and performed on model plants, therefore, in some cases they may refer to phytotoxicity or toxic interactions rather than allelopathy in its "pure form." Nevertheless, we hope that the presented data referring to particular allelochemicals may be useful to create a general, global view that allows to link plant allelopathic interaction to metabolism and regulatory function of ROS.

2 Allelochemicals Impact on ROS Production and Scavenging, Gene Expression, and Signal Transduction Is Accompanied by Deterioration of Cellular Structures of Acceptor Plants

From the beginning of XXI century there is an increasing amount of publications referring to induction of oxidative stress in plants subjected to allelopathic interactions. Some reports were focused on prooxidant action of mixture of allelochemicals present in extracts, leachates, exudates, while the others pointed on particular isolated compounds exhibiting strong phytotoxicity.

In the literature, there is an increasing amount of examples of induction of ROS production in plants suffering from their neighborhood. In this chapter some examples of the link between plant morphophysiological response to allelopathic plant or extract/compound and ROS production/metabolism are described.

Sunflower (*Helianthus annuus* L.) belongs to the group of crop plants of high allelopathic potential, due to variety of chemicals synthesized in its tissue (Macias et al. 1999). Water extract from sunflower leaves inhibited seed germination and growth of mustard (*Sinapis alba* L.) (Oracz et al. 2007), due to over-accumulation of ROS, which were not sufficiently scavenged by activated antioxidant enzymatic system (CAT, SOD and glutathione reductase—GR). Allelochemical stress caused by *Sicyos deppei* G. Don, an invasive annual tendril-bearing vine, endemic to Mexico, was specified by producing an oxidative imbalance evidenced by alteration in ROS generation and modification of antioxidant cellular system in roots of tomato (*Solanum lycopersicum* L.) (Lara-Nuñez et al. 2006; Cruz-Ortega et al. 2007). Despite well-described overproduction of ROS in tomato plants, observed mainly in short-term reaction to *S. deppei* leachates, the authors stated that they believe that it is not an oxidative damage responsible for toxicity of the plant (Cruz-Ortega et al. 2007), so it is possible that generated ROS may only act as signals activating cascade of other events leading to cell malformations. On the other hand, studies of Lara-Nuñez et al. (2006) showed decline inactivity of NADPH oxidase in roots of tomato exposed to leachate of *S. deppei*. As activity of this enzyme is known to control root development and growth by generation of ROS (Foreman et al. 2003), it could be possible that the inhibition of tomato root elongation by allelochemicals of *S. deppei* might be due to the inhibition of NADPH oxidase, and the subsequent reduction in root ROS level in prolonged treatment.

Although, the precise molecular target of ROS generated in plants responding to allelochemicals is not fully recognized, there is no doubt that several allelochemicals act as prooxidants. Allelopathy is not limited only to terrestrial ecosystems, but can also exist in aquatic ones, and what is more, it is a promising tool in water bloom control (Qian et al. 2009 and references therein). *N*-phenyl-2-naphthylamine, which has been shown to exist in exudates of water hyacinth (*Eichhornia crassipes* (Mart.) Solms), is an antialgal allelochemical inhibiting growth of unicellular green alga *Chlorella vulgaris* at concentration range of

0.5–5 mg L^{-1} (Qian et al. 2009). Activities of SOD and POX increased in lower (2.5 mg L^{-1}) concentration and decreased at higher (4 mg L^{-1}) concentration of *N*-phenyl-2-naphthylamine, while CAT activity exhibited almost opposite pattern. The constantly high level of malondialdehyde (MDA) in cells subjected to the allelochemical suggested that the antioxidant enzymatic system although induced, did not completely eliminate generated ROS. Exposure of the alga to *N*-phenyl-2-naphthylamine at high concentration did not increase the activities of the antioxidants, so the authors speculated that in this case, the allelochemical could directly inhibit antioxidizing enzymes in some way, leaving the plant under the risk to oxidative damage (Qian et al. 2009).

Berberine produced in Chinese herb golden threat (*Coptis chinensis* Franch) is also known to act as allelochemical in aquatic ecosystems as it inhibits growth of cyanobacteria *Microcystis aeruginosa* Kützing responsible for harmful algal blooms. Berberine (0.005–0.03 %) enhanced superoxide anion ($O_2^{\bullet -}$) level, upregulated SOD activity at low concentration, while downregulated it at higher concentration. In the opposite, reduced glutathione (GSH) level increased dependently on concentration of allelochemical and duration of treatment (Zhang et al. 2011). Changes in oxidative damage (MDA content) and antioxidant responses in *M. aeruginosa* cells under berberine stress suggested that the algal antioxidant defense system might be susceptible to berberine, as it was observed in case of other allelochemicals, e.g., previously mentioned *N*-phenyl-2-naphthylamine, and oxidative damage might be one of the mechanisms of mode of action of berberine in inhibition of *M. aeruginosa* growth. More detailed investigation have been done on ethyl 2-methylacetoacetate (EMA) isolated from common reed (*Phragmites communis* Trin.) (Li and Hu 2005). It exhibited allelopatic activity on *Chlorella pyrenoidosa* and *M. aeruginosa* at concentration range 0.25–4 mg L^{-1}. The algae responded to EMA at lower concentration by increasing activity of SOD and POX, however, higher concentration of EMA led to decreased activity of the enzymes (Li and Hu 2005). A peak in CAT activity in *M. arunginosa* cells treated with EMA occurred after dissolution of a medium concentration of allelochemical (1 mg L^{-1}) and generally declined upon increasing concentration and longer than 2 days exposure to allelochemical (Hong et al. 2008). Transient increase in CAT activity coexisted with peaks in content of reduced form of ascorbic acid (ASA) and GSH. It seems that in this case activation of CAT and high availability of ASA and GSH were the most important reactions in algae defense against oxidative stress induced by mild allelopathy stress, but the protective function would be exhausted with extended duration and dose of treatment with phytotoxin. Induction of oxidative stress by EMA was detected also in marine diatom *Phaeodactylum tricornutum* Bohlin (Yang et al. 2011). In this organism EMA enhanced SOD, CAT, and GR activities in dose-dependent manner, similarly as induced generation of ROS. Other GSH depending enzymatic activities such as glutathione peroxidase (GPX) and glutathione *S*-transferase (GST) increased more remarkably under EMA stress, suggesting that they might serve as more effective defense tool than CAT and GR in avoidance of oxidative damage in algal cells. In contrast to described above experiments focused mostly on modification of ROS detoxifying mechanisms,

Rudrappa et al. (2007) pointed on molecular target of gallic acid (2,3,5-dihydroxy benzoic acid)—a phenolic compound exuded by roots of common reed. They demonstrated elevated levels of ROS in roots of *A. thaliana* treated with 50 μM gallic acid or root exudates of common reed. The generated ROS were completely quenched when the roots were exposed to ascorbic acid (ASA). With both root exudates and gallic acid treatment, the microtubule assembly appeared to be dismantled and formed localized aggregates of microtubule fibers. This effect was reversed by antioxidant in a similar manner as was detected for ROS production. The key role of ROS in plant response to gallic acid was confirmed in *A. thaliana*. Several ROS-related gene products, such as ubiquinol—cytochrome c reductase complex, Cu–Zn SOD, POX, and cytochrome P450, were upregulated in plants exposed to gallic acid (20 mg kg^{-1}) (Golisz et al. 2008). These results suggest that gallic acid which is also a potent antioxidant (Soobrattee et al. 2005) generated elevated level of ROS. Moreover, more than twofold increase in transcription level of gene of putative GST was noticed (Golisz et al. 2008).

α-Tubulins as molecular targets in allelochemicals mode of action was under investigation also in onion (*Allium cepa* L.) and maize (*Zea mays* L.) roots subjected to cyanamide, synthesized by hairy vetch (*Vicia villosa* Roth.) (Soltys et al. 2011, 2014). In cells of root tips of both plant species treated with cyanamide some disappearance of the characteristic cytoskeleton patterns were detected, together with induction of ROS production, suggesting a direct link between ROS and cytoskeleton action in cyanamide-stressed organs.

Juglone (5-hydroxy-1,4-naphthoquinone) is one of the most commonly known allelochemical, produced by walnut (*Juglans nigra* L.). It represses seed germination, growth, and development of various plant species and acts as inhibitor of photosynthesis, respiration, or nutrient uptake. The phytotoxic effects of juglone may be attributed to its ability to disrupt function of electron transport chain both in chloroplasts and mitochondria (Hejl et al. 1993). Induction of oxidative stress by juglone treatment was investigated in a range of plants, e.g., maize (Mylona et al. 2007), rice (*Oryza sativa* L.) (Chi et al. 2011), or soybean (*Glycine max* L.) (Böhm et al. 2006). In germinating maize kernels juglone (0.1–10 mM) induced enzymatic activity of CAT, mainly CAT2 isoform, with no effect on SOD activity. In contrast to enzymatic activity of SOD, relative expression of *CAT* and *SOD* gene transcripts was upregulated by juglone (0.1–1 mM). Authors stated that existing data revealed that ROS can induce enzymatic activities of CAT in maize kernels by possible involvement of ROS in the signaling cascade affecting CAT responses (Mylona et al. 2007). Moreover, the promoter architecture of *CAT1*, *CAT2*, *CAT3*, *SOD4*, *SOD4A* genes encoding antioxidant enzymes was examined to search for the occurrence of 70 different stress-related motifs. It was found that more than 30 ROS and stress-related *cis*-elements were present in the 5′ upstream regions of these genes. Selected promoter elements have been recognized as regulatory components conferring ROS-regulated gene expression of maize antioxidant genes, that could be critical in mediating induction after treatment with ROS-producing chemicals such as allelopathins, e.g., juglone (Mylona et al. 2007). Chi et al. (2011) examined involvement of ROS and calcium (Ca^{2+}) in allelochemical signal transduction pathway induced by juglone in roots of rice. Juglone (10 μM) led to increased generation of ROS after 1 and 3 h of treatment, which was accompanied by enhanced cytoplasmic Ca^{2+} level. Transcriptomic analysis showed upregulation of 31 genes related to both ROS production and detoxification, including among them alternative oxidase (AOX), GR, GST, GPX, glutaredoxins (GRX), and Trx. Transcripts of genes encoding ROS scavenging enzymes dependent on GSH increased more than fivefold in juglone-treated roots, suggesting key role of GSH in juglone tolerance. Similarly, abundance of several transcripts related to signal transduction pathways was also elevated. Juglone increased transcription of genes encoding calmodulin CAM (CML31) and calcium-regulated kinases [CBL interacting protein kinase, CIPK (CIPK07) and calcium-dependent protein kinase, CDPK (CPK4, CPK20)]. Activation of these elements of signal transduction cascade is typical for plants subjected to different abiotic stresses (cold, salinity, drought), that are known to induce secondary oxidative stress. In addition to overexpression of genes, the activation of CDPK-like kinases and MAP kinases (MAPK) was also observed (Chi et al. 2011). The coincidence of ROS and Ca^{2+} signaling seems to be a characteristic in plant cells. It was suggested that in roots of *Arabidopsis* (±)-catechin triggers a wave of ROS initiated at the root meristem, which leads to a Ca^{2+}-signaling cascade responsible for initialization of genomewide changes in gene expression, and leading finally to death of the root system (Bais et al. 2003). This observation was discussed by demonstrating rather antioxidant properties of (−)-catechin (Duke et al. 2009). Chobot et al. (2009) have proved that catechin can act as an efficient scavenger of ROS, antioxidant effect was pronounced especially when iron was chelated directly by the allelochemical. After using ethylenediaminetetraacetic acid (EDTA) to chelate iron, catechin acted as pro-oxidative agent, as it reduced molecular oxygen and iron to reagents required by the Fenton reaction to produce hydroxyl radicals ($^{\bullet}OH$). Therefore, it is still questionable whether ROS is involved in the mode of action of catechin. Weir et al. (2006) demonstrated that prevention of production of ROS by oxalate declines phytotoxicity of (−)-catechin. Moreover, after a long-term discussion, induction of ROS production by racemic catechin (100 μg mL^{-1}) was described again in *Arabidopsis* roots. It was followed by elevated expression of cell death-associated genes: *acd2*, *acdl* (Kaushik et al. 2010) confirming suggestion of oxidative stress as a mode of action of (±)-catechin (Bais et al. 2003). Independent research group demonstrated quite recently that (+)-catechin, allelochemical causing inhibition of growth of cyanobacteria and green algae cells, acts *via* induction of overproduction of ROS (Wang et al. 2011). *M. aeruginosa* and *Pseudokirchneriella subcapitata* Korshikov responsed to catechin (25–100 mg L^{-1}) by enhanced formation of ROS, which was significantly stimulated in light conditions probably due to continuous formation of $O_2^{\bullet-}$ and NADPH in photosynthetic light reaction. Both components are necessary for redox cycling of catechin (formation of semiquinone radical and *o*-quinone). It amplifies generation of $O_2^{\bullet-}$, which is dismutated to hydrogen peroxide (H_2O_2)—a substrate for Fenton reaction.

We should underline that in case of some allelochemicals, e.g., catechin effect of hormesis, defined as stimulatory, the beneficial action of nontoxic doses of phytotoxins ought to be analyzed. A hormetic effect was exhibited also by juglone at lower (6–12 μg mL^{-1}) concentration in mustard (*Sinapis alba* L.) seedlings, but only when plants were stressed with 10 % methanol (Chobot and Hadacek 2009). Thus, juglone may act as a scavenger of $^{\bullet}OH$ in organisms subjected to higher oxidative challenge, while prooxidative activities, accelerated by iron ions chelatation may be responsible for toxicity of this allelochemical (Chobot and Hadacek 2009) as it was described earlier in the chapter.

The question on pro- or antioxidative properties of some allelochemicals is not so rare, as it raised in connection with toxicity of L-3,4-dihydroxyphenylalanine (L-DOPA), non-protein amino acid, synthesized in velvet bean (*Mucuna pruriens* L.). It was hypothesized that the harmful action of L-DOPA might be due to generation of ROS during oxidation of amino acid to melanin in reaction that requires polyphenol oxidase and POX activity (Hachinohe and Matsumoto 2007b). In suspension culture of carrot (*Daucus carota* L.) cells or in roots of lettuce (*Lactuca sativa* L.) it was shown that L-DOPA was metabolized to melanin, and it was accompanied by oxidative damage expressed as increased lipid peroxidation (Hachinohe and Matsumoto 2007a). Authors proved that antioxidants such as ASA and α-tocopherol reduced the oxidative damage, removed inhibition of lettuce root growth by L-DOPA, and prevented the accumulation of melanin. However, they did not determine the level of ROS in L-DOPA-treated plants. Completely contradicting data were shown on roots of soybean (Soares et al. 2011) in which the levels of ROS ($O_2^{\bullet-}$, H_2O_2) decreased after treatment with L-DOPA. It may be explained by differences in sensitivity of various plant species to L-DOPA, but also to putative antioxidant activity of L-DOPA and melanin. In addition in roots of soybean, activities of SOD and POX increased following allelochemical exposure while level of ROS and lipid peroxidation declined (Soares et al. 2011). The transcriptomic analysis of L-DOPA action in *Arabidopsis* revealed that after 6 h of treatment with the allelochemical, the expression of 110 genes was significantly upregulated, and the expression of 69 genes was significantly downregulated (Golisz et al. 2011). Among significantly downregulated genes there were three ferritins crucial for iron homeostasis. In the absence of ferritin, plants have higher levels of ROS and increased activity of enzymes involved in their detoxification, but in the examined model among upregulated genes there were no genes encoding elements of antioxidant cellular system (Golisz et al. 2011). In addition, six genes of POX were also downregulated.

3 Essential Oils as Inducers of Oxidative Stress. Essential Oils Lead to Typical Oxidative Deterioration in Acceptor Plants

Aromatic plants have been popular and used by humans in medicine and cosmetology since antiquity due to their flavor and fragrance. But essential oils, mainly volatile terpenes, produced by aromatic plants, exhibit a variety of functions in plant kingdom. They act as pollinator attractants, provide an important defense strategy against herbivores and pathogenic fungi, and function as allelochemicals in plant–plant interactions. This chapter is focused mostly on biochemical alterations that may originate from disturbances in ROS production or scavenging induced by various essential oils recognized as strong allelochemicals.

The most frequently used essential oils such as α-β-pinene, eucalyptol, camphene, camphor, thymol, geraniol, and 1,8-cineole are known to act as inhibitors of seed germination and suppressors of seedling growth (Abrahim et al. 2000, 2003; Romagni et al. 2000; Zunino and Zygadlo 2004). Exposure of coffee senna (*Cassia occidentalis* L.) roots to α-pinene (5–10 mM) or seedlings of various weed species to β-pinene (0.02–0.8 mg mL^{-1}) led to rapid increase in H_2O_2 production (Singh et al. 2006; Chowhan et al. 2013). It was accompanied by enhanced electrolyte leakage, and high level of MDA, observed even at lower concentration of α-β-pinene indicating lipid peroxidation. In general, after exposition to 2.5–10 mM α-pinene activities of the antioxidant enzymes SOD, CAT, GPX, ascorbate peroxidase (APX), and GR were significantly elevated (Singh et al. 2006). α-Pinene (0.5 mM) decreased germination rate of little bell (*Ipomea triloba* L.) seeds and inhibited seedling growth (Pergo and Ishii-Iwamoto 2011). At this concentration α-pinene did not significantly enhance activity of SOD, CAT, APX in little bell but resulted in almost doubled, as compared to control, accumulation of MDA and enhanced lipoxygenase (LOX) activity accompanied by high activity of POX and GR. In addition, α-pinene increased level of GSH. This suggests that the tested allelochemical at 0.5 mM concentration may induce oxidative stress, as activation of LOX is believed to be one of the rapid responses to deterioration of cell membrane structures resulting from ROS overproduction. Therefore, in seedlings of little bell subjected to α-pinene, the activation of antioxidant enzymatic system is probably not sufficient to scavenge ROS, but it is possible that other detoxification mechanisms may be rather induced. α-Pinene greatly stimulated oxygen consumption by AOX (Pergo and Ishii-Iwamoto 2011), which is suggested as a compensatory mechanism that reduces ROS generation by mitochondrial electron transport chain (Clifton et al. 2005). Data presented by Mucciarelli et al. (2001) implied similar action of (+)-pulegon and menthone, compounds of peppermint (*Mentha piperita* L.) on mitochondrial respiration of roots of cucumber (*Cucumis sativus* L.) seedlings. Experiment with simultaneous treatment with essential oils of peppermint and salicylhydroxamic acid (inhibitor of AOX) indicated that two tested monoterpenes exerted maximum inhibition on oxygen consumption when AOX was active. Therefore, it is probable that volatile allelochemicals may increase partitioning of the nonphosphorylating pathway of mitochondrial electron transport chain, which leads to decrease of ubiquinone reduction and results in avoidance of ROS generation in this cellular compartment.

Some researchers investigated impact of other monoterpenes (1,8-cineole, menthol, thymol, geraniol, camphor) on lipid peroxidation in maize roots (Zunino and Zygadlo 2004). Exposure of maize seedlings to monoterpenes resulted in high level of MDA and increased values of conjugated dienes during the first 24 h. However, at prolonged toxicity stress oxidation values led to a lower state than in the control. Kriegs et al. (2010) studied influence of camphor and menthol fumigation on expression of several genes encoding proteins involved in abiotic stress in seedlings of *A. thaliana*. They demonstrated a rapid increase in the abundance of *LOX2* transcript, and upregulation (3–6-fold) of transcription of gene encoding MAP kinase 3 (MPK3) after 4–6 h of plant fumigation with monoterpenes. MPK3 is known to be activated in response to H_2O_2 and abscisic acid (ABA) and plays an important role in stomata closure. Alterations in stomata closure after *A. thaliana* treatment with camphor and menthol increased transpiration rate due to extreme swelling and a final breakdown of the protoplasts leading to wilting of the seedlings (Schulz et al. 2007). More detailed data, focused on microtubular cytoskeleton formation in stomata guard cells in *Arabidopsis* plants exposed to tested monoterpenes, indicated reorganization and partial disruption of F-actin filaments (Kriegs et al. 2010). It can correspond to induction of not only oxidative stress suggested by increased level of *MPK3* and *LOX2* transcripts, but also may indicate induction of nitrosative stress due to a strong correlation between nitric oxide (NO) biosynthesis and structure of cytoskeleton (Yemets et al. 2011).

Among some aromatic plants that exhibit high allelopathic potential representatives of the genus *Artemisia* (commonly wormwoods; family *Asteraceae*) are under investigation, as they are popular weeds inhibiting seed germination, seedling emergence, and growth of neighboring plants. They are also well recognized due to a characteristic vegetation patterning (death zone) formed around them due to emission of volatile terpenes. Analyses of *Artemisia* oil revealed it to be a mixture of monoterpenoids, sesquiterpenes, aromatic compounds, aliphatic esters, ketones, and alcohols. A total of 33 chemical constituents representing 99.83 % of the essential oil were identified (Singh et al. 2009). β-Myrcene was the main constituent, accompanied by (+)-limonene, (Z)-β-ocimene, γ-terpinene, other authors indicated also presence of β-pinene, limonene, and (E)-β-ocimene methyl eugenol or thujone, camphor, and 1,8-cineole in oil from aerial parts of *Artemisia* plants. Fumigation with the essential oil from leaves of redstem wormwood (*Artemisia scoparia* Waldst. & Kit.) and its major monoterpene (β-myrcene) suppressed seed germination and seedling growth of nut grass (*Cyperus rotundus* L.) and correlated with cell viability, membrane integrity, and ROS concentration (Singh et al. 2009). Membrane damage upon exposure to *Artemisia* oil or β-myrcene was confirmed by a significant decline in the content of conjugated dienes. Similarly to MDA accumulation, the level of H_2O_2 increased in roots of the seedlings exposed to wormwood oil or β-myrcene. Accumulation of H_2O_2 in nut grass roots enhanced lipid peroxidation, resulting in induction of oxidative stress, and leading to disruption of metabolic activities in the cell. Recent experiments have proved that monoterpene induced transcriptome changes in *Arabidopsis* (Godard et al. 2008). Microarray gene expression profiling indicated alterations in the abundance of many transcripts in *Arabidopsis* plants treated with myrcene or a mixture of ocimene volatiles consisting of (E)-b-ocimene, (Z)-b-ocimene, and allo-ocimene (acyclic monoterpenes that are released from the leaves in some plants when damaged by insect herbivores and acts an attractant to predators of the herbivores). Many of the monoterpene-induced transcripts were annotated as either transcription factors or as stress or defense genes including several steps in the octadecanoid pathway with *LOX2* transcript encoding lipoxygenase. In general, myrcene volatiles resulted in a significant change of *Arabidopsis* transcript abundance for 986 oligo micro arrays, of which 730 (74 %) were up- and 256 (26 %) were downregulated. Moreover, transcripts associated with abiotic and biotic stress and transcripts associated with general stress were significantly over-represented. Similarly, transcripts of the genes encoding proteins associated with membranes were significantly over-represented in *Arabidopsis* plants treated with myrcene. An increased abundance of 10 genes related to response to oxidative stress was detected in plants fumigated with myrcene (Godard et al. 2008).

Essential oils (0.01 %) from the aerial parts of another aromatic plant—catmint (*Nepeta meyeri* Benth.) inhibited seed germination of various agricultural weeds [redroot pigweed (*Amaranthus retroflexus* L.), pigweed (*Portulaca oleracea* L.), oat brome (*Bromus danthoniae* L.), prickly lettuce (*Lactuca serriola*), goose foot (*Chenopodium album* L.)]. This effect was associated with over-accumulation of H_2O_2 and enhancement in lipid peroxidation, manifested as elevated MDA level (Mutlu et al. 2011). In the majority of tested weed plants, an increased CAT activity was noticed, but in contrast neither POX nor SOD was stimulated, likewise reported for α-pinene (Singh et al. 2006). The experiments performed quite recently by Maketç et al. (2013) showed that the essential oils from catmint inhibited also seed germination of crop plants such as canola (*B. napus*), and maize to a similar extent as observed for prickly lettuce and oat brome. In addition, it was linked to damage of DNA expressed as changes in RAPD band profiles, nondirectly suggesting ROS dependent harmful action of catmint allelochemicals on nucleic acids.

4 Autotoxicity (Interspecific Allelopathy) Is Due to Autotoxin-Dependent Enrichment in ROS Level

The phenomenon of allelopathy is usually interspecific, but it also occurs within the same species, and is called autotoxicity. It is defined as the deleterious allelopathic effect among the individuals of the same species and has been documented in both natural and agro-ecosystems. The problem of autotoxicity is common in monocultures being the main cause of yield reduction. In cultivation, alfalfa (*Medicago sativa* L.) and cucumber are known as species displaying the strongest autotoxic potential. Phytotoxicity of cinnamic acid, the most important after benzoic acid allelochemical, in root exudates and root extracts of cucumber, is associated with alteration in ROS metabolism (Yu et al. 2003; Ye et al. 2006; Ding et al. 2007; Li et al. 2011a). Cinnamic acid is considered to act as an autotoxic agent leading to soil sickness, which is a serious problem in horticultural crop production. In roots of cucumber seedlings cinnamic acid (0.05–0.25 mM) generated high level of ROS (both $O_2^{\bullet-}$ and H_2O_2) (Ding et al. 2007; Ye et al. 2006). It was accompanied by stimulation of activity of antioxidant enzymatic cellular system. Activities of CAT, SOD, GPX, and APX in cucumber roots increased in dose dependent manner after cinnamic acid treatment (Ding et al. 2007); SOD and POX also increased after treatment with exudates or root extracts (Yu et al. 2003). Analogous reactions were detected in leaves of cucumber seedlings exposed to cinnamic acid (0.05 mM) (Li et al. 2011a). Similarly to other allelochemicals, cinnamic acid induced also membrane deterioration resulting in enhanced MDA content. A comparison between the visualization of the pattern of ROS production, cell viability, and changes in Ca^{2+} level in roots of cucumber seedlings indicated strong correlation (Ding et al. 2007; Yu et al. 2009). A significant increase in cytosolic Ca^{2+} level was found in the elongation zone of the cucumber root tip. More pronounced ROS formation and cell mortality was detected in the elongation zone than in the root apex.

While the data indicating induction of oxidative stress and alteration in antioxidant enzymes as a general reaction to allelochemicals are up-and-coming, the knowledge on the mechanisms by which allelochemicals induce ROS formation is only fragmentary. It is suggested that the increased activity of plasma membrane NADPH oxidase is associated with increased ROS production observed in plants under biotic and abiotic stresses (Foreman et al. 2003). The key role of this enzyme in ROS generation in roots of cucumber seedlings subjected to cinnamic acid was proposed by Ding et al. (2007), as exposition of roots to this allelochemical resulted in twofold stimulation of NADPH oxidase activity.

The apple (*Malus domestica* Borkh.) autotoxicity is a widespread agricultural problem documented in all of the major fruit-growing regions of the world. Inhibition of growth of young seedlings of apple is caused by the chemicals released into the soil in root exudates or after decomposition of residues. Phthalic acid acts as a potent allelochemical inhibiting growth of apple rootstock (*Malus prunifolia* Borkh.) (Bai et al. 2009). Phthalic acid (1 mM) induced $O_2^{\bullet-}$ and H_2O_2 generation in apple roots. The enhanced activity of SOD indicated that excessive generation of $O_2^{\bullet-}$ has been triggered by phthalic acid treatment, and consequently, that SOD activity was upregulated to mitigate the oxidative damage. Stimulation of activities of CAT, POX was parallel to the accumulation of MDA and ROS level. Moreover, phthalic acid also stimulated activities of enzymes of the ascorbate–gluthatione pathway, such as dehydroascorbate reductase (DHAR) and monodehydroascorbate reductase (MDHAR) (Bai et al. 2009). Authors suggested that activation of antioxidant enzymes could reflect a defensive response to the cellular damage provoked by phthalic acid *via* ROS generation, but this increase was not strong enough to eliminate all injurious effects provoked by phthalic acid.

5 Allelochemicals Impact on ROS-Related Modification of Cell Wall Flexibility

Plant growth and development depends on cell expansion (Cosgrove 1997). Size and shape as well as number of the cells are essential for organ morphology. Cell enlargement is driven by two simultaneous processes as water uptake that increases turgor and cell wall relaxation. Cell wall loosening is crucial for all stages of plant development starting from endosperm weakening and embryo growth to fruit ripening. Mechanical properties of cell wall may be modified by proteins and ROS (Dolan and Davies 2004). Among cell wall proteins are those whose activity directly enhance cell wall extensibility, e.g., endoglucanases, xyloglucan endotransglycosylases, pectinases, pectin esterases, debranching enzymes, and nonenzymatic proteins such as expansins (Cosgrove 1999). These proteins lead to cell enlargement by rearranging cell wall matrix polymers and spreading the cellulose and/or matrix network. The second group of agents affecting cell wall extensibility are ROS, derived by spontaneous reaction or produced/consumed by cell wall associated proteins such as apoplastic POX, NADPH oxidase, and SOD.

Two major ROS playing opposite roles in cell wall extensibility are $^{\bullet}OH$ and H_2O_2. $^{\bullet}OH$ can be provided by Fenton reaction from NADPH oxidase-derived $O_2^{\bullet -}$, or by peroxidases supplied with O_2 and NADH (Liszkay et al. 2004). Additionally, $^{\bullet}OH$ may be generated by peroxidases coupled with Mn-SOD. The substrate of this reaction, H_2O_2, is derived from the autocatalytic reaction of O_2 with cell wall hydroxycinnamic acids or NADH as a reductant (Liszkay et al. 2003; Kukavica et al. 2009). $^{\bullet}OH$ is capable of non-specific breakdown of polysaccharides in cell wall matrix (Liszkay et al. 2004). This reaction mediates cell wall loosening by splitting covalent bonds and, in this way, promotes extension growth. However, $^{\bullet}OH$ is a short-lived intermediate and its reactivity is limited to the side of its formation (Schopfer 2001). As outlined above, the apoplastic H_2O_2 may be generated by an autocatalytic reaction, however, it can also be formed by a spontaneous dismutation, often accelerated by SOD. H_2O_2 may be consumed in the peroxidative cycle by POX (Liszkay et al. 2003). Peroxidative function of the apoplastic POX allows utilization of H_2O_2 to H_2O *via* conversion of phenolic compounds into reactive phenoxy radicals that can polymerize into, e.g., lignin. Such cross-linking of phenolic compounds contributes to cell wall stiffening and inhibition of cell growth (Liszkay et al. 2003).

Allelochemicals are well-known inhibitors of germination and plant growth. Its mode of action often proceeds *via* modification of cell redox status. Any changes in redox homeostasis in apoplast have far-reaching repercussion in cell growth due to alteration in cell wall extension. Myrigalone A (MyA) is a flavonoid of sweet gale (*Myrica gale* L.) (Oracz et al. 2012; Voegele et al. 2012). This compound strongly inhibits growth of seedlings. Lately, it has been shown that MyA also affects embryo growth due to induction of atypical endosperm rupture (increased potential of embryo growth without endosperm weakening). Embryo growth of cress (*Lepidium sativum* L.) after MyA treatment was inhibited in specific, radical/

hypocotyl region (RAD) particularly in the second or third zone of RAD (Oracz et al. 2012). This RAD region is responsible for elongation growth of radicle during endosperm rupture due to accumulation of $O_2^{\bullet-}$ allowing cell extension. MyA significantly decreased concentration of $O_2^{\bullet-}$ and $^{\bullet}OH$ in the apoplast of second RAD region. MyA probably directly acts as ROS-scavenger (Oracz et al. 2012; Voegele et al. 2012). However, it may indirectly down-regulate production of $O_2^{\bullet-}$ and $^{\bullet}OH$ in apoplast by affecting gibberellins (GA) metabolism and/or signaling that decreases GA-dependent cell extension (Oracz et al. 2012).

Decrease in ROS accumulation under allelochemical treatment may be observed also during post-germination plant growth. Proper distribution of ROS in growing root is an important factor allowing transition of cells from proliferation to differentiation (Tsukagoshi et al. 2010). ROS regulates redox status and cell cycle progression by acting as a signal at the transcriptional and/or posttranscriptional level. Accumulation of $O_2^{\bullet-}$ in meristematic zone regulates proliferation while H_2O_2 induces elongation in root elongation zone (Tsukagoshi et al. 2010). Simple amide, cyanamide is recognized to be an allelopathic agent. Tomato (*Solanum lycopersicum* L.) seedlings treated with cyanamide (1.2 mM) characterized inhibition of growth (Soltys et al. 2012). Localization studies of $O_2^{\bullet-}$ in roots confirmed its accumulation in meristematic zone; however, in case of cyanamide-treated roots, $O_2^{\bullet-}$ was less abundant (Soltys et al. 2013). This phenomenon was accompanied by higher than in the control frequency of differentiated cells and overexpression of the expansins genes *LeExpA4* and *LeExpA5*, which are directly related to the process of cell wall loosening (Soltys et al. 2013).

Allelochemicals, especially phenolic compounds, may contribute to cell wall stiffening. Cinnamic acid derivatives (e.g., ferulic, caffeic, and *p*-coumaric acids) usually induce premature lignification of root cell wall (dos Santos et al. 2008; Zanardo et al. 2009). Phenolics can be channeled into the phenylpropanoid pathway contributing to the excess synthesis of lignin monomers and in this way increasing lignification (dos Santos et al. 2008).

In the lignification process, two pivotal enzymes are involved. Phenylalanine ammonia-lyase (PAL) induces polymerization of monolignols at the early stage of the lignification process (Boerjan et al. 2003). The second enzyme POX, acts at the last step of phenypropanoid pathway forming H, G and S lignin units. Exogenously applied caffeic acid (1 and 2 mM), ferulic acid (1 mM), and *p*-coumaric acid (at $\geq$0.5 mM) significantly inhibited soybean root growth (Bubna et al. 2011; Zanardo et al. 2009; dos Santos et al. 2008). These cinnamic acid derivatives have similar modes of action in acceptor plants. Influx of more phenolic compounds into phenylpropanoid pathway increased lignin production *via* enlarged amount of lignin monomer H, G and S in case of caffeic acid; H, G for *p*-coumaric acid and G for ferulic acid. In spite of modified lignin monomer composition, PAL activity has been reduced (Bubna et al. 2011; Zanardo et al. 2009). The activity of POX significantly increased after caffeic, ferulic and *p*-coumaric acid treatment with simultaneous decline in H_2O_2 concentration (Bubna et al. 2011; dos Santos et al. 2008; Zanardo et al. 2009). Cinnamic acid derivatives accumulate in cell wall which supports the idea that apoplastic POX may be directly engaged in cell wall lignification (Kovácik et al. 2009). During cell wall lignification, POX consumes H_2O_2 to build cross-linkages between cell wall polymers. Lignification mainly occurred in root vascular cylinder tissues, especially in metaxylem vessels. After phenolic treatment, lignification of cell walls resulted in enhanced diameter of the vascular cylinder and differentiated xylem vessels as well as formation of lateral roots (dos Santos et al. 2008). On the one hand, the increased influx of caffeic acid into cell wall may increase apoplastic POX activity and consumption of H_2O_2. On the other hand, it may downregulate PAL activity (dos Santos et al. 2008).

Other phenolic compound, benzoic acid derivative—gallic acid also shows phytotoxic activity on root growth (Rudrappa et al. 2007). However, its mode of action is distinct from that proposed for cinnamic acid derivatives. Exogenously applied gallic acid (50 μM) not only inhibited root growth but also altered its morphology by inducing its collapse. Gallic acid-induced ROS wave on the root surface 10 min after its application, which directly reorganized cortical microtubules assembly in cells. Microtubules were diffused and single fibers were gathered into local aggregates (Rudrappa et al. 2007). Cortical microtubules determine proper cellulose microfibrils deposition during cell elongation (Verbelen et al. 2003). Such undirected reorientation of microtubules as was observed in case of gallic acid led to disorder in cellulose deposition, disturbed proper elongation growth and resulted in collapse of roots (Rudrappa et al. 2007).

6 Allelochemicals-Induced ROS Level Alterations Provoke Phytohormonal Imbalance Resulting in Abnormal Growth and Development of Susceptible Plants

Plants that are subjected to sublethal level of abiotic conditions exhibit a broad range of morphogenic response. They include inhibition of cell elongation, localized stimulation of cell division, alterations in cell differentiation status and are known as stress-induced morphogenic responses (SIMRs) (Chobot et al. 2009). It was postulated by Potters et al. (2007, 2009), that similarities in plant morphological response to various stressors are due to elevated ROS generation and distribution/content of plant hormones. The close interplay between ROS and phytohormones is out of any doubt. Several years ago Kwak et al. (2006) reviewed the function of ROS acting as second messengers in physiological reactions governed by plant hormones, e.g., it was proven that gravitropism correlates with organ asymmetric generation of ROS and reactive nitrogen species (RNS) formation and corresponds to asymmetric concentration of auxins (Hu et al. 2005; Kwak et al. 2006 and references therein). Similarly, root hair growth seems to depend on ROS production maintained by NADPH oxidase and ethylene (Foreman et al. 2003). Root growth is under control of auxins and ABA playing in orchestra with ROS and calcium signaling (Jiao et al. 2013), as a close dependence of

ROS/calcium signal in cell to cell communication, is necessary for long distance signal transduction (Steinhorst and Kudla 2013).

There are many data on induction of morphological signs typical for SIMRs after treatment of various plants with the range of allelochemicals. Such response has been shown for maize and tomato seedlings stressed with cyanamide (Soltys et al. 2012, 2014), or non-protein amino acid *m*-tyrosine (Olechowicz et al. 2013), as well as *Arabidopsis* plants subjected to citral (Graña et al. 2013), or cumarine derivative, 4-methylumbelliferone (4-MU) (Li et al. 2011b). In most of these studies auxin—indole-3-acetic acid (IAA), ethylene emission and ROS concentration increased after treatment with the allelochemical leading to disruption in hormonal and redox homeostasis. In *Arabidopsis* plants exposed to 4-MU, transcription of genes encoding PIN2 and PIN3 was upregulated, in a similar way as *IAA14/SLR1* one of the most important auxin responsive AUX/IAA genes involved in lateral root initiation (Li et al. 2011b), suggesting that this allelochemical affects auxin distribution more than synthesis. Non-protein amino acids such as canavanine and 5-hydroxy-tryptophane acted as inhibitors of root growth of tomato seedlings (Krasuska et al. 2013). We demonstrated just recently that this was accompanied by increase of IAA level and induction of nitro-oxidative stress expressed as overproduction of ROS and alterations in NO level (unpublished results). Similar data were obtained also for *Arabidopsis* roots subjected to farnesene, the allelochemical isolated from lesser calamint (*Calamintha nepeta* Mill.), that induced loss of gravitropism and restriction in root growth (F. Araniti, personal communication). Farnesene caused an accumulation of auxin, and transient burst in ROS/NO production, which resulted in strong alteration of cell ultrastructure. Moreover, root malformation and loss of gravitropism, together with a lack in root hair production and a significant deformation of the root, were observed. These data are in agreement with the impact of ROS on root hair development. The inhibitory action of allelochemical on root hair development and graviperception was shown in lettuce seedlings after treatment with rabdosin B [*ent*-kaurene diterpenoid isolated from *Isodon japonicus* (Burm. f.) H. Hara)] (Ding et al. 2010) or leaf leachates from false yellowhead (*Dittrichia viscose* (L.) Greuter) (Levizou et al. 2002). Growth and architecture of roots is determined by crosstalk between auxins, ROS, and NO. The general summary of this orchestra was presented by Yu et al. (2014). It is possible that close connection between key regulatory agents, ROS and auxins, is due to NO-dependent modifications, mainly *S*-nitrosylation of enzymes or elements of signal transduction pathways. Therefore, any smooth influence of allelochemicals on NO level could lead to important molecular, biochemical, and morphological consequences.

Plant growth and development is regulated also by other molecules than classical phytohormones. Polyamines (PAs) are aliphatic amines of low molecular weight acting as regulators of plant growth and development. They are also a putative source of NO in reaction mechanism that has not been resolved yet (Yu et al. 2014). It is commonly known that PAs stabilize membranes by inhibiting transbilayer phospholipids movement, and are positively related to plant stress tolerance (Martin-Tonguy 2001), although integration of PAs and anti-allelochemical response was not widely investigated. In the literature there are only rare data on impact of allelochemical on PAs level and metabolism. Cinnamic acid (0.1 mM) disturbed PAs homeostasis in leaves of cowpea (*Vigna unguiculata* (L.) Walp.), by decreasing spermine (Spm) and spermidine (Spd) level and increasing putrescine (Put) content (Huang and Bie 2010). The mechanism of decline in the level of both tetra- and triamine PAs was not investigated but, it may include activation of polyamine oxidase (PAO), leading to production of H_2O_2, which may confirm the described above induction of oxidative stress in plants treated with cinnamic acid. Harmful effect of this allelochemical exhibited as increased lipid peroxidation may be due to the fact that Spd and Spm seem to act as more effective scavengers of ROS than Put; therefore, their lack favors oxidative damage of biomembranes. Similar pattern of changes in PAs content was detected also in lettuce roots after exposure to 2 (3H)-benzoxazolinone (BOA) (Sánchez-Moreiras and Reigosa 2005).

Conclusions

The survival of plants in natural or agro-ecosystems depends on the outcome of various positive and negative interactions with neighboring organisms; therefore, one of the ways of controlling plant–plant coexistence involves communication by chemical messages. We have demonstrated that the impact of various allelochemicals on plant growth and development may occur *via* ROS and probably RNS. It should be emphasized that most of described data come from laboratory experiments; therefore, although they are valid for *in vitro* studies, some of them are far from being realistic from an ecological point of view.

To date, it is clear that not all allelochemicals induce oxidative stress, as some of them may even act as antioxidants. There is no doubt that pro- or antioxidative character of allelochemical depends on its chemical structure, concentration, and sensitivity of acceptor plant and may differ depending on the type of tissue. Supposing that allelochemical belongs to the group of prooxidative agents we may define its action as leading to modification of ROS production and metabolism, associated with degradation of cell components (proteins, nucleic acids) and structures (mainly biomembranes) (for overview see Fig. 1). Although at the current moment it is hard to point on molecular target of most of examined allelochemicals and to generalize their mode of action, based on presented data we can assume that in many cases the action of allelochemicals via ROS includes alteration in cell wall stiffness and imbalance of level of phytohormones. Thus, in certain environmental conditions plant response to allelopathy stress may be compared to SIMR reactions, characteristic for some biotic stresses. Overproduction of ROS accompanied by modification of cell wall structure in plants suffering from allelopathic neighbors may implicate a role of NADPH oxidase as the most important component of internal signal transduction pathway, as the ROS

(continued)

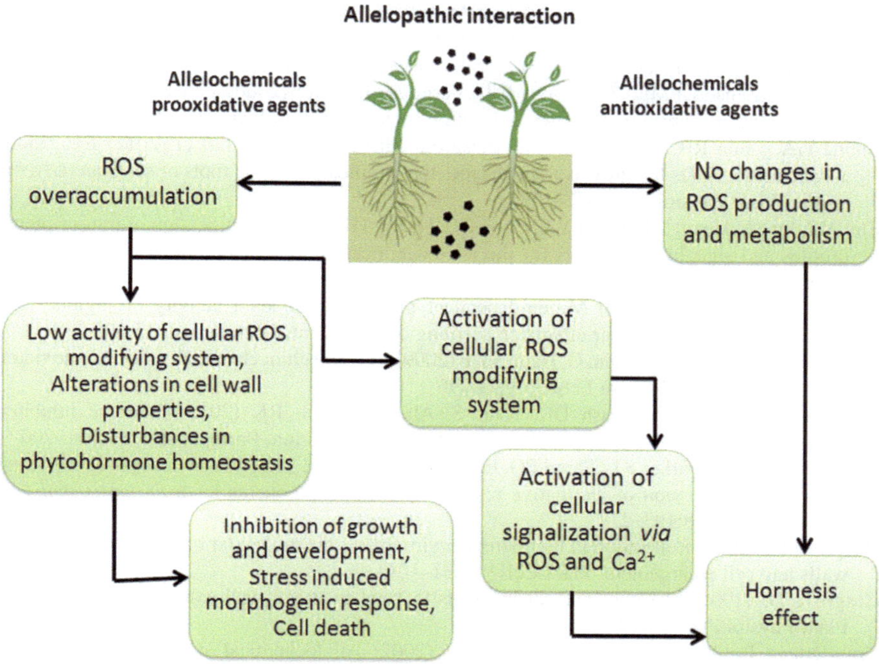

Fig. 1 Action of allelochemicals (*black pentagons*) as ROS inducing or ROS scavenging agents in acceptor plants. Allelochemicals may scavenge ROS resulting in decreasing intracellular ROS level. At low concentration they may act as growth stimulators leading in hormetic effect. At high concentration or in susceptible acceptor plants allelochemicals may induce overproduction of ROS and alteration in operation of antioxidant cellular system. Plant reaction to allelochemical depends on signaling pathway activated by ROS and calcium ions

wave maintained by this enzyme seems to be crucial for the pattern of plant development. In future perspectives much more attention should be paid on correlation between harmful effect of allelochemicals on plant growth and alterations in cellular NO metabolism, as now it is becoming apparent that NO exhibits various biological functions in every aspect of plant physiology.

References

Abrahim D, Braguini WL, Kelmer-Bracht AM, Ishii-Iwamoto EL (2000) Effects of four monoterpenes on germination, primary root growth, and mitochondrial respiration of maize. J Chem Ecol 26:611–624

Abrahim D, Francischini AC, Pergo EM, Kelmer-Bracht AM, Ishii-Iwamoto EL (2003) Effects of α-pinene on the mitochondria respiration of maize seedlings. Plant Physiol Biochem 41:985–991

Bai R, Ma F, Liang D, Zhao X (2009) Phthalic acid induces oxidative stress and alters the activity of some antioxidant enzymes in roots of *Malus prunifolia*. J Chem Ecol 35:488–494

Bais HP, Vepachedu R, Gilroy S, Callaway RM, Vivanco JM (2003) Allelopathy and exotic plant invasion: from molecules and genes to species interaction. Science 301:1377–1380
Boerjan W, Ralph J, Baucher M (2003) Lignin biosynthesis. Annu Rev Plant Biol 54:519–546
Böhm PAF, Zanado FML, Ferrarese MLL, Ferrarese-Filho O (2006) Peroxidase activity and lignification in soybean root growth-inhibition by juglone. Biol Plant 50:315–317
Bubna GA, Lima RB, dos Santos WD, Ferrarese MLL, Ferrarese-Filho O (2011) Exogenous caffeic acid inhibits the growth and enhances the lignification of the roots of soybean (*Glycine max*). J Plant Physiol 168:1627–1633
Chi WC, Fu SF, Huang TL, Chen YA, Chen CC, Huang HJ (2011) Identification of transcriptome profiles and signaling pathways for the allelochemical juglone in rice roots plant. Mol Biol 77:591–607
Chobot V, Hadacek F (2009) Milieu-dependent pro-and antioxidant activity of juglone may explain linear and nonlinear effects on seedling development. J Chem Ecol 35:383–390
Chobot V, Huber C, Trettebhahn G, Hadacek F (2009) (±)-Catechin: chemical weapon, anioxidant or stress regulator? J Chem Ecol 35:980–996
Chowhan N, Singh HP, Batish DR, Kaur S, Ahja N, Kohli RK (2013) β-Pinene inhibited germination and early growth involves membrane peroxidation. Protoplasma 250:691–700
Clifton R, Lister R, Parker KL, Sappl PG, Elhafez D, Millar AH, Day DA, Whelan J (2005) Stress induced co-expression of alternative respiratory chain components in *Arabidopsis thaliana*. Plant Mol Biol 58:193–212
Cosgrove DJ (1997) Relaxation in high-stress environment: the molecular bases of extensible cell walls and cell enlargement. Plant Cell 9:1031–1041
Cosgrove DJ (1999) Enzymes and other agents that enhance cell wall extensibility. Annu Rev Plant Physiol Plant Mol Biol 50:391–417
Cruz-Ortega R, Ayala-Cordero G, Anaya AL (2002) Allelochemical stress produced by the aqueous leachate of *Callicarpa acuminata*: effects on roots of bean, maize, and tomato. Physiol Plant 116:20–27
Cruz-Ortega R, Lara-Nunez A, Anaya AL (2007) Allelochemical stress can trigger oxidative damage in receptor plants. Plant Signal Behav 2:269–270
Ding J, Sun Y, Xiao CL, Shi K, Zhou YH, Yu JQ (2007) Physiological basis of different allelopathic reactions of cucumber and fig leaf gourd plants to cinnamic acid. J Exp Bot 58:3765–3773
Ding L, Jing H, Qin B, Qi L, Li J, Wang T, Liu G (2010) Regulation of cell division and growth in roots of *Lactuca sativa* L. seedlings by the Ent-Kaurene diterpenoid rabdosin B. J Chem Ecol 36:553–563
Dolan L, Davies J (2004) Cell expansion in roots. Curr Opin Plant Biol 7:33–39
dos Santos WD, Ferrarese MLL, Nakamura CV, Mourão KSM, Mangolin CA, Ferrarese-Filho O (2008) Soybean (*Glycine max*) root lignifications induced by ferulic acid. The possible mode of action. J Chem Ecol 34:1230–1241
Duke SO, Blair AC, Dayan FE, Johnson RD, Meepagala KM, Cook D, Bajsa J (2009) Is (−)-catechin a novel weapon of spotted knapweed (*Centaurea stoebe*)? J Chem Ecol 35:141–153
Foreman J, Demidchik V, Bothwell JH, Mylona P, Miedema H, Torres MA, Linstead P, Costa S, Brownlee C, Jones JD, Davies JM, Dolan L (2003) Reactive oxygen species produced by NADPH oxidase regulate plant cell growth. Nature 27:442–446
Foyer CH, Noctor G (2005) Redox homeostasis and antioxidant signaling: a metabolic interface between stress perception and physiological responses. Plant Cell 17:1866–1875
Gapper C, Dolan L (2006) Control of plant development by reactive oxygen species. Plant Physiol 141:341–345
Gniazdowska A, Bogatek R (2005) Allelopathic interaction between plants. Multi site action of allelochemicals. Acta Physiol Plant 27:395–408
Godard KA, White R, Bohlmann J (2008) Monoterpene-induced molecular responses in *Arabidopsis thaliana*. Phytochemistry 69:1838–1849

Golisz A, Sugano M, Fujii Y (2008) Microarray expression profile of *Arabidopsis thaliana* L. in response to allelochemcials identified in buckwheat. J Exp Bot 59:3099–3109

Golisz A, Sugano M, Hiradate S, Fujii Y (2011) Microarray analysis of *Arabidopsis* plants in response to allelochemical L-DOPA. Planta 233:231–240

Graña E, Sotelo T, Díaz-Tielas C, Araniti F, Krasuska U, Bogatek R, Reigosa MJ, Sánchez-Moreiras AM (2013) Citral induces auxin and ethylene-mediated malformations and arrests cell division in *Arabidopsis thaliana* roots. J Chem Ecol 39:271–282

Hachinohe M, Matsumoto H (2007a) Involvement of melanin synthesis and reactive oxygen species in phytotoxic action of L-DOPA in carrot cells. Crop Prot 26:294–298

Hachinohe M, Matsumoto H (2007b) Mechanism of selective phytotoxicity of L-3,4-dihydroxyphenylalanine (L-DOPA) in barnyardglass and lettuce. J Chem Ecol 33:1919–1926

Hejl AM, Einhellig FA, Rasmussen J (1993) Effects of juglone on growth, photosynthesis and respiration. J Chem Ecol 19:559–568

Hong Y, Hu HY, Xie X, Li FM (2008) Responses of enzymatic antioxidants and non-enzymatic antioxidants in the cyanobacterium *Microcystis aeruginosa* to the allelochemical ethyl 2-methyl acetoacetate (EMA) isolated from reed (*Phragmites communis*). J Plant Physiol 165:1264–1273

Hu X, Neill SJ, Tang Z, Cai W (2005) Nitric oxide mediates gravitropic bending in soybean roots. Plant Physiol 137:663–670

Huang X, Bie Z (2010) Cinnamic acid-inhibited ribulose-1,5-bisphosphate carboxylase activity is mediated through decreased spermine and changes in the ratio of polyamines in cowpea. J Plant Physiol 167:47–53

Jiao Y, Sun L, Song Y, Wang L, Liu L, Zhang L, Liu B, Li N, Miao C, Hao F (2013) AtrbohD and AtrbohF positively regulate abscisic acid-inhibited primary root growth by affecting Ca^{2+} signalling and auxin response of roots in *Arabidopsis*. J Exp Bot 64:4183–4192

Kacperska-Lewak A (2004) Sensor types in signal transduction pathways in plant cells responding to abiotic stressors: do they depend on stress intensity? Physiol Plant 122:159–168

Kaushik S, Bais HP, Biedrzycki M, Venkatachalam L (2010) Catechin is a phytotoxin and pro-oxidant secreted from the roots of *Centaurea stoebe*. Plant Signal Behav 5:1088–1098

Kekeç G, Mutlu S, Alpsoy L, Sakçali MS, Atici Ö (2013) Genotoxic effects of catmint (*Nepeta meyeri* Benth.) essential oils on some weed and crop plants. Toxicol Ind Health 29:504–513

Kovácik J, Grúz J, Hedbavny J, Klejdus B, Strnad M (2009) Cadmium and nickel uptake are differentially modulated by salicylic acid in *Matricaria chamomilla* plants. J Agric Food Chem 57:9848–9855

Krasuska U, Staszek P, Antosik A, Gniazdowska A (2013) Modification of basic processes of plant growth and development by non protein amino acids. In: Book of abstracts of 56th Meeting of Polish Society of Botany, Olsztyn, pp 330–331

Kreslavski VD, Los DA, Allakhverdiev SI, Kuznetsov VV (2012) Signaling role of reactive oxygen species in plants under stress. Russ J Plant Physiol 59:141–154

Kriegs B, Jansen M, Hahn K, Peisker H, Šamajová O, Beck M, Braun S, Ulbrich A, Baluška F, Schulz M (2010) Cyclic monoterpene mediated modulations of *Arabidopsis thaliana* phenotype: effects on cytoskeleton and on the expression of selected genes. Plant Signal Behav 5:832–838

Kukavica B, Mojović M, Vučinić Ž, Maksimović V, Takahama U, Veljović Jovanović S (2009) Generation of hydroxyl radical in isolated pea root cell wall, and the role of cell wall-bound peroxidase. Mn-SOD and phenolics in their production. Plant Cell Physiol 50:304–317

Kwak JM, Nguyen V, Schroeder JI (2006) The role of reactive oxygen species in hormonal responses. Plant Physiol 141:323–329

Lara-Nuñez A, Romero-Romero T, Ventura JL, Blancas V, Anaya AL, Cruz-Ortega R (2006) Allelochemical stress causes inhibition of growth and oxidative damage in *Lycopersicon esculentum* Mill. Plant Cell Environ 29:2009–2016

Levizou E, Karageorgou P, Psaras GK, Manetas Y (2002) Inhibitory effects of water soluble leaf leachates from *Dittrichia viscosa* on lettuce root growth, statocyte development and graviperception. Flora 197:152–157

Li FM, Hu HY (2005) Isolation and characterization of a novel antialgal allelochemical from *Phragmites communis*. Appl Environ Microbiol 71:6545–6553

Li Q, Yu B, Gao Y, Dai AH, Bai JG (2011a) Cinnamic acid pretreatment mitigates chilling stress of cucumber leaves through altering antioxidant enzyme activity. J Plant Physiol 168:927–934

Li X, Gruber MY, Hegedus DD, Lydiate DJ, Gao MJ (2011b) Effects of a coumarin derivative, 4-methylumbelliferone, on seed germination and seedling establishment in *Arabidopsis*. J Chem Ecol 37:880–890

Liszkay A, Kenk B, Schopfer P (2003) Evidence for the involvement of cell wall peroxidase in the generation of hydroxyl radicals mediating extension growth. Planta 217:658–667

Liszkay A, Van der Zalm E, Schopfer P (2004) Production of reactive oxygen intermediates ($O_2^{\bullet -}$ H_2O_2, and $^{\bullet}OH$) by maize roots and their role in wall loosening and elongation growth. Plant Physiol 136:3114–3123

Macias FA, Oliva RM, Varela RM, Torres A, Molinillo JMG (1999) Allelochemicals from sunflower leaves cv Peredovick. Phytochemistry 52:613–621

Martin-Tonguy J (2001) Metabolism and function of polyamines in plants: recent development (new approaches). Plant Growth Regul 34:135–148

Miller G, Shulaev V, Mittler R (2008) Reactive oxygen signaling and abiotic stress. Physiol Plant 133:481–489

Mucciarelli M, Camusso W, Bertea CM, Maffei M (2001) Effect of (+)-pulegone and other oil components of Mentha x Piperita on cucumber respiration. Phytochemistry 57:91–98

Mutlu S, Atici Ö, Esim E, Mete E (2011) Essential oils of catmint (*Nepeta meyeri* Benth.) induce oxidative stress in early seedlings of various weed species. Acta Physiol Plant 33:943–951

Mylona PV, Polidoros AN, Scandalios JG (2007) Antioxidant gene responses to ROS-generating xenobiotics in developing and germinated scutella of maize. J Exp Bot 58:1301–1312

Nanda AK, Andrio E, Marino D, Pauly N, Dunand C (2010) Reactive oxygen species during plant-microorganism early interactions. J Integr Plant Biol 52:195–204

Olechowicz J, Gniazdowska A, Krasuska U, Bogatek R (2013) Phytotoxic effects of *m*-tyrosine on tomato (*Solanum lycopersicum* L.) root growth—alterations in ROS production and hormonal balance. BioTechnologia 94:345

Oracz K, Bailly C, Gniazdowska A, Come D, Corbineau F, Bogatek R (2007) Induction of oxidative stress by sunflower phytotoxins in germinating mustard seeds. J Chem Ecol 33:251–264

Oracz K, Voegele A, Tarkowská D, Jacquemoud D, Turečková V, Urbanová T, Strnad M, Sliwinska E, Leubner-Metzger G (2012) Myrigalone A inhibits *Lepidium sativum* seed germination by interference with gibberellin metabolism and apoplastic superoxide production required for embryo extension growth and endosperm rupture. Plant Cell Physiol 53:81–95

Pergo EM, Ishii-Iwamoto EL (2011) Changes in energy metabolism and antioxidant defense systems during seed germination of the weed species *Ipomoea triloba* L. and the responses to allelochemicals. J Chem Ecol 37:500–513

Potters G, Pasternak TP, Guisez Y, Palme KJ, Jansen MAK (2007) Stress-induced morphogenic responses: growing out of trouble? Trends Plant Sci 12:98–105

Potters G, Pasternak TP, Guisez Y, Jansen MA (2009) Different stresses, similar morphogenic responses: integrating a plethora of pathways. Plant Cell Environ 32:158–169

Qian H, Xu X, Chen W, Jiang H, Jin Y, Liu W, Fu Z (2009) Allelochemical stress causes oxidative damage and inhibition of photosynthesis in *Clorella vulgaris*. Chemosphere 75:368–375

Romagni JG, Allen SN, Dayan FE (2000) Allelopathic effects of volatile cineoles on two weedy plant species. J Chem Ecol 26:303–313

Rudrappa T, Bonsall J, Galladher JL, Seliskar DM, Bais HP (2007) Root-secreted allelochemical in the noxious weed *Phragmites australis* deploys a reactive oxygen species response and microtubule assembly distruption to execute rhizotoxicity. J Chem Ecol 33:1898–1918

Sánchez-Moreiras AM, Reigosa MJ (2005) Whole plant response of lettuce after root exposure to BOA (2(3H)benzoxazolinone). J Chem Ecol 31:2689–2703

Schopfer P (2001) Hydroxyl radical-induced cell-wall loosening in vitro and in vivo: implications for the control of elongation growth. Plant J 28:679–688

Schulz M, Kussmann P, Knop M, Kriegs B, Gregens F, Eichert T, Ulbrich A, Marx F, Fabrycius H, Goldbach H, Noga G (2007) Allelopathic monoterpene interfere with *Arabidopsis thaliana* cuticula waxes and enchance transpiration. Plant Signal Behav 2:231–239

Sharma P, Jha AB, Dubey RS, Pessarakli M (2012) Reactive oxygen species, oxidative damage, and antioxidative defense mechanism in plants under stressful conditions. J Bot 2012:1–26

Singh HP, Batish DR, Kaur S, Arora K, Kohli RK (2006) α-Pinene inhibits growth and induces oxidative stress in roots. Ann Bot 98:1261–1269

Singh HP, Kaur S, Mittal S, Batish DR, Kohli RK (2009) Essential oil of *Artemisia scoparia* inhibits plant growth by generating reactive oxygen species and causing oxidative damage. J Chem Ecol 35:154–162

Soares AR, Ferrarese MLL, Siqueira-Soares RC, Marchiosi R, Finger-Teixeira A, Ferrarese-Filho O (2011) The allelochemical L-DOPA increases melanin production and reduces reactive oxygen species in soybean roots. J Chem Ecol 37:891–898

Soltys D, Rudzińska-Langwald A, Kurek W, Gniazdowska A, Sliwinska E, Bogatek R (2011) Cyanamide mode of action during inhibition of onion (*Allium cepa* L.) root growth involves disturbances in cell division and cytoskeleton formation. Planta 234:609–621

Soltys D, Rudzinska-Lagwald A, Wiśniewska A, Gniazdowska A, Bogatek R (2012) Inhibition of tomato (*Solanum lycopersicum* L.) root growth by cyanamide is due to altered cell division, phytohormone balance and expansion gene expression. Planta 236:1629–1638

Soltys D, Gniazdowska A, Bogatek R (2013) Inhibition of tomato (*Solanum lycopersicum* L.) root growth by cyanamide is not always accompanied with enhancement of ROS production. Plant Signal Behav 8:e23994

Soltys D, Rudzińska-Langwald A, Kurek W, Szajko K, Sliwinska E, Bogatek R, Gniazdowska A (2014) Phytotoxic cyanamide affects maize (*Zea mays*) root growth and root tip function: from structure to gene expression. J Plant Physiol 171:565–575

Soobrattee MA, Neergheen VS, Luximon-Ramma A, Aruoma OI, Bahorun T (2005) Phenolics as potential antioxidant therapeutic agents: mechanism and actions. Mutat Res 579:200–213

Steinhorst L, Kudla J (2013) Calcium and reactive oxygen species rule the waves of signaling. Plant Physiol 163:471–485

Torres MA (2010) ROS in biotic interactions. Physiol Plant 138:414–429

Tsukagoshi H, Busch W, Benfey PN (2010) Transcriptional regulation of ROS controls transition from proliferation to differentiation in the root. Cell 143:606–616

Verbelen JP, Le J, Vissenberg K (2003) The cytoskeleton, elongation and the control of elongation. Cell Biol Int 27:287–288

Voegele A, Graeber K, Oracz K, Tarkowská D, Jacquemoud D, Turečková V, Urbanová T, Strnad M, Leubner-Metzger G (2012) Embryo growth, testa permeability, and endosperm weakening are major targets for the environmentally regulated inhibition of *Lepidium sativum* seed germination by myrigalone A. J Exp Bot 63:5337–5350

Wang J, Zhu J, Liu S, Liu B, Gao Y, Wu Z (2011) Generation of reactive oxygen species in cyanobacteria and green algae induced by allelochemicals of submerged macrophytes. Chemosphere 85:977–982

Weir T, Park SW, Vivanco JM (2004) Biochemical and physiological mechanisms mediated by allelochemicals. Curr Opin Plant Biol 7:472–479

Weir TL, Bais HP, Stull VJ, Callaway RM, Thelen GC, Redenhour WM, Bhamidi S, Stermitz FR, Vivanco JM (2006) Oxalate contributes to the resistance of *Gaillardia grandiflora* and *Lupinus sericeus* to a phytotoxin produced by *Centaurea maculosa*. Planta 223:785–795

Yang CY, Liu SJ, Zhou SW, Wu HF, Yu JB, Xia CH (2011) Allelochemical ethyl 2-methyl acetoacetate (EMA) induces oxidative damage and antioxidant responses in *Phaeodactylum tricornutum*. Pestic Biochem Physiol 100:93–103

Ye SF, Zhou YH, Sun Y, Zou LY, Yu JQ (2006) Cinnamic acid causes oxidative stress in cucumber roots, and promotes incidence of *Fusarium wilt*. Environ Exp Bot 56:255–262

Yemets AI, Krasylenko YA, Lytvyn DI, Sheremet YA, Blume YB (2011) Nitric oxide signalling *via* cytoskeleton in plants. Plant Sci 181:545–554

Yu JQ, Ye SF, Zhang MF, Hu WH (2003) Effects of root exudates, aqueous root extracts of cucumber (*Cucumis sativus* L.) and allelochemicals on photosynthesis and antioxidant enzymes in cucumber. Biochem Syst Ecol 31:129–139

Yu J, Sun Y, Zhang Y, Ding J, Xia X, Xiao C, Shi K, Zhou Y (2009) Selective trans-cinnamic acid uptake impairs [Ca^{2+}]cyt homeostasis and growth in *Cucumis sativus* L. J Chem Ecol 35:1471–1477

Yu M, Lamattina L, Spoel SH, Loake G (2014) Nitric oxide function in plant biology: redox cue in deconvolution. New Phytol 202:1142–1156

Zanardo DIL, Lima RB, Ferrarese MLL, Bubna GA, Ferrarese-Filho O (2009) Soybean root growth inhibition and lignification induced by p-coumaric acid. Environ Exp Bot 66:25–30

Zhang S, Zhang B, Dai W, Zhang X (2011) Oxidative damage and antioxidant responses in *Microcystis aeruginosa* exposed to the allelochemical berberine isolated from golden thread. J Plant Physiol 168:639–643

Zunino MP, Zygadlo JA (2004) Effect of monoterpenes on lipid oxidation in maize. Planta 219:303–309

The Role of Reactive Oxygen and Nitrogen Species in Bioenergetics, Metabolism, and Signaling During Seed Germination

Natalia V. Bykova, Junjie Hu, Zhenguo Ma, and Abir U. Igamberdiev

Abbreviations

ABA	Abscisic acid
GA	Gibberellic acid
GC	Guanylate cyclase
Hb	Hemoglobin
NO	Nitric oxide
PCD	Programmed cell death
RNS	Reactive nitrogen species
ROS	Reactive oxygen species

N.V. Bykova (✉)
Cereal Research Centre, Agriculture and Agri-Food Canada, Morden, MB, Canada R6M 1Y5
e-mail: Natalia.Bykova@agr.gc.ca

J. Hu
Cereal Research Centre, Agriculture and Agri-Food Canada, Morden, MB, Canada R6M 1Y5

Department of Biology, Memorial University of Newfoundland, St. John's, NL, Canada A1B 3X9

Z. Ma • A.U. Igamberdiev
Department of Biology, Memorial University of Newfoundland, St. John's, NL, Canada A1B 3X9
e-mail: igamberdiev@mun.ca

K.J. Gupta, A.U. Igamberdiev (eds.), *Reactive Oxygen and Nitrogen Species Signaling and Communication in Plants*, Signaling and Communication in Plants 23,
DOI 10.1007/978-3-319-10079-1_9

1 Introduction

Seed germination is a critical stage in the lifecycle of plants. It is framed by the time starting from the events which succeed the uptake of water by dry seed and completing by the emergence of embryonic tissues from the seed coat (Bewley 1997). After radicle protrusion, the early (pre-autotrophic) stage of seedling development (postgerminative growth), when the plant is dependent on mobilization of storage compounds, physiologically represents a direct continuation of the germination stage and is often considered as a part of it, despite of recommendations to separate it from the germination stage taken *sensu stricto* (Bewley and Black 1994). Germination is distinguished by rapid mobilization of storage reserves and initiation of growth and metabolic activity within the embryo. Rapid resumption of metabolic and respiratory activity following seed imbibition indicates that ATP synthesis due to initiation and/or repair of mitochondrial respiration is essential for supporting high energetic demands of germinating seeds and early (pre-autotrophic) stages of seedling growth (Bewley and Black 1994; Bewley 1997; Logan et al. 2001).

Both late seed maturation (Rolletschek et al. 2004; Weber et al. 2005) and early seed germination (Taylorson and Hendricks 1977; Bewley 1997) events encounter low in vivo oxygen levels, anticipated to limit respiration. The oxygen environment in tissues of dry seeds is rapidly depleted upon imbibition. Such a scenario has physiological consequences in that oxygen deficiency may be limiting for oxidative respiration and result in depleted energy status of developing or germinating seeds (Geigenberger 2003; Rolletschek et al. 2004; Weber et al. 2005). As a consequence, fermentative and/or other adaptive mechanisms are initiated during these critical stages of plant development. Although low in vivo oxygen should lower seed energy status by limiting respiration, this is generally not the case, as seed ATP levels and energy charge remain elevated (Duff et al. 1998; Logan et al. 2001; Benamar et al. 2003), suggesting that alternative mechanisms for energy generation could be in effect. We will discuss below how the varying oxygen levels of seed from imbibition to seedling development trigger formation of reactive oxygen species (ROS) and reactive nitrogen species (RNS) and how the cross-talk between ROS, RNS, hormones, and gene expression regulates bioenergetics and metabolism of seeds and triggers major physiological processes related to seed germination.

2 Reactive Oxygen Species, Dormancy Alleviation, and Germination

Upon imbibition, the immediate activation of metabolism results in the increase of redox level resulting in rapid consumption of oxygen in the reaction of oxidative metabolism. This leads to the formation of reactive oxygen species (ROS) targeting proteins, membranes, and nucleic acids. The available oxygen is rapidly consumed due to low permeability of seed coat, which turns the conditions inside the seed to highly hypoxic. In these conditions, respiration is suppressed and bioenergetics of seeds becomes mainly anaerobic. Later, after radicle protrusion, oxygen supply becomes restored and oxygen availability stimulates aerobic respiration in mitochondria and flavin-dependent oxidation in peroxisomes and other compartments. Therefore, in regard to oxygen availability, we can distinguish between three phases: (1) early oxygen consumption during the first hours after imbibition resulting in ROS production; (2) the following anaerobic stage before radicle protrusion characterized by fermentation and NO production in anaerobic reduction of nitrate and nitrite; (3) aerobic metabolism after radicle protrusion involving fully developed mitochondria and peroxisomes.

In the first phase the accumulation of ROS takes place, and this accumulation is related to breaking dormancy at this stage. A significant amount of data has been accumulated indicating that seeds of plant species from different taxonomic groups during imbibition produce ROS such as H_2O_2, O_2^-, and hydroxyl radicals, and reactive nitrogen species (RNS), such as nitric oxide (NO) (Cakmak et al. 1993; Caliskan and Cuming 1998; Wojtyla et al. 2006; Oracz et al. 2009; Kranner et al. 2010). ROS formation occurs primarily due to the restarting of metabolism and also ROS are generated by NADPH oxidase or peroxidases as part of an oxidative burst during rehydration (Colville and Kranner 2010). Most ROS are scavenged by specific antioxidant systems and directly by proteins that scavenge an estimated 50–75 % of reactive radicals (Davies et al. 1999; Møller et al. 2007). Seeds generally activate their antioxidant systems immediately upon rehydration (De Gara et al. 1997; Tommasi et al. 2001; Müller et al. 2010). The accumulating data suggest that the ability of seeds to germinate is linked to the accumulation of a critical level of H_2O_2 and that this ROS is produced at a higher level in nondormant imbibed seeds than in dormant imbibed seeds (Oracz et al. 2007; Bailly et al. 2008). The balance between ROS-producing and ROS-scavenging system plays a key role in seed germination and dormancy alleviation (Oracz et al. 2009). The effect may be direct or can be mediated by ROS interaction with the phytohormones that control dormancy (Kwak et al. 2006; Finkelstein et al. 2008; Bailly et al. 2008).

It has been demonstrated that in Arabidopsis seeds, dormancy is associated with low ROS levels, while germination is mediated by accumulation of O_2^- and H_2O_2 in the radicle triggering GA signaling activation (Leymarie et al. 2012). Higher ROS levels in non-dormant seeds appeared first in cytoplasm, then in nucleus and finally in cell wall and were correlated with low catalase activity as compared to dormant seeds. In barley (*Hordeum vulgare* L.), H_2O_2 treatment upregulated the expression of GA-induced gene *HvExpA11*, inhibited the expression of *HvGA2ox3* involved in GA catabolism and enhanced the expression of *HvGA20ox1* implicated in GA synthesis, while ABA treatment upregulated catalase activity through transcriptional activation of the gene *HvCAT2* (Bahin et al. 2011). In sunflower (*Helianthus annuus* L.) embryos, the stimulating HCN effect on germination was shown to be associated with a marked increase in hydrogen peroxide and superoxide anion generation in the embryonic axes. This increase results from an inhibition of catalase and superoxide dismutase activities and also involves activation of

NADPH oxidase (Oracz et al. 2009). When sunflower seeds become able to fully germinate during dry after-ripening, they accumulate high amount of hydrogen peroxide and exhibit a low detoxifying ability through catalase, resulting from the decrease in *CATA1* transcript (Bouteau et al. 2007). The role of catalase expression in regulating H_2O_2 levels during dormancy and in the course of dormancy breakage was shown also in grapevines (*Vitis vinifera* L.) by Perez and Lira (2005). One of possible targets of H_2O_2 is ABA catabolism. It was shown that H_2O_2 upregulated ABA catabolism genes in Arabidopsis seeds, the action that required also NO. H_2O_2 is also involved in upregulation of GA biosynthesis (Liu et al. 2010). Stimulating effects of H_2O_2 on breaking dormancy and germination were demonstrated also for several C4 grasses (Sarath et al. 2007). Inhibition of NADPH oxidase by diphenylene-iodonium suppressed germination, and it was also shown that H_2O_2 stimulated endogenous production of NO, and scavengers of NO abolished the peroxide-responsive stimulation of seed germination.

It has been suggested that the antioxidant-rich environment in fruits protects the developing plant embryos from the damage caused by ROS and prevents early germination. After fruit maturation and seed dispersal, a prolonged exposure to oxygen and moisture enables the seeds to produce the ROS needed for seed germination (Grover 2013). In apples (*Malus domestica* Borkh.) a biphasic pattern of variation in H_2O_2 concentration in the embryos during cold stratification was detected. H_2O_2 content increased markedly upon imbibition. NO production by the embryos was low during entire period of stratification, but increased significantly when seeds started to germinate. It was concluded that the increase of H_2O_2 and protein carbonylation are important in seed dormancy alleviation by cold stratification, while NO becomes necessary for germination (Debska et al. 2013). A key role of H_2O_2 in the coordination of pea seed germination, acting as a priming factor that involves specific changes at the proteome, transcriptome, and hormonal levels was demonstrated, indicating, in particular, the role of activation of one of MAP kinase genes by H_2O_2 and of carbonylation of storage proteins in dormancy alleviation (Barba-Espin et al. 2011).

Upon imbibition, drastic changes in redox level are observed not only at the metabolite level but also at the protein level. The protein thiol redox state is drastically increased upon imbibition and treatment with GA. In imbibed dormant seeds higher abundance of antioxidant proteins and enzymes important for redox control, ROS scavenging, and detoxification was observed, which indicates that more intensive ROS scavenging prevents dormancy alleviation (Bykova et al. 2011a, b). These observations resulted in development of the concept of "oxidative window" in dormancy alleviation (Bailly et al. 2008), i.e., of the necessary range of ROS concentrations to break seed dormancy. Antioxidant defense mechanisms protect seeds during a dormancy stage and prevent their germination, while the break of dormancy is achieved under conditions of after-ripening when ROS accumulation modifies redox status and breaks these control mechanisms by altering protein functions through oxidative modifications, elevated levels of which may eventually lead to irreversible cell damage (Bykova et al. 2011a, b; Bykova and Rampitsch 2013). As cysteine often has catalytic or

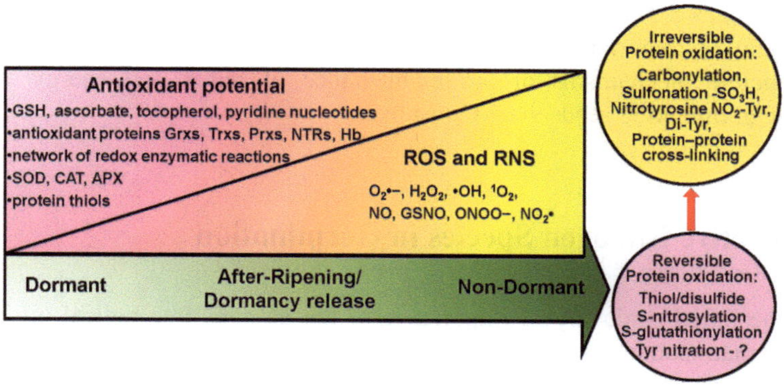

Fig. 1 Interconnection between ROS accumulation, dormancy release, and modification of amino acid residues in proteins. Grxs, glutaredoxins; Trxs, thioredoxins; Prxs, peroxyredoxins; NTRs, nitroreductases; Hb, hemoglobin; SOD, superoxide dismutase; CAT, catalase; APX, ascorbate peroxidase

structural roles in proteins, such modifications profoundly influence protein function (Wouters et al. 2010). This is schematically presented in Fig. 1, showing the interconnection between ROS accumulation, dormancy release and reduction/oxidation of cysteine residues in proteins.

The redox signaling via biochemical communication between ROS-driven carbonylation and RNS-driven nitrosylation was demonstrated in a number of recent redox proteomics studies with photosynthetic and nonphotosynthetic plant tissues and in response to abiotic stresses indicating their importance in redox homeostasis (Bykova and Rampitsch 2013, and references therein). Protein carbonylation is a permanent marker of oxidative stress, and it has been implicated in physiological transitions of seeds from dormancy to high germination potential and aging (Lounifi et al. 2013, and references therein). This irreversible protein modification was shown to be concomitantly increased with ROS accumulation and involved in breaking dormancy and/or the loss of germination ability due to aging. The physiological functions of RNS-driven protein modifications in seed biology have not been clearly demonstrated. However, reversible nitrosylation can possibly have an intermediate protective role (similar to glutathionylation) to preserve critical protein thiols. The multistage Cys oxidation resulting in irreversible oxidative damage and tyrosine nitration by peroxynitrite is likely to take place during transition between the state in which ROS prevail and the state in which RNS prevail, i.e., under conditions when both ROS and RNS are present at high levels. S-nitrosylation of NADPH oxidase, the enzyme responsible for ROS production, was recently shown to follow high level of NO production during a biotic stress response, thus limiting ROS production (Yun et al. 2011). It is important to note that the NADPH oxidase AtrbohB was shown to play a role in ROS production during Arabidopsis seeds after-ripening (Müller et al. 2009). The role of RNS becomes prevailing when oxygen in the seed covered by coat is depleted and NO starts to generate in anaerobic nitrite reduction. The bioenergetics of this anaerobic stage includes fermentation but it is also related to major fluxes of NO turnover which may be important for regulating redox state and energy. We discuss below the concrete mechanisms of the maintenance of redox and energy level of hypoxic germinating seeds via NO turnover.

3 Reactive Nitrogen Species in Germination

After the initial discovery of importance of the gaseous free radical nitric oxide (NO) in mammalian systems, its multiple cellular functions in plant systems have been revealed (Neill et al. 2003; Lamattina et al. 2003; Wendehenne et al. 2004; Guo and Crawford 2005). Among these functions is its proposed role as an endogenous dormancy-breaking agent (Bethke et al. 2004a, b, 2006, reviewed in Arc et al. 2013). However, later it was realized that the role of NO in dormancy breakage is rather secondary, resulting from its inhibitory effect on the electron transport proteins followed by stimulation of ROS production (Hancock et al. 2011). The endogenous production of NO starts likely not immediately upon but rather shortly after seed imbibition and corresponds to the condition when the internal oxygen becomes depleted. NO production from germinating seeds, similarly to other hypoxic plant tissues, coincides with expression of class-1 nonsymbiotic plant hemoglobin, which appears within 2 h after imbibition and which function was shown to be related to NO scavenging (Taylor et al. 1994; Nie and Hill 1997; Duff et al. 1998; Hunt et al. 2001; Ross et al. 2001; Igamberdiev and Hill 2004). In the reaction, which involves class 1 hemoglobin and corresponding methemoglobin reductase, NO is metabolized in a NADH-dependent reaction which yields nitrate (NO_3^-) (Igamberdiev et al. 2004, 2006a, b, reviewed in Gupta et al. 2011). Seed germination is a unique stage in the lifecycle of plant where both signaling and metabolic roles of NO as well as the expression level of class 1 nonsymbiotic plant hemoglobin are spatially and temporally integrated.

The sequence of events leading to NO formation and its role in metabolism and signaling in germinating seeds is presented in Fig. 2. It shows that the hypoxic environment leads to the formation of NO, which in reaction with hemoglobin restores the redox balance and maintains energy homeostasis of the hypoxic cell. Figure 3 depicts different pathways of NO formation operating in plants. Although several pathways (reductive and oxidative) are possible, the most powerful NO formation is associated with anoxic mitochondria (Gupta and Igamberdiev 2011). Other reductive reactions include side reaction of nitrate reductase with nitrite (Harper 1981; Dean and Harper 1988; Yamasaki and Sakihama 2000), putative plasma membrane-bound nitrite: NO reductase (Stöhr et al. 2001), other nitrite-reducing proteins containing hemes or molybdocofactors (Igamberdiev et al. 2010). Apart from enzymatic NO synthesis in plants, under acidic ($pH < 6$) conditions, a series of reactions ensue during which nitrous acid (HNO_2) and nitrogen dioxide (NO_2) generate NO (Bethke et al. 2004a). However, the low pH requirements of this reaction discourage cytoplasmic operation. Nonetheless, this particular nonenzymatic reaction bears significance for seed physiologists, as NO generation

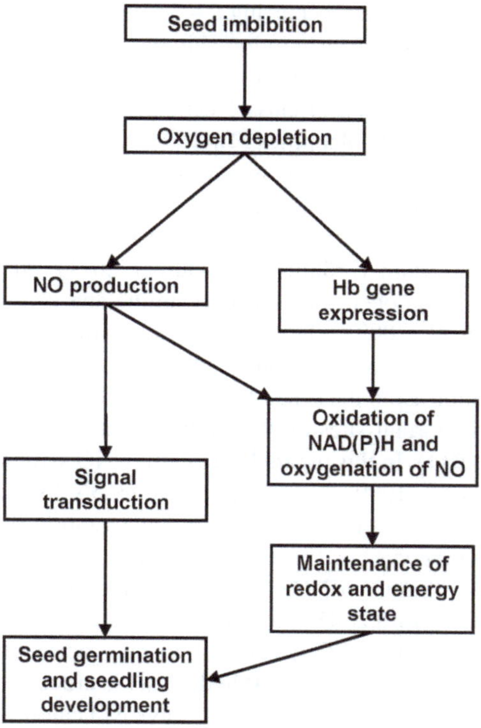

Fig. 2 NO formation and its role in metabolism and signaling in germinating seeds

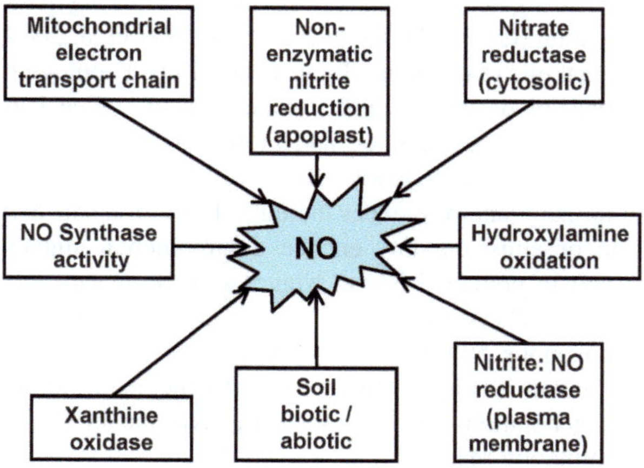

Fig. 3 Potential sources of nitric oxide for germinating seeds

via nitrite reduction was reported to occur in the apoplast of barley aleurone layers (Bethke et al. 2004a). The EPR spectra of NO emission in germinating barley and sorghum seeds revealed both enzymatic and nonenzymatic sources of NO generation during early seed germination events (Bethke et al. 2004a; Simontacchi et al. 2004). Moreover, in sorghum seeds, both nitrate reductase- and NO synthase-mediated NO generation appeared to be substrate-limited with maximal

NO emission occurring 24–30 h postimbibition and preceding any rapid increases in oxygen consumption (Simontacchi et al. 2004). Coincidentally, both the initial and maximal points of NO generation in germinating seeds appear to parallel with the induction of class 1 hemoglobin, which subsequently extends into developing roots and coleoptiles (Taylor et al. 1994; Duff et al. 1998).

The oxidative pathways usually imply the involvement of L-arginine-dependent NO synthase, but while NO production from arginine has been demonstrated in plants with certainty (Corpas et al. 2009), the nature of a plant NO synthase remains uncertain (Fröhlich and Durner 2011). Another oxidative pathway for NO formation occurs via hydroxylamine and is associated with ROS production (Rümer et al. 2009). During seed germination, the involvement of oxidative pathways demanding availability of oxygen is unlikely at early stages but after radicle protrusion these pathways may deliver NO for signaling purposes. One of possible important sites of NO formation in the oxidative pathway is peroxisome (del Río 2011). In this organelle a putative NO synthase activity was demonstrated (Barroso et al. 1999). However, the attempts to identify the NO synthase gene remain unsuccessful, except the characterization of NO synthase sequence in green alga *Ostreococcus* (Foresi et al. 2010). An abscisic-acid (ABA)-activated nitric oxide synthase (AtNOS1) initially identified in *Arabidopsis thaliana* (Guo and Crawford 2005) was shown to be a cyclic GTPase (Moreau et al. 2008). Its indirect relation to NO accumulation may occur at the level of ribosome assembly and operation where GTP hydrolysis is crucial for functioning.

While in oxidative pathways NO can be produced in small quantities sufficient for signaling function, the function of NO produced in reductive pathways spans that of signal molecule, as in the case during pathogen attack and hormonal signaling (Guo and Crawford 2005; Zeidler et al. 2004), to one of substrate during a metabolic reaction involving a class 1 plant hemoglobin (Igamberdiev et al. 2004, 2006a, b) or *S*-nitrosoglutathione reductase (Feechan et al. 2005). In terrestrial ecosystems, both abiotic and biotic processes operating external to the seed or seedling contribute toward fluxes of gaseous nitrogen oxides, including NO (Giba et al. 2003 and references therein). Therefore, soil-derived nitrogen oxide production or emission may encode environmental signals that integrate with endogenously produced NO to trigger seed dormancy release and germination (Giba et al. 2003; Alboresi et al. 2005). Several potential sources of exogenous and endogenous NO generation are presented in Fig. 3. For seed physiologists in particular, it is imperative to consider the manifold of mechanisms contributing to NO generation and catabolism in vivo and in the soil environment external to germinating seeds.

Due to the role of nitrate reductase in NO turnover, the production of NO is enhanced upon germination in soils and media containing nitrate. Nitrate accelerated testa rupture of hedge mustard [*Sisymbrium officinale* (L.) Scop.] seeds, modified expression pattern of the cell-wall remodeling mannanase proteins (Carrillo-Barral et al. 2013) and induced expression of key genes belonging to metabolism and signaling of ABA. It has been suggested that the nitrate signaling is already operative during imbibition of seeds increasing the favorable molecular conditions that trigger germination. NO acts metabolically to provide oxidation of reducing equivalents and contributing to ATP synthesis (Stoimenova et al. 2007), and it also acts as a signaling agent, where its effect is mediated via its interaction with hormones. Evidence has recently been presented that blue light inhibits germination in dormant grain and that jasmonate and NO counteract this effect by reducing dormancy in wheat (*Triticum aestivum* L.) via interdendent control of ABA level (Jacobsen et al. 2013). The stimulation of ethylene biosynthesis by NO was demonstrated in the course of dormancy breaking in apples (Gniazdowska et al. 2007). In Arabidopsis seeds the cross-talk between NO, H_2O_2, and ethylene was demonstrated in dormancy breaking (Lin et al. 2013).

Several reports provide evidence for overlapping functions of NO, UV radiation, light (via phytochrome) and gibberellins in seed germination events (Giba et al. 1998; Batak et al. 2002) and plant development (Beligni and Lamattina 2000; An et al. 2005). Collectively, these studies imply that the stimulatory roles of NO, if hormone-mediated, are more likely to occur through GA, as compared to ABA, signal transduction pathways. Additional roles for NO in seed germination include the stimulatory effect of NO upon α or β-amylase activity (Beligni et al. 2002; Zhang et al. 2005) and the potential for NO to improve the redox status of seeds (Beligni et al. 2002). Further studies may reveal that NO is capable of circumventing the classical abscisic acid–gibberellin antagonism by preferentially stimulating β-amylase activity, which, in comparison to α amylase, is unresponsive to gibberellin (Wang et al. 1996; Zhang et al. 2005).

4 Bioenergetics in Germinating Seeds and NO Production

The seed coat, acting as a diffusion barrier to oxygen and the effectiveness of mitochondria at the onset of imbibition may impose an influence on germination events (Mayer and Shain 1974). Maize seed mitochondria are capable of synthesizing ATP as soon as the seeds are rehydrated (Attucci et al. 1991). Similarly, in germinating barley seed, ATP increases after 1 h of imbibition (Duff et al. 1998). Changes in energy charge suggest an involvement of respiratory processes in ATP production (Bewley and Black 1982). The development of mitochondrial activity during seed germination is depicted in Fig. 4. Mitochondria in dry seeds usually

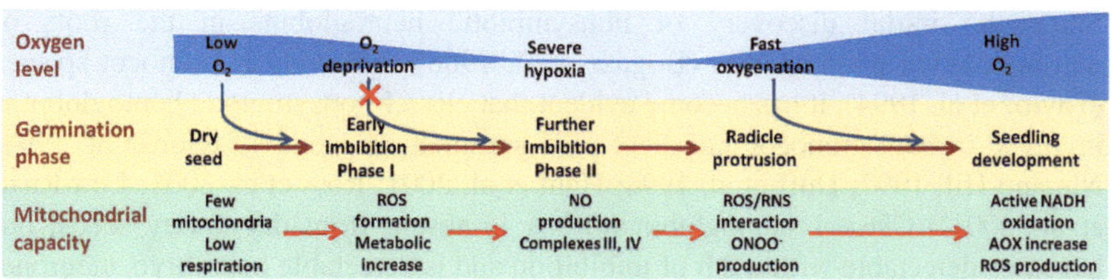

Fig. 4 Development of the mitochondrial capacity in seeds during germination. The scheme is based on data from Wilson and Bonner (1971), Ching and Ching (1972), Logan et al. (2001) and Benamar et al. (2003). AOX—alternative oxidase

have poor internal structure with a few cristae, although they can carry out respiratory activity. Dry seed mitochondria exhibit marked NADH-dependent oxygen uptake with lower activity of the citric acid cycle enzymes than mitochondria from germinated seeds (Logan et al. 2001). Upon imbibition, they develop higher rates of oxidation of NADH and the citric acid cycle substrates. This is further accompanied by the development of cristae structure and higher capacity for ATP synthesis (Benamar et al. 2003; Logan et al. 2001). The high cytosolic concentrations of NADH generated by glycolysis may permit the operation of the external mitochondrial NADH dehydrogenases facing the cytosolic site of the inner mitochondrial membrane (Logan et al. 2001). This may have a particular relevance under the hypoxic conditions that exist during early germination. When oxygen availability becomes higher, the development of the alternative oxidase (AOX) activity takes place.

Mitochondria are now considered as the primary site of NO generation in plants in the reductive pathway (Gupta and Igamberdiev 2011), although other sources also contribute (Fig. 3). The main role in anaerobic nitrite reduction to NO belongs to complex IV (cytochrome c oxidase), although the involvement of complex III (bc_1) and AOX is also plausible (Planchet et al. 2005). The poorly developed mitochondria that are present during seed imbibition can be fully capable of NO production, because the reported low level of the citric acid cycle enzymes in them does not have influence on this process. NO can be scavenged by mitochondria at the level of complex IV (back to nitrite) and in the reaction with superoxide following peroxynitrite formation and its conversion (e.g., by peroxiredoxin) only when oxygen concentration increases. In the conditions close to anaerobic, NO is diffused into the cytosol where it is scavenged by the non-symbiotic hemoglobin having affinity to oxygen two orders of magnitude higher than that of complex IV (Gupta and Igamberdiev 2011). The class 1 hemoglobin has likely a dual localization (in cytosol and in nucleus) (Seregélyes et al. 2000; Hebelstrup et al. 2008), and it also protects genome from the toxic (and mutagenic) effect of NO.

5 Hemoglobin-Dependent Nitric Oxide Turnover in Germinating Seeds

Since the initial discovery of nonsymbiotic hemoglobins in the roots of non-nodulating plant species (Bogusz et al. 1988) and seeds of monocot species (Taylor et al. 1994), it has become evident that class 1 nonsymbiotic hemoglobin is involved in both monocot and dicot seed germination events (Taylor et al. 1994; Nie and Hill 1997; Duff et al. 1998; Hunt et al. 2001; Ross et al. 2001; Lira-Ruan et al. 2001). Class 1 hemoglobin mRNA is absent from dry barley seeds, but becomes detectable within 2 h of imbibition and is detectable in embryo, aleurone, roots, and coleoptile (Andersson et al. 1996; Duff et al. 1998; Ross et al. 2001; Guy et al. 2002). In barley and rice, class 1 hemoglobin expression persists for 2–3 days postimbibition and is further enhanced by exposing these tissues to hypoxic (low oxygen) stress (Guy et al. 2002). Maize seedling Hb reaches concentrations of 0.04 % of total soluble seedling protein, whereas in wheat, wild oat, and *Echinochloa crus-galli* Hb concentrations are in the order of 0.01 %. Hb amounts are the highest in the root and aleurone layer of barley seedlings, with very low concentrations in the coleoptile (Duff et al. 1998).

Hemoglobin gene expression during hypoxic stress has proven essential for enhancing cellular energy status in maize cell culture (Sowa et al. 1998) and alfalfa roots (Dordas et al. 2003; Igamberdiev et al. 2004). The expression of the Hb gene during early seed germination and postgermination, therefore, may reflect its participation in maintaining the energy and redox status within embryo and aleurone layer of the seed and in developing oxygen-limited hypoxic organs (primarily roots) of the young seedling. ATP levels in barley embryos reach plateau within 2 h of imbibition (Duff et al. 1998) and that is roughly coincident with increases in Hb transcripts in aleurone layer tissue (Taylor et al. 1994).

A strong evidence has been accumulated that the primary function of class 1 Hb is related to NO metabolism, which involves not just NO scavenging but also the establishment of a connection between NO turnover and the maintenance of redox and energy levels in the plant cell (Sowa et al. 1998; Dordas et al. 2003; Igamberdiev et al. 2004, 2006a, b). The sequence of reactions involved in NO scavenging under hypoxic conditions such as during the early imbibition of seeds, designated the Hb/NO cycle (Igamberdiev and Hill 2004; Igamberdiev et al. 2005), is shown in Fig. 5. While NO may be formed by any of the processes described in Fig. 3, the reductive formation of NO from nitrite via nitrate in mitochondria is

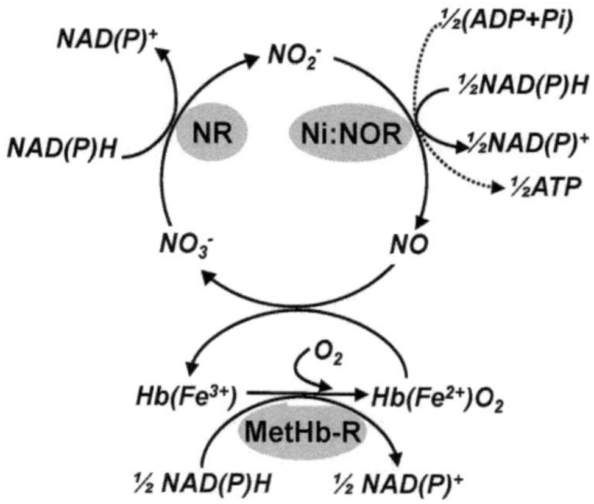

Fig. 5 The hemoglobin–nitric oxide (Hb/NO) cycle, a sequence of reactions of NO scavenging by oxyhemoglobin followed by its regeneration via methemoglobin reductase (MetHb-R) protein. The cycle is linked to the oxidation of NADH and NADPH in MetHb-R, nitrate reductase (NR), and nitrite: NO reductase (Ni:NOR) reactions. The latter can be attributed to the mitochondrial electron transport chain (Planchet et al. 2005; Gupta et al. 2005) where it is putatively linked to ATP synthesis, or to the side reaction of the cytosolic NR (Rockel et al. 2002), or to a special nitrite: NO reductase enzyme associated with plasma membrane (Stöhr et al. 2001)

suggested here. The steps of NO turnover in the Hb/NO cycle involve oxidation of NAD(P)H, contributing to the maintenance of redox status in a reductive environment. The mitochondrial reaction, if effectively coupled, will lead to the rates of ATP synthesis comparable to glycolysis (Stoimenova et al. 2007). This reaction indicates that Hb may be responsible for the diversification of metabolism from fermentation during seedling development. The NO produced in cells can be oxygenated to nitrate with the tightly bound oxygen of class 1 Hb, which, in the process, is oxidized to the ferric Hb (Fe^{3+}) state. To maintain the cycle, Hb(Fe^{2+}) must be regenerated. A special enzyme possessing ferric Hb (metHb) reductase activity is involved in the ascorbate-mediated reduction of ferric Hb (Igamberdiev et al. 2006a, 2011); other reductases or free reduced flavins may also be involved (Sainz et al. 2013). All necessary components for the operation of the Hb/NO cycle during seed germination and postgermination are therefore present. The involvement of the cycle in the maintenance of the redox and energy balance in germinating seeds may explain how at early stages of germination, high metabolic rates and ATP concentrations are achieved in the hypoxic environment of the seed.

6 Aerobic Metabolism After Radicle Protrusion

After breakage of seed coat and radicle protrusion (during postgerminative growth), the supply of oxygen drastically increases and the processes dependent on oxygen supply are greatly enhanced. One of these processes is mitochondrial respiration and the other is peroxisomal oxidation, which in seeds provides mobilization of stored fats via β-oxidation followed by conversion of acetyl-CoA to succinate in the glyoxylate cycle and metabolism of succinate to carbohydrates in gluconeogenesis (Igamberdiev and Lea 2002). In fact, the appearance of seed plants was directly connected to the acquirement by storage tissues of a peroxisomal fatty acid oxidation function linked to the glyoxylate cycle, which is induced during seed germination and maturation. The type of peroxisome performing β-oxidation and glyoxylate cycle is known as glyoxysome. The autonomy of seed metabolism, which is a key factor in higher plants evolution, is provided by the acquisition of glyoxysomal enzymes. Glyoxysomal metabolism in higher plant seeds is located mostly in the cotyledons, but in some cases it operates in the endosperm (e.g., castor bean).

A special type of peroxisome is the glyoxysome in the scutellum of cereal plants. The scutellum has to a certain extent similar functions to the cotyledon of dicots but reveals a homology to the third leaf of seedlings of plants from the order of Nymphaeales (Burger 1998). Besides a fat-storing function, the scutellum together with the aleurone layer has "digestive" functional activities involved in supplying organic acids for the hydrolysis of starch in the endosperm (Drozdowicz and Jones 1995). During germination of cereal seeds, carbohydrates are provided by starch hydrolysis, while glyoxysomes participate in acidification of endosperm, promoting operation of amylases. Operation of the glyoxylate cycle in the scutellum also provides carbon skeletons for amino acid synthesis, which was proven by the study showing incorporation of label from ^{14}C-acetate by the scutellum mostly into amino acids, but not in carbohydrates (Igamberdiev and Rodionova 1991). Metabolism of succinate occurs primarily in mitochondria via the TCA cycle succinate dehydrogenase, however under conditions favoring its intensive formation it can be oxidized in glyoxysomes by the special enzyme system forming malate from succinate. This system is membrane-bound, has low affinity for succinate, contains flavin and forms hydrogen peroxide (Igamberdiev et al. 1995). Hence, the function of glyoxysome-like organelles in the scutellum of monocots may be the interconnection of fatty acid degradation with amino acid synthesis and with the synthesis of organic acids resulting in acidification of the endosperm. Fumarate formed in the succinate dehydrogenase reaction can be metabolized by the cytosolic form of fumarase highly expressed in scutellum and mostly absent in leaves of cereal plants (Eprintsev et al. 2014). This form of fumarase can be specifically associated with the synthesis of amino acids (Pracharoenwattana et al. 2010).

After the radicle protrusion, production of NO in the reductive pathways is ceased due to the availability of oxygen, which also correlates with a decrease in hemoglobin expression (Duff et al. 1998). NO may be produced in oxidative pathways, in particular in glyoxysomes, and participate in signaling functions, however, we do not have sufficient information on the role of NO during this stage of plant development. On the other hand, the production of hydrogen peroxide in flavin-dependent oxidation of fatty acids and other compounds increases drastically, and the role of ROS at this stage again becomes prevailing over RNS. Catalase keeps H_2O_2 at physiological levels, and enzymes such as ascorbate peroxidase and other peroxidases can also be involved. The reaction of catalase, which produces half of O_2 consumed in flavin-dependent oxidation, prevents oxygen depletion in the proximity of flavin-containing oxidases such as fatty acyl-CoA oxidase. When the storage compounds in seeds are consumed, ROS may play a primary role in the processes related to apoptosis and resulting in degradation of corresponding tissues after fulfilling their functions (Bailly et al. 2008; He and Kermode 2010). This process completes the function of ROS (and possibly RNS) during germination and postgermination and facilitates the transition of plants to the next stage of development (seedling growth).

Conclusion

Seed germination and postgerminative growth are accompanied by intensive production of reactive oxygen (superoxide, hydrogen peroxide, etc.) and nitrogen (NO and its derivatives) species (ROS and RNS). Intensive ROS accumulation during imbibition approaching dangerous levels is critically important condition for breaking dormancy. Upon depletion of oxygen under the seed coat, NO becomes produced anaerobically in the reductive pathway associated mainly with mitochondria and participates in energy metabolism

(continued)

of the seed before radicle protrusion. Its turnover involves nitrate reduction to nitrite in the cytosol, nitrite reduction to NO in mitochondria, and NO oxygenation in the cytosol in the reaction involving hypoxically induced (class 1) hemoglobin. After radicle protrusion, an important organelle participating in ROS and RNS formation becomes peroxisome, which functions in utilization of seed storage compounds including fatty acids. When storage compounds are consumed, ROS become involved in degradation of seed tissues. ROS and RNS participate in posttranslational modification of proteins, in mediation of hormonal responses and in other signaling events, which are essential in regulation and integration of cellular processes in germinating seeds.

References

Alboresi A, Gestin C, Leydecker MT, Bedu M, Meyer C, Truong HN (2005) Nitrate, a signal relieving seed dormancy in Arabidopsis. Plant Cell Environ 28:500–512

An LZ, Liu YH, Zhang MX, Chen T, Wang XL (2005) Effects of nitric oxide on growth of maize seedling leaves in the presence or absence of ultraviolet-B radiation. J Plant Physiol 162: 317–326

Andersson CR, Jensen EO, Llewellyn DJ, Dennis ES, Peacock WJ (1996) A new hemoglobin gene from soybean: a role for hemoglobin in all plants. Proc Natl Acad Sci USA 93:5682–5687

Arc E, Galland M, Godin B, Cueff G, Rajjou L (2013) Nitric oxide implication in the control of seed dormancy and germination. Front Plant Sci 4:346

Attucci S, Carde JP, Raymond P, Saint-Ges V, Spiteri A, Pradet A (1991) Oxidative phosphorylation by mitochondria extracted from dry sunflower seeds. Plant Physiol 95:390–398

Bahin E, Bailly C, Sotta B, Kranner I, Corbineau F, Leymarie J (2011) Crosstalk between reactive oxygen species and hormonal signalling pathways regulates grain dormancy in barley. Plant Cell Environ 34:980–993

Bailly C, El-Maarouf-Bouteau H, Corbineau F (2008) From intracellular signaling networks to cell death: the dual role of reactive oxygen species in seed physiology. C R Biol 331:806–814

Barba-Espin G, Diaz-Vivancos P, Job D, Belghazi M, Job C, Hernandez JA (2011) Understanding the role of H_2O_2 during pea seed germination: a combined proteomic and hormone profiling approach. Plant Cell Environ 34:1907–1919

Barroso JB, Corpas FJ, Carreras A, Sandalio LM, Valderrama R, Palma JM, Lupiáñez JA, del Río LA (1999) Localization of nitric-oxide synthase in plant peroxisomes. J Biol Chem 274: 36729–36733

Batak I, Dević M, Giba Z, Grubišić D, Poff KL, Konjević R (2002) The effects of potassium nitrate and NO-donors on phytochrome A- and phytochrome B-specific induced germination of *Arabidopsis thaliana* seeds. Seed Sci Res 12:253–259

Beligni MV, Lamattina L (2000) Nitric oxide stimulates seed germination and de-etiolation, and inhibits hypocotyl elongation, three light-inducible responses in plants. Planta 210:215–221

Beligni MV, Fath A, Bethke PC, Lamattina L, Jones RL (2002) Nitric oxide acts as an antioxidant and delays programmed cell death in barley aleurone layers. Plant Physiol 129:1642–1650

Benamar A, Tallon C, Macherel D (2003) Membrane integrity and oxidative properties of mitochondria isolated from imbibing pea seeds after priming or accelerated ageing. Seed Sci Res 13:35–45

Bethke PC, Gubler F, Jacobsen JV, Jones RL (2004a) Dormancy of Arabidopsis seeds and barley grains can be broken by nitric oxide. Planta 219:847–855

Bethke PC, Badger MR, Jones RL (2004b) Apoplastic synthesis of nitric oxide by plant tissues. Plant Cell 16:332–341

Bethke PC, Libourel IGL, Reinöhl V, Jones RL (2006) Sodium nitroprusside, cyanide, nitrite, and nitrate break Arabidopsis seed dormancy in a nitric oxide-dependent manner. Planta 223: 805–812

Bewley JD (1997) Seed germination and dormancy. Plant Cell 9:1055–1066

Bewley JD, Black M (1982) Physiology and biochemistry of seeds in relation to germination. Springer, New York

Bewley JD, Black M (1994) Seeds: physiology of development and germination. Plenum Press, New York

Bogusz D, Appleby CA, Landsmann J, Dennis ES, Trinick MJ, Peacock WJ (1988) Functioning hemoglobin genes in non-nodulating plants. Nature 331:178–180

Bouteau HEM, Job C, Job D, Corbineau F, Bailly C (2007) ROS signaling in seed dormancy alleviation. Plant Signal Behav 2:362–364

Burger WC (1998) The question of cotyledon homology in angiosperms. Bot Rev 64:356–371

Bykova NV, Rampitsch C (2013) Modulating protein function through reversible oxidation: redox-mediated processes in plants revealed through proteomics. Proteomics 13:579–596

Bykova NV, Hoehn B, Rampitsch C, Banks T, Stebbing JA, Fan T, Knox R (2011a) Redox-sensitive proteome and antioxidant strategies in wheat seed dormancy control. Proteomics 11: 865–882

Bykova NV, Hoehn B, Rampitsch C, Hu J, Stebbing JA, Knox R (2011b) Thiol redox-sensitive seed proteome in dormant and non-dormant hybrid genotypes of wheat. Phytochemistry 72: 1162–1172

Cakmak I, Strbac D, Marschner H (1993) Activities of hydrogen peroxide scavenging enzymes in germinating wheat seeds. J Exp Bot 44:127–132

Caliskan M, Cuming AC (1998) Spatial specificity of H_2O_2-generating oxalate oxidase gene expression during wheat embryo germination. Plant J 15:165–171

Carrillo-Barral N, Matilla AJ, Iglesias-Fernandez R, Rodriguez-Gacio MD (2013) Nitrate-induced early transcriptional changes during imbibition in non-after-ripened *Sisymbrium officinale* seeds. Physiol Plant 148:560–573

Ching TM, Ching KK (1972) Content of adenosine phosphates and adenylate energy charge in germinating ponderosa pine seeds. Plant Physiol 50:536–540

Colville L, Kranner I (2010) Desiccation tolerant plants as model systems to study redox regulation of protein thiols. Plant Growth Regul 62:241–255

Corpas FJ, Palma JM, del Río LA, Barroso JB (2009) Evidence supporting the existence of L-arginine-dependent nitric oxide synthase activity in plants. New Phytol 184:9–14

Davies MJ, Fu S, Wang H, Dean RT (1999) Stable markers of oxidant damage to proteins and their application in study of human disease. Free Radic Biol Med 27:1151–1161

De Gara L, de Pinto MC, Arrigoni O (1997) Ascorbate synthesis and ascorbate peroxidase activity during the early stage of wheat germination. Physiol Plant 100:894–900

Dean JV, Harper JE (1988) The conversion of nitrite to nitrogen oxide(s) by the constitutive NAD(P)H-nitrate reductase enzyme from soybean. Plant Physiol 88:389–395

Debska K, Krasuska U, Budnicka K, Bogatek R, Gniazdowska A (2013) Dormancy removal of apple seeds by cold stratification is associated with fluctuation in H_2O_2, NO production and protein carbonylation level. J Plant Physiol 170:480–488

del Río LA (2011) Peroxisomes as a cellular source of reactive nitrogen species signal molecules. Arch Biochem Biophys 506:1–11

Dordas C, Hasinoff BB, Igamberdiev AU, Manac'h N, Rivoal J, Hill RD (2003) Expression of a stress-induced hemoglobin affects NO levels produced by alfalfa root cultures under hypoxic stress. Plant J 35:763–770

Drozdowicz YM, Jones RL (1995) Hormonal regulation of organic and phosphoric acid release by barley aleurone layers and scutella. Plant Physiol 108:769–776

Duff SMG, Guy PA, Nie X, Durnin DC, Hill RD (1998) Hemoglobin expression in germinating barley. Seed Sci Res 8:431–436

Eprintsev AT, Fedorin DN, Starinina EV, Igamberdiev AU (2014) Expression and properties of the mitochondrial and cytosolic forms of fumarase in germinating maize seeds. Physiol Plant 152:231–240

Feechan A, Kwon E, Yun BW, Wang Y, Pallas JA, Loake GJ (2005) A central role for S-nitrosothiols in plant disease resistance. Proc Natl Acad Sci USA 102:8054–8059

Finkelstein R, Reeves W, Ariizumi T, Steber C (2008) Molecular aspects of seed dormancy. Annu Rev Plant Biol 59:387–415

Foresi N, Correa-Aragunde N, Parisi G, Caló G, Salerno G, Lamattina L (2010) Characterization of a nitric oxide synthase from the plant kingdom: NO generation from the green alga *Ostreococcus tauri* is light irradiance and growth phase dependent. Plant Cell 22:3816–3830

Fröhlich A, Durner J (2011) The hunt for plant nitric oxide synthase (NOS): is one really needed? Plant Sci 181:401–404

Geigenberger P (2003) Response of plant metabolism to too little oxygen. Curr Opin Plant Biol 6: 247–256

Giba Z, Grubišić D, Todorović S, Sajc L, Stojaković Đ, Konjević R (1998) Effect of nitric oxide-releasing compounds on phytochrome-controlled germination of Empress tree seeds. Plant Growth Regul 26:175–181

Giba Z, Grubišić D, Konjević R (2003) Nitrogen oxides as environmental sensors for seeds. Seed Sci Res 13:187–196

Gniazdowska A, Dobrzynska U, Babanczyk T, Bogatek R (2007) Breaking the apple embryo dormancy by nitric oxide involves the stimulation of ethylene production. Planta 225: 1051–1057

Grover AK (2013) Why fruits are rich in antioxidants? An opinion review. Can J Physiol Pharmacol 91:191–197

Guo FQ, Crawford NM (2005) Arabidopsis nitric oxide synthase1 is targeted to mitochondria and protects against oxidative damage and dark-induced senescence. Plant Cell 17:3436–3450

Gupta KJ, Igamberdiev AU (2011) The anoxic plant mitochondrion as a nitrite: NO reductase. Mitochondrion 11:537–543

Gupta KJ, Stoimenova M, Kaiser WM (2005) In higher plants, only root mitochondria, but not leaf mitochondria reduce nitrite to NO, in vitro and in situ. J Exp Bot 56:2601–2609

Gupta KJ, Hebelstrup KH, Mur LA, Igamberdiev AU (2011) Plant hemoglobins: important players at the crossroads between oxygen and nitric oxide. FEBS Lett 585:3843–3849

Guy PA, Sidaner J-P, Schroeder S, Edney M, MacGregor AW, Hill RD (2002) Embryo phytoglobin gene expression as a measure of germination in cereals. J Cereal Sci 36:147–156

Hancock JT, Neill SJ, Wilson ID (2011) Nitric oxide and ABA in the control of plant function. Plant Sci 181:555–559

Harper JE (1981) Evolution of nitrogen oxide(s) during in vivo nitrate reductase assay of soybean leaves. Plant Physiol 68:1488–1493

He X, Kermode AR (2010) Programmed cell death of the megagametophyte during post-germinative growth of white spruce (*Picea glauca*) seeds is regulated by reactive oxygen species and the ubiquitin-mediated proteolytic system. Plant Cell Physiol 51:1707–1720

Hebelstrup KH, Østergaard-Jensen E, Hill RD (2008) Bioimaging techniques for subcellular localization of plant hemoglobins and measurement of hemoglobin-dependent nitric oxide scavenging in planta. Methods Enzymol 437:595–604

Hunt PW, Watts RA, Trevaskis B, Llewelyn DJ, Burnell J, Dennis ES, Peacock WJ (2001) Expression and evolution of functionally distinct hemoglobin genes in plants. Plant Mol Biol 47:677–692

Igamberdiev AU, Hill RD (2004) Nitrate, NO and hemoglobin in plant adaptation to hypoxia: an alternative to classic fermentation pathways. J Exp Bot 55:2473–2482

Igamberdiev AU, Lea PJ (2002) The role of peroxisomes in the integration of metabolism and evolutionary diversity of photosynthetic organisms. Phytochemistry 60:651–674

Igamberdiev AU, Rodionova MI (1991) Role of the glyoxylate cycle in metabolism of acetate and other organic acids. Sov Plant Physiol 38:360–365
Igamberdiev AU, Popov VN, Falaleeva MI (1995) Alternative system of succinate oxidation in glyoxysomes of higher plants. FEBS Lett 367:287–290
Igamberdiev AU, Seregélyes C, Manac'h N, Hill RD (2004) NADH-dependent metabolism of nitric oxide in alfalfa root cultures expressing barley hemoglobin. Planta 219:95–102
Igamberdiev AU, Baron K, Manac'h-Little N, Stoimenova M, Hill RD (2005) The hemoglobin/nitric oxide cycle: involvement in flooding stress and effects on hormone signalling. Ann Bot 96:557–564
Igamberdiev AU, Bykova NV, Hill RD (2006a) Scavenging of nitric oxide by barley hemoglobin is facilitated by monodehydroascrbate reductase-mediated ascorbate reduction of methemoglobin. Planta 223:1033–1040
Igamberdiev AU, Stoimenova M, Seregélyes C, Hill RD (2006b) Class-1 hemoglobin and antioxidant metabolism in alfalfa roots. Planta 223:1041–1046
Igamberdiev AU, Bykova NV, Shah JK, Hill RD (2010) Anoxic nitric oxide cycling in plants: participating reactions and possible mechanisms. Physiol Plant 138:393–404
Igamberdiev AU, Bykova NV, Hill RD (2011) Structural and functional properties of class 1 ant hemoglobins. IUBMB Life 63:146–152
Jacobsen JV, Barrero JM, Hughes T, Julkowska M, Taylor JM, Xu Q, Gubler F (2013) Roles for blue light, jasmonate and nitric oxide in the regulation of dormancy and germination in wheat grain (*Triticum aestivum* L.). Planta 238:121–138
Kranner I, Minibayeva FV, Beckett RP, Seal CE (2010) What is stress? Concepts, definitions and applications in seed science. New Phytol 188:655–673
Kwak JM, Nguyen V, Schroeder JI (2006) The role of reactive oxygen species in hormonal responses. Plant Physiol 141:323–329
Lamattina L, Garcia-Mata C, Graziano M, Pagnussat G (2003) Nitric oxide: the versatility of an extensive signal molecule. Annu Rev Plant Biol 54:109–136
Leymarie J, Vitkauskaite G, Hoang HH, Gendreau E, Chazoule V, Meimoun P, Corbineau F, El-Maarouf-Bouteau H, Bailly C (2012) Role of reactive oxygen species in the regulation of Arabidopsis seed dormancy. Plant Cell Physiol 53:96–106
Lin YC, Yang L, Paul M, Zu YG, Tang ZH (2013) Ethylene promotes germination of Arabidopsis seed under salinity by decreasing reactive oxygen species: evidence for the involvement of nitric oxide simulated by sodium nitroprusside. Plant Physiol Biochem 73:211–218
Lira-Ruan V, Sarath G, Klucas RV, Arredondo-Peter R (2001) Synthesis of hemoglobins in rice (*Oryza sativa* var. Jackson) plants growing in normal and stress conditions. Plant Sci 161: 279–287
Liu YG, Ye NH, Liu R, Chen MX, Zhang JH (2010) H_2O_2 mediates the regulation of ABA catabolism and GA biosynthesis in Arabidopsis seed dormancy and germination. J Exp Bot 61:2979–2990
Logan DC, Millar AH, Sweetlove LJ, Hill SA, Leaver CJ (2001) Mitochondrial biogenesis during germination in maize embryos. Plant Physiol 125:662–672
Lounifi I, Arc E, Molassiotis A, Job D, Rajjou L, Tanou G (2013) Interplay between protein carbonylation and nitrosylation in plants. Proteomics 13:568–578
Mayer AM, Shain Y (1974) Control of seed germination. Annu Rev Plant Physiol 25:167–193
Møller IM, Jensen PE, Hansson A (2007) Oxidative modifications to cellular components in plants. Annu Rev Plant Biol 58:459–481
Moreau M, Lee GI, Wang Y, Crane BR, Klessig DF (2008) AtNOS/AtNOA1 is a functional *Arabidopsis thaliana* cGTPase and not a nitric-oxide synthase. J Biol Chem 283:32957–32967
Müller K, Carstens AC, Linkies A, Torres MA et al (2009) The NADPH-oxidase AtrbohB plays a role in Arabidopsis seed after-ripening. New Phytol 184:885–897
Müller K, Job C, Belghazi M, Job D, Leubner-Metzger G (2010) Proteomics reveal tissue-specific features of the cress (*Lepidium sativum* L.) endosperm cap proteome and its hormone-induced changes during seed germination. Proteomics 10:406–416

Neill SJ, Desikan R, Hancock JT (2003) Nitric oxide signalling in plants. New Phytol 159:11–35
Nie XZ, Hill RD (1997) Mitochondrial respiration and hemoglobin gene expression in barley aleurone tissue. Plant Physiol 114:835–840
Oracz K, El-Maarouf Bouteau H, Farrant JM, Cooper K, Belghazi M, Job C, Job D, Corbineau F, Bailly C (2007) ROS production and protein oxidation as a novel mechanism for seed dormancy alleviation. Plant J 50:452–465
Oracz K, El-Maarouf-Bouteau H, Kranner I, Bogatek R, Corbineau F, Bailly C (2009) The mechanisms involved in seed dormancy alleviation by hydrogen cyanide unravel the role of reactive oxygen species as key factors of cellular signaling during germination. Plant Physiol 150:494–505
Perez FJ, Lira W (2005) Possible role of catalase in post-dormancy bud break in grapevines. J Plant Physiol 162:301–308
Planchet E, Gupta KJ, Sonoda M, Kaiser WM (2005) Nitric oxide emission from tobacco leaves and cell suspensions: rate limiting factors and evidence for the involvement of mitochondrial electron transport. Plant J 41:732–743
Pracharoenwattana I, Zhou WX, Keech O, Francisco PB, Udomchalothorn T, Tschoep H, Stitt M, Gibon Y, Smith SM (2010) Arabidopsis has a cytosolic fumarase required for the massive allocation of photosynthate into fumaric acid and for rapid plant growth on high nitrogen. Plant J 62:785–795
Rockel P, Strube F, Rockel A, Wildt J, Kaiser WM (2002) Regulation of nitric oxide (NO) production by plant nitrate reductase in vivo and in vitro. J Exp Bot 53:103–110
Rolletschek H, Weschke W, Weber H, Wobus U, Borisjuk L (2004) Energy state and its control on seed development: starch accumulation is associated with high ATP and steep oxygen gradients within barley grains. J Exp Bot 55:1351–1359
Ross EJH, Shearman L, Mathiesen M, Zhou YJ, Arredondo-Peter R, Sarath G, Klucas RV (2001) Nonsymbiotic hemoglobins in rice are synthesized during germination and in differentiating cell types. Protoplasma 218:125–133
Rümer S, Gupta KJ, Kaiser WM (2009) Plant cells oxidize hydroxylamines to NO. J Exp Bot 60: 2065–2072
Sainz M, Pérez-Rontomé C, Ramos J, Mulet JM, James EK, Bhattacharjee U, Petrich JW, Becana M (2013) Plant hemoglobins may be maintained in functional form by reduced flavins in the nuclei, and confer differential tolerance to nitro-oxidative stress. Plant J 76:875–887
Sarath G, Hou GC, Baird LM, Mitchell RB (2007) Reactive oxygen species, ABA and nitric oxide interactions on the germination of warm-season C-4-grasses. Planta 226:697–708
Seregélyes C, Mustárdy L, Ayaydin F, Sass L, Kovács L, Endre G, Lukács N, Kovács I, Vass I, Kiss GB, Horváth GV, Dudits D (2000) Nuclear localization of a hypoxia-inducible novel non-symbiotic hemoglobin in cultured alfalfa cells. FEBS Lett 482:125–130
Simontacchi M, Jasid S, Puntarulo S (2004) Nitric oxide generation during early germination of sorghum seeds. Plant Sci 167:839–847
Sowa AW, Duff SMG, Guy PA, Hill RD (1998) Altering hemoglobin levels changes energy status in maize cells under hypoxia. Proc Natl Acad Sci USA 95:10317–10321
Stöhr C, Strube F, Marx G, Ullrich WR, Rockel P (2001) A plasma membrane-bound enzyme of tobacco roots catalyses the formation of nitric oxide from nitrite. Planta 212:835–841
Stoimenova M, Igamberdiev AU, Gupta KJ, Hill RD (2007) Nitrite-driven anaerobic ATP synthesis in barley and rice root mitochondria. Planta 226:465–474
Taylor ER, Nie XZ, MacGregor AW, Hill RD (1994) A cereal hemoglobin gene is expressed in seed and root tissues under anaerobic conditions. Plant Mol Biol 24:853–862
Taylorson RB, Hendricks SB (1977) Dormancy in seeds. Annu Rev Plant Physiol 28:331–354
Tommasi F, Paciolla C, de Pinto MC, De Gara L (2001) A comparative study of glutathione and ascorbate metabolism during germination of *Pinus pinea* L. seeds. J Exp Bot 52:1647–1654
Wang SM, Lue WL, Eimert K, Chen J (1996) Phytohormone-regulated beta-amylase gene expression in rice. Plant Mol Biol 31:975–982

Weber H, Borisjuk L, Wobus U (2005) Molecular physiology of legume seed development. Annu Rev Plant Biol 56:253–279

Wendehenne D, Durner J, Klessig DF (2004) Nitric oxide: a new player in plant signalling and defence responses. Curr Opin Plant Biol 7:449–455

Wilson SB, Bonner WD Jr (1971) Studies of electron transport in dry and imbibed peanut embryos. Plant Physiol 48:340–344

Wojtyla Ł, Garnczarska M, Zalewski T, Bednarski W, Ratajczak L, Jurga S (2006) A comparative study of water distribution, free radical production and activation of antioxidative metabolism in germinating pea seeds. J Plant Physiol 163:1207–1220

Wouters MA, Fan SW, Haworth NL (2010) Disulfides as redox switches: from molecular mechanisms to functional significance. Antioxid Redox Signal 12:53–91

Yamasaki H, Sakihama Y (2000) Simultaneous production of nitric oxide and peroxynitrite by plant nitrate reductase: in vitro evidence for the NR-dependent formation of active nitrogen species. FEBS Lett 468:89–92

Yun BW, Feechan A, Yin M, Saidi NB et al (2011) S-nitrosylation of NADPH oxidase regulates cell death in plant immunity. Nature 478:264–268

Zeidler D, Zahringer U, Gerber I, Dubery I, Hartung T, Bors W, Hutzler P, Durner J (2004) Innate immunity in *Arabidopsis thaliana*: lipopolysaccharides activate nitric oxide synthase (NOS) and induce defense genes. Proc Natl Acad Sci USA 101:15811–15816

Zhang H, Shen WB, Zhang W, Xu LL (2005) A rapid response of β-amylase to nitric oxide but not gibberellin in wheat seeds during the early stage of germination. Planta 220:708–716

ROS Signalling in Plant Embryogenesis

Mohamed Elhiti and Claudio Stasolla

1 Introduction

Reactive oxygen species (ROS) are chemically reactive molecules containing oxygen with unpaired electrons. These partially reduced or activated derivatives of oxygen [singlet oxygen $^1O_2$, superoxide anion O^{2-}, hydrogen peroxide H_2O_2, and hydroxyl radical $O^{\bullet}$] are highly reactive and toxic and can lead to the oxidative destruction of cells. In planta, ROS are continuously produced as by-products of various metabolic pathways activated in different cellular organelles such as chloroplasts, mitochondria, and peroxisomes (Foyer and Harbinson 1994). Consequently, plants have evolved efficient ROS-scavenging mechanisms (Mittler et al. 2004). Beside their harmful effects, ROS also act as signal molecules regulating a variety of biological processes (Kovtun et al. 2000). Perception of the cellular or intracellular ROS levels relies on specific sensors triggering signaling cascades culminating in developmental responses. The best examples of ROS-controlled processes include programmed cell death (PCD) (Jacobson 1996; Tan et al. 1998), cell cycle control (Fehér et al. 2008), response to pathogen attack (Liu et al. 2010), plant defence responses through hormonal signalling (Sakamoto et al. 2008; Miller et al. 2010), biotic and abiotic stress responses (Fujita et al. 2006), and embryogenesis (Mantiri et al. 2008).

Plant embryogenesis results in the formation of a plant embryo from a fertilized ovule through elaborated and accurate cell division and differentiation patterns. This process can also occur without fertilization through the generation of asexual or apomictic embryos, generated from the maternal tissue or the unfertilized egg

M. Elhiti
Faculty of Science, Department of Botany, Tanta University, Tanta 31527, Egypt

C. Stasolla (✉)
Department of Plant Science, University of Manitoba, Winnipeg, MB, Canada R3T 2 N2
e-mail: stasolla@ms.umanitoba.ca

K.J. Gupta, A.U. Igamberdiev (eds.), *Reactive Oxygen and Nitrogen Species Signaling and Communication in Plants*, Signaling and Communication in Plants 23,
DOI 10.1007/978-3-319-10079-1_10

cell (Nogler 1984). Recapitulation of embryogenesis is not an in vivo prerogative as embryos can be produced in vitro through two main processes: gametophytic and somatic embryogenesis. Gametophytic embryogenesis is the generation of haploid embryos from the male or female gametophyte. When using male gametophytic tissue the embryogenic process is called androgenesis and involves the utilization of microspores and pollen grains as the initial explants (Maraschin et al. 2005). While androgenesis often results in the formation of haploid embryos, somatic embryogenesis is initiated from somatic cells, i.e. cells other than gametes and produces diploid embryos (Raghavan 2000). A key event during embryogenesis in culture is the acquisition of embryogenic potential. The competent cells within the explants must be able to respond to inductive signals, usually provided by specific changes in culture conditions and medium supplements (Rose and Nolan 2006; Elhiti et al. 2010). Following this induction period, cells start to de-differentiate into embryogenic cells which can be further directed to generate embryos.

Stress, plant growth regulators (PGRs) or a combination of both, have been identified as critical factors during the acquisition of the embryogenic potential (Feher et al. 2003; Nolan et al. 2006). While in some systems wounding following the dissection of the explant is often sufficient to trigger the formation of embryogenic cells, other systems require additions of PGRs (Zavattieri et al. 2010). Furthermore, response of the tissue to these factors is often variable and depends on the physiological status of the explants (Zavattieri et al. 2010). In several instances stress not only promotes the de-differentiation step leading to the formation of embryogenic cells but can also induce the development of embryos. This is the case in pine, canola, wheat, rice, and carrot.

The relevance of stress during plant embryogenesis is tightly linked to ROS production and homeostasis (Kreslavski et al. 2012; Elhiti et al. 2013). Investigation of ROS signalling during in vivo embryogenesis is a very difficult task as zygotic embryos are embedded in the maternal tissue and difficult to isolate. This poses some serious challenges in collecting a large amount of tissue to analyse. These problems, however, can be overcome using in vitro embryogenesis which allows the production of a large number of embryos, often synchronized in their development, in a short period of time. Furthermore, the ability to manipulate both culture conditions and medium composition at specific stages of embryogenesis represents another advantage in studying ROS signalling in vitro.

2 ROS and the Stress-Mediated Induction of Plant Embryogenesis

Conversion of oxygen ground state to reactive oxygen species (ROS) can occur either by energy transfer or by electron transfer reactions (Apel and Hirt 2004). As shown in Fig. 1, energy transfer leads to the formation of singlet oxygen, while sequential reduction by electron transfer generates superoxide, hydrogen peroxide,

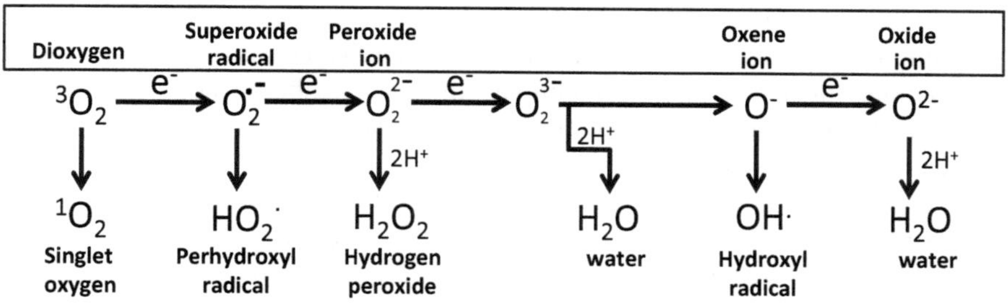

Fig. 1 Generation of different reactive oxygen species (ROS) in plants. Singlet oxygen is produced by energy transfer, while sequential univalent reduction of ground state triplet oxygen generates ROS including superoxide, hydrogen peroxide, and hydroxyl radicle

and hydroxyl radical. In plants, ROS are continuously produced not only as by-products of various metabolic processes but also in response to specific perturbation in environmental conditions which activates various peroxidase and oxidases. The rapid increase of endogenous ROS is referred to as "oxidative burst" (Apostol et al. 1989). External conditions that adversely affect the plants, i.e. biotic and abiotic stress conditions, often trigger an oxidative burst. During wounding responses, for example a rapid and transient increase in hydrogen peroxide and superoxide is observed in the damaged tissue (Doke et al. 1991; Orozco-Cardenas and Ryan 1999). It has been shown that hydrogen peroxide reaches a peak after 4–6 h of wounding, while superoxide accumulation occurs after a few minutes (Doke et al. 1991; Orozco-Cardenas and Ryan 1999). A model for hydrogen peroxide synthesis in response to wounding has been proposed by Orozco-Cárdenas et al. (2001) and involves the activation of a cytoplasmic phospholipase that releases linolenic acid from the cell membranes. Linolenic acid is then converted to jasmonic acid (JA), which through downstream signalling regulates transcription resulting in modifications of cellular metabolism leading to production of hydrogen peroxide (Orozco-Cárdenas et al. 2001).

Wounding is only one example of ROS-inducing stimuli, as accumulations of ROS follows a broad range of stress responses including heat stress, UV radiation stress, photoinhibition, heavy metal stress and anoxia, as well as pathogen attacks, and herbivore damage (Apel and Hirt 2004).

In vitro culture of cells, tissues, and organs is accompanied by a combination of stresses. Many embryogenic systems are initiated by culturing immature explants dissected from a host plant. During the dissection process, wounding is inevitable and stress-responses and production of ROS occur, as documented by the induction of stress-related genes (Dudits et al. 1995). Once dissected, the explants are cultured in media which, even when optimized to yield high regeneration frequencies, might lack basic nutrients. This sub-optimal nutritional requirement also induces stress and possibly results in the accumulation of ROS. Finally, the culture environment, i.e. temperature, humidity, photoperiod, etc., can impose substantial stress on the explant also leading to the generation of ROS (Ikeda-Iwai et al. 2003).

Being inherent to tissue culture, stress conditions are often considered one of the most important inductive factors initiating plant morphogenesis (Stasolla 2010a, b).

Examples of abiotic stress conditions triggering embryogenesis even in the absence of hormonal supplementations are abundant. In Brassica, for example, production of microspore-derived embryos can be initiated by a heat shock treatment for 72 h at 37 °C in the dark (Yeung 2002; Stasolla 2010a). Cold treatments are also used to initiate embryogenesis in other species, including *Pinuspatula* where excised shoot apices exposed to 2 °C for 3 days can be used as explants to generate somatic embryos (Malabadi and Van Staden 2005).

Other ROS-inducing conditions such as heavy metal accumulation and water stress are also often used to trigger the embryogenic process. In *Daucus carota* applications of 0.6 mM cadmium chloride (CdCl) is utilized to produce somatic embryos (Kiyosue et al. 1990), and a similar effect is achieved in wheat by exposing leaf explants to high levels of cadmium (Patnaik et al. 2005).

Osmotic and dehydration stresses are known to elevate ROS levels in vivo (reviewed by Stasolla 2010a, b) and their application in culture is widespread. Application of 5 % of the osmotic agent polyethylene glycol-8000 (PEG-8000) significantly improved date palm somatic embryogenesis along with 60 g/L sucrose. A similar improvement was also observed by Heringer et al. (2013) on the regeneration of papaya. These results confirm the importance of PEG in reducing the osmoticum of the culture medium and promoting somatic embryogenesis (Stasolla and Yeung 2003). Another effective, but less used osmoticum agent is betaine which enhances the induction of somatic embryos in tea (Akula et al. 2000). Water availability in culture medium can also be limited by the right choice and concentration of the gelling agent (Ladyman and Girard 1992).

The most common inductive signal for in vitro embryogenesis is auxin. Among all the auxins tested, 2,4-D is the most effective, an observation suggesting that its mode of action is more extensive than that of other auxins. It has been proposed that the efficiency of 2,4-D is related to the induction of stress responses in the explants (Raghavan et al. 2006). In support of this notion, Pasternak et al. (2002) showed that addition of Fe^{+3} as ferric ethylenediaminetetraacetic acid (Fe-EDTA) to alfalfa protoplast cultures mimicked the effects of exogenous applications of 2,4-D. Another possible effect of 2,4-D is to initiate embryogenesis by creating internal auxin maxima from which morphogenesis is initiated (Elhiti and Stasolla 2011).

A few studies showed that direct applications of ROS in the culture medium also promote embryo production. Addition of hydrogen peroxide (200 μM) to *Lycium barbarum* embryogenic cultures significantly enhanced the formation of somatic embryos (Kairong et al. 1999). The same work showed that compared to non-embryogenic lines, embryogenic cells have higher levels of hydrogen peroxide. In agreement with the beneficial effects of ROS, suppression of the embryogenic process often occurs in media depleted of ROS. Rose et al. (2010) demonstrated that diphenyleneiodonium (DPI), an inhibitor of flavoprotein-dependent ROS production, inhibits embryogenesis in a variety of species, including alfalfa, Arabidopsis, and tobacco.

Collectively, these studies show that stress conditions associated to ROS production favour in vitro embryogenesis in a variety of species.

3 ROS Scavenging Mechanisms During Embryogenesis: Ascorbate and Glutathione

Plant cells have a variety of mechanisms responsible for lowering the level of ROS under stress conditions. Removal of superoxide and hydrogen peroxide, for example, can be achieved through antioxidant systems involving glutathione and ascorbic acid, as well as tocopherol, flavonoids, alkaloids, and carotenoids (reviewed by Stasolla 2010a). The best characterized antioxidant system is the Halliwell–Asada cycle, which is composed of the glutathione and ascorbate redox pairs (see Potters et al. 2002). In this cycle removal of hydrogen peroxide occurs through the oxidation of reduced ascorbic acid (ASA) and glutathione (GHS) which are converted into the oxidized forms [(monodehydroascorbate (MDA), dehydroascorbate (DHA) and reduced glutathione (GSSG)]. Recycling mechanisms reconvert these oxidized forms back to their ASA and GSH (see Potters et al. 2002). While intimately linked and interconnected, the glutathione and ascorbate redox pairs will be discussed separately.

3.1 The Glutathione Redox Pair and In Vitro Embryogenesis

Glutathione is a ubiquitous thiol which is present in many organisms, including plants and animals. Its biosynthesis occurs through two distinct steps: The first requires the conversion of glutamate and cysteine to γ-glutamylcysteine by the enzyme γ-glutamylcysteine synthase, while the second involves the formation of glutathione from γ-glutamylcysteine (Noctor and Foyer 1998). Feedback mechanisms are also present to prevent the over-production of glutathione. Within the cell the glutathione pool is comprised of GSH and GSSG and their relative proportion is regulated by glutathione reductase responsible for the reduction of GSSG to GSH. Rapid conversion of one form to another, especially in the presence of ROS, alters the overall glutathione pool and induces cellular responses. Noctor et al. (1998) suggested that during stress responses and production of ROS, changes in the GSH/GSSG ratio are more important than absolute variations in total glutathione (GSH + GSSG) pool.

Over the past few years glutathione has emerged as an important determinant for plant embryogenesis in culture (Stasolla 2010a). Marre and Arrigoni (1957) were first to demonstrate that the effect of auxin on tissue grown in culture was influenced by glutathione levels. Specifically the auxin-promotion of tissue proliferation correlated with high levels of GSH relative to GSSG, while a switch of this ratio was observed when growth was suppressed. Earnshaw and Johnson (1985) further confirmed this observation demonstrating that during carrot somatic embryogenesis the proliferating embryogenic tissue cultured in the presence of auxin accumulated high GSH levels. Embryo development was stimulated by the removal of auxin which was associated to an increase in cellular GSSG relative to GSH. Furthermore the inhibition of GSH synthesis by L-buthionesulfoximine (BSO) mimicked the removal of auxin in stimulating embryo formation. The notion that fluctuations in the GSH/GSSG ratio modulate in vitro embryogenesis, with a reduced environment promoting embryogenic tissue proliferation and an oxidized environment embryo development, was demonstrated in many later studies.

By manipulating the GSH/GSSG ratio, Belmonte et al. (2005) was able to double the number of white spruce somatic embryos and increase their frequency of conversion, i.e. ability to regenerate viable plants. The authors showed that the imposition of a reduced glutathione environment (by GSH applications) induced mitotic activity of the embryogenic tissue and increased its fresh weight by 25 %. Once produced, the ability of the embryogenic tissue to generate somatic embryos was increased by a switch of the glutathione pool towards its oxidized form (GSSG). This can be achieved either by direct GSSG treatments, or by using BSO. By applying GSSH, Belmonte et al. (2005) were able to lower the endogenous GSH/GSSG ratio from 17 to 8 and increased the number of fully developed embryos. These embryos displayed improved post-embryonic growth due to "structurally" organized shoot meristems. The requirement of an oxidized glutathione environment for successful embryo production was also demonstrated in Brassica where BSO supplementations enhanced the conversion frequency of the embryos by more than three times (Belmonte et al. 2006). The physiological roles of an oxidized glutathione environment during in vitro morphogenesis have been investigated. Henmi et al. (2001) reported that by altering cellular ROS levels GSSG regulates PCD, an obligatory event in the formation of somatic embryos (discussed in the next section).

Another role attributed to GSSG is to inhibit the cell cycle progression (de Pinto et al. 1999). This function might be important in reducing proliferation of the embryogenic tissue, thereby inducing the formation of somatic embryos. Growth characteristics of spruce embryogenic tissue following the imposition of an oxidized environment confirm this notion. It has also been speculated that alterations in ROS levels by manipulations in the oxidized/reduced forms of glutathione trigger transcriptional changes beneficial for embryogenesis. Microarray studies showed that the middle and late stages of BSO-treated canola microspore embryos express ABA-responsive proteins and late embryogenesis abundant proteins (LEA) at high levels (Stasolla et al. 2008). Expression of these genes was associated with the enhanced embryogenic performance of BSO-treated tissue. Other genes up-regulated by the imposition of an oxidized glutathione environment, were related to the enhanced development of the shoot meristems following BSO applications. The transcription of CLAVATA 3 ESR-related (CLE27) and ARGONAUTE1 were highly induced by BSO. In Arabidopsis both genes participate in the maintenance of the shoot meristem. This is especially true for

ARGONAUTE1 which besides promoting normal meristems, as its suppression generated pin-shaped structures (Cerrutti et al. 2000), is required for the proper localization of SHOOTMERISTEMLESS and CUPSHAPED COTYLEDONS 1 and 2. These genes regulate the function of the meristematic cells (reviewed by Thair and Stasolla 2005). Activation of SHOOTMERISTEMLESS was also observed during the middle-late phases of Brassica microspore-derived embryos treated with BSO (Stasolla et al. 2008), thus confirming that a switch of the glutathione pool towards an oxidized state acts as a molecular signal required for proper tissue patterning in the embryo body.

As well as affecting transcription, the imposition of an oxidized environment influences hormone synthesis and metabolism. Additions of BSO during Brassica microspore-embryo development increased the synthesis of ABA, as well as its degradation to phaseic and dehydrophaseic acid via 8′hydroxylation (Belmonte et al. 2006). These changes were interpreted by the authors as a rapid turnover of ABA which enhanced embryogenesis. Of interest, exogenous ABA mimicked the effects of BSO applications by inducing similar structural and physiological responses (Belmonte et al. 2006).

Another hormone affected by the oxidized environment is ethylene, which has deleterious effects on the formation of the shoot apical meristem produced in vitro and therefore reduces the ability of the embryos to regenerate plants (reviewed by Stasolla 2010a). Using white spruce somatic embryogenesis as a model, Belmonte et al. (2005) documented a significant repression in ethylene biosynthesis following an experimental switch of the glutathione pool towards an oxidized state. This observation was also supported by later studies showing an increase in expression of 1-aminocyclopropane-1-oxidase, an ethylene biosynthetic enzyme, in tissue culture with GSH (Stasolla et al. 2004).

Collectively, these studies suggest that alterations in the glutathione pool affect in vitro embryogenesis through complex regulatory mechanisms involving transcriptional changes and hormone responses.

3.2 The Ascorbate Redox Cycle and In Vitro Embryogenesis

Ascorbic acid is an important metabolite tightly linked to glutathione in the removal of ROS. Plant cells are able to synthesise ascorbate through a de novo pathway in which fructose 6-phosphate is converted to 1-galactono-1,4-lactone, the precursor of ASC. Once formed, reduced ascorbate (ASC) contributes to the removal of ROS through oxidation mechanisms requiring ascorbate oxidase, ascorbate peroxidase, and ascorbate-dependent dioxygenases. The oxidized forms, monodehydroascorbate (MDA) and dehydroascorbate (DHA), can be reduced back to ASC by NAD(P)H-dependent MDA reductase and the glutathione-dependent DHA reductase (Arrigoni and De Tullio 2002).

Ascorbic acid metabolism affects many basic cellular processes such as cell division and differentiation. High levels of ASC are often observed in rapidly growing and young tissues, while low levels occur in older and differentiated tissues (De Gara and Tommasi 1999). The requirement of ASC in mitotic activity has been well documented in animals, where supplementation of ASC amplified chemically induced forms of cancer (Shibata et al. 1992). In plants, increased cell division rates were observed in tobacco, corn, and onion root cells following ASC treatments (de Pinto et al. 1999; Kerk and Feldman 1995), whereas reduced proliferation was observed in tissues in which ASC was depleted with lycorine, a biosynthetic inhibitor (Kerk and Feldman 1995). Potters et al. (2002) reviewed mechanisms by which ASC influences the cell cycle machinery by facilitating the G1-S and G2-S transitions.

Several independent experiments showed that ASC is also a regulator of somatic embryogenesis. Changes in the ascorbate redox pool delineate key events in the embryogenic pathway of white spruce: a reduced ascorbate pool correlates with the proliferation of the embryogenic tissue while a switch of the pool towards its oxidized state occurs at the onset of embryo development (Stasolla and Yeung 2001). Furthermore, exogenous applications of ASC promote the regeneration frequency of the embryos into plants by inducing cell division in the embryonic meristems (reviewed in Stasolla and Yeung 2003), while a depletion of the ASC content causes the differentiation of the meristematic cells leading to meristem and embryo abortion (Stasolla and Yeung 2007). Of note, an experimental reduction in ASC level within the meristems of somatic embryos increased the activity of several peroxidases, including those cross-linking cell wall components and inhibiting the division of the meristematic cells (Stasolla and Yeung 2007). This observation was explained by the ability of ASC to scavenge hydrogen peroxide a substrate of peroxidases, by activating ASC peroxidase (De Gara and Tommasi 1999). High levels of ASC and ASC peroxidase activityoccur in those tissues and organs depleted in hydrogen peroxide and with reduced peroxidase levels (Stasolla and Yeung unpublished). In addition, Stasolla et al. (2004) were able to identify one ASC peroxidase (*At1g07890*) as one of the abundant transcripts expressed during the middle stages of somatic embryogenesis in *Norway spruce*. Based on the above it is suggested that by scavenging hydrogen peroxide and decreasing peroxidase activity, ASC might promote the relaxation of wall component in the meristematic cells and induce cell division. These events favour the reactivation of the meristems at germination and the regeneration of embryos into viable plants.

3.3 Other ROS Scavenging Mechanisms and In Vitro Embryogenesis

Two key enzymes involved in the removal of ROS are superoxide dismutase (SOD), which catalyses the dismutation of superoxide producing hydrogen peroxide, and catalase (CAT), a hydrogen peroxide scavenger (Dat et al. 2000; Mittler 2002). Biochemical studies showed that the activity of both enzymes can be used as markers to delineate the developmental stages of somatic and zygotic embryogenesis in *Aesculus hippocastanum*. Furthermore, the authors also recognised the occurrence of oxidative stress conditions as one of the main factors responsible for the regeneration of embryos into plantlets. Most recently, comprehensive expression profile analysis indicated that different SOD types exhibited different spatial and temporal expression patterns consistent with their involvement during the middle and late stages of *Dimocarpus longan* somatic embryo development (Lin and Lai 2013).

4 ROS Mode of Action During Embryogenesis

Establishment of the embryo body in culture is a very complex process due to the absence of maternal cues. The precise cell division and differentiation patterns characterizing in vivo embryogenesis, especially the early phases of embryo development, are not visible in culture. Once de-differentiated, cells of the explants must reorganize into embryogenic clusters which under appropriate conditions develop further into embryos. While the participation of ROS during this transition is not well characterized, it cannot be excluded. The following sections deal with the participation of ROS in cell death mechanisms influencing in vitro embryogenesis.

4.1 ROS and PCD

Together with cell division and differentiation, PCD is fundamental in shaping tissues and organs during development and/or stress conditions, by eliminating unwanted or supernumerary cells. The suicide mode is initiated and sustained by complex molecular mechanisms with ROS as key signals. Specifically, their site of production and homeostasis, as well as their interaction with other components of the PCD machinery, have been show to control the death/survival decision (reviewed by Stasolla 2010a).

The importance of PCD during embryogenesis is well established in animal systems, where a disordered morphogenic pattern often leading to embryo death occurs if the PCD pathway is experimentally blocked (Kuida et al. 1996). Compared to animals, the plant embryo body contains fewer cell, tissue and organ types, and based on this structural simplification it would be expected that PCD plays a less pronounced role. This is not the case during in vivo embryogenesis. The suicide mode is manifested in many seed and embryonic compartments, including suspensor and nucellus (Bozhkov et al. 2005). In extreme cases PCD is necessary for the removal of entire embryos, a process referred to as monozygotic polyembryony, which ensures the growth of a dominant embryo among the several formed (Bozhkov et al. 2005)

The requirement of PCD during in vitro embryogenesis was initially demonstrated in *Norway spruce*. In this species somatic embryogenesis is delineated by two waves of PCD, the first implicated in the transition from pre-embryonic masses to somatic embryos, while the second in the elimination of the suspensor (Filonova et al. 2000). The importance of the initial suicide wave was confirmed by the observation that factors abrogating PCD reduced the number of somatic embryos (Bozhkov et al. 2005). Our recent work (Huang et al. 2014) confirmed the importance of PCD during *Zea mays*somatic embryogenesis and proposed a model with NO as the initiator of the suicide mode and hydrogen peroxide and superoxide as the final executors. By using antisense-mediated suppression of two maize hemoglobins (*ZmHb1* and *ZmHb2*), as modulators of NO, we showed that accumulation of NO in many embryonic cells (by suppression of *ZmHb1*) induced massive PCD leading to embryo abortion, while elevation of NO in cells connecting the developing embryos to the embryogenic tissue (by suppression of *ZmHb2*) increased embryogenesis. The induction of PCD by NO was mediated by an initial increase in Zn^{2+} level followed by the activation of the mitogen-activated protein kinase (MAPK) cascade and NADPH oxidase leading to the production of hydrogen peroxide and superoxide in cells committed to die. This model clearly shows the relevance of ROS during cell death and somatic embryogenesis and provides a framework for manipulating embryo production in vitro through selective and targeted alterations in ROS levels.

5 ROS Signalling Pathways

Independent molecular and biochemical studies conducted on Arabidopsis suggest that perception of ROS can occur through at least three different pathways involving unknown receptors proteins, redox-sensitive transcription factors, and/or phosphatases (Fig. 2) (Mittler 2002; Neill et al. 2002; Vranová et al. 2002a, b; Apel and Hirt 2004). The produced ROS in the plant cell could be sensed by unknown ROS receptor protein(s) which triggers Ca^{+2} influxes. Ca^{+2} ions promote the activity of serine/threonine protein kinase (OXI1) and phospholipase C/D (PLC/PLD) which generate phosphatidic acid (PA). PA triggers phosphoinositide-dependent kinase 1 (PDK1) which consequently activates the serine/threonine protein kinase OXI1. OXI1 has been shown to play a central role in ROS sensing through the activation of mitogen-activated-protein kinases (MAPKs) 3 and 6 by Ca^{+2} (Rentel et al. 2004). The activation of MAPK3/6 triggers different defense mechanisms in response to ROS (Kovtun et al. 2000; Apel and Hirt 2004). Alternative ROS sensing pathways involve the participation of redox sensitive transcription factors which directly promote OXI1, and the suppression of phosphatases, negative regulators of OXI1 (Apel and Hirt 2004). Pharmacological and genetic studies showed the possible existence of positive amplification loops involving NADPH oxidases in ROS signalling (Dat et al. 2003; Rizhsky et al. 2004). These loops might be promoted by low levels of ROS and result in enhanced production and amplification of the

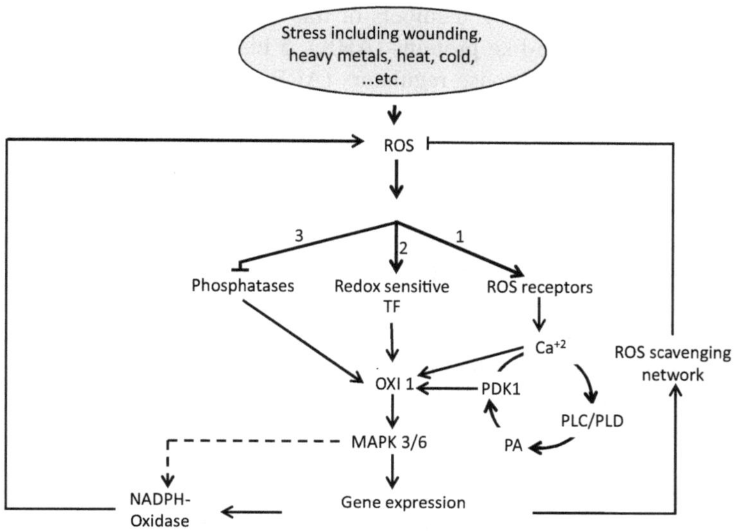

Fig. 2 Model showing reactive oxygen species (ROS) signal transduction pathways in plant cells. Stress-induced ROS can be detected by at least three different mechanisms. (1) Detection of ROS by unknown receptor(s) initiate Ca^{+2} signals which activate the protein kinase OXI1 either directly or indirectly via phosphoinositide-dependent kinase 1 (PDK1). Activation of OXI1 induces the Mitogen Activated Protein Kinase (MAPK) cascade and consequently the transcriptional regulation of downstream factors which fine tune ROS homeostasis through their production (NADPH oxidase) or scavenging. (2) Stress-produced ROS may activate the redox sensitive transcription factors (TF) which could directly activate OXI1. (3) Inhibition of phosphatases by ROS might result in the activation of several kinases including OXI1

ROS signals (Fig. 2). The accumulation of ROS in cells might trigger the ROS scavenging pathways and leads to suppression of ROS. While these proposed signal pathways have been established using in vivo systems, emerging evidence suggest they might also operate during in vitro morphogenesis. The sections below will highlight possible links.

5.1 *Do ROS interact with Cytokinins Through a Two-Component Signalling System?*

Several signal transduction systems employ phosphotransfer schemes including the two-component signalling mechanism, comprising two elements: a histidine (His) kinase and a response regulator (Lohrmann and Harter 2002). A signal-induced activation by the input domain causes the autophosphorylation of a His residue. The phosphoryl group is then transferred to an invariant asparagine (Asp) within a conserved domain of the receiver. Through these phosphorylation events the stimulus is converted into a response (Stock et al. 2000). The Arabidopsis genome contains many genes encoding members of this signalling mechanism, including 17 His and hybrid kinase-like proteins (AHK), 5 histidine phosphotransfer (HPt) proteins (AHP), and 23 response regulators (ARR) (Lohrmann and Harter 2002). Independent studies have shown that the two component signalling system functions as a redox sensing mechanism (Whistler et al. 1998; Quinn et al. 2002) integrated in the ROS response and involved in the regulation of redox-sensitive transcription factors. While there is no information of similar mechanisms operating during in vitro embryogenesis, cytokinin sensing is also mediated by a two-component signalling system (To and Kieber 2008) and its effects influence the ROS response (Miller 1980).

Cytokinins play crucial roles during the initial phases of embryogenesis (Sagare et al. 2000). In some plants such as *Dendrobium*, applications of thidiazuron (TDZ) in the induction medium promote direct somatic embryos (Chung et al. 2005). This finding was supported by Kuo et al. (2005) who found clusters of directly induced somatic embryos originating from leaf segments of Phalaenopsis cultured in the presence of benzyl adenine (BA) and TDZ (Canhoto and Cruz 1994). In the same study it was shown that the number of somatic embryos produced was proportional to the amount of cytokinins added in the medium. While the mode of action of cytokinin in culture is debatable, several reports indicate that this plant growth regulator is tightly linked to ROS. First, applications of cytokinin often trigger cell death, a process commonly induced by ROS. Carimi et al. (2005) showed that 6-benzylaminopurine (BAP) accelerates senescence in Arabidopsis and carrot tissues leading to PCD, as revealed by structural and ultrastructural features. A similar effect was also observed in tobacco epidermal and sub-epidermal leaves following BAP treatments. These effects are most likely due to an increase in ROS production, as reported by Mlejnek (2013) who showed that cytokinin applications cause the death of tobacco BY-2 cells through an oxidative burst elevating ROS levels. An indirect link between ROS, an oxidized environment, and cytokinin signalling occurs in the transcriptional regulation of genes. As indicated in the previous section, a switch of the glutathione redox state towards its oxidized forms, a process modulated by ROS, increases the expression of WUSCHEL, a gene required for the formation of the shoot meristem and the specification of embryogenic cells (Belmonte et al. 2006). The expression of this gene in vivo is tightly linked to cytokinin (Laux et al. 1996).

5.2 *Do ROS Interact with Auxin?*

The mitogen-activated protein kinase (MAPK) cascade participates in many signal transduction pathways connecting external stimuli with internal responses. Activated by ROS-mediated oxidative stress (Fig. 2), MAPKs regulate morphogenesis through a variety of mechanisms ranging from the control of cell division to cell death (Zhang et al. 2007).

Increasing evidence suggests that the MAPK cascade might also have a role during auxin signalling. Mizogushi et al. (1994) showed that supplementations of auxin increased both the myelin basic protein (MBP) phosphorylation activity and the activity responsible for phosphorylating MAPK activity in vitro. A rapid elevation in MAPK activation was also documented in Arabidopsis seedling treated with auxin (Mockaitis and Howell 2000), although the authors interpreted this response as being "genetically separable" from the MAPK activation resulting from the imposition of abiotic stress.

While a direct link between ROS and auxin in the regulation of MAPK has not been identified it cannot be excluded, especially in vitro. The initial phases of somatic embryogenesis are in fact favoured by a possible elevation in ROS levels produced by the wounding responses in the explant, and a high auxin environment.

Another possible link between ROS and auxin is MYC2, a basic helix–loop–helix (bHLH) domain-containing transcription factor (Dombrecht et al. 2007). MYC2 is a downstream regulator of the JA respons, in many abiotic and biotic stress responses involving production of ROS. Recent studies have identified MYC2 as an important regulator of somatic embryogenesis in Arabidopsis. Factors suppressing MYC2 expression favours the formation of somatic embryos while an experimental increase in MYC2 expression has deleterious effects on embryogenesis biosynthetic genes including, ANTHRANILATE SYNTHASE (α subunit) (ASA1), CYTOCHROME P79B2 (CYP79B2) and AMIDASE1 (AMI1), causing an increase in auxin within the embryogenic tissue which encourages the production of somatic embryos. While still preliminary, these results suggest a possible connection between stress and auxin signalling during in vitro embryogenesis.

Conclusions

Reactive oxygen species (ROS) are ubiquitous molecules in plant systems, produced during development and especially in response to stress factors. Increasing evidence suggests that they might function as signal molecules transducing external stimuli into cellular responses. Formation of embryos in culture occurs under pronounced stress due to the initial dissection and wounding of the explant, and the sub-optimal culture conditions and medium composition. The inevitable generation of ROS might be one of the many signals directing embryogenesis and affecting embryo development. While many studies have documented how alterations in ROS-detoxification mechanisms, such as the glutathione and ascorbate systems, modulate embryogenesis, no information is available on a more direct involvement of ROS. Future efforts should be directed in the identification of possible interactions between ROS and hormone responses influencing embryogenesis and, to a larger extent, morphogenesis.

References

Akula A, Akula C, Bateson M (2000) Betaine a novel candidate for rapid induction of somatic embryogenesis in tea (Camellia sinensis (L.) O. Kuntze). Plant Growth Regul 30:241–246

Apel K, Hirt H (2004) Reactive oxygen species: metabolism, oxidative stress, and signal transduction. Annu Rev Plant Biol 55:373–399

Apostol I, Heinstein PF, Low PS (1989) Rapid stimulation of an oxidative burst during elicitation of cultured plant cells: role in defense and signal transduction. Plant Physiol 90:109–116

Arrigoni O, De Tullio MC (2002) Ascorbic acid: much more than just an antioxidant. Biochim Biophys Acta 1569:1–9

Belmonte MF, Donald G, Reid DM, Yeung EC, Stasolla C (2005) Alterations of the glutathione redox state improve apical meristem structure and somatic embryo quality in white spruce (*Picea glauca*). J Exp Bot 56:2355–2364

Belmonte MF, Ambrose SJ, Ross ARS, Abrams S, Stasolla C (2006) Improved development of microspore-derived embryo cultures of *B. napus* cv Topas following changes in glutathione metabolism. Physiol Plant 127:690–700

Bozhkov PV, Filonova LH, Suarez MF (2005) Programmed cell death in plant embryogenesis. Curr Top Dev Biol 67:135–179

Canhoto JM, Cruz GS (1994) Improvement of somatic embryogenesis in *Feijoa sellowiana* berg (Myrtaceae) by manipulation of culture media composition. In Vitro Cell Dev Biol Plant 30:21–25

Carimi F, Zottini M, Costa A, Cattelan I, De Michele R, Terzi M, Lo Schiavo F (2005) NO signalling in cytokinin-induced programmed cell death. Plant Cell Environ 28:1171–1178

Cerrutti L, Mian N, Bateman A (2000) Domains in gene silencing and cell differentiation proteins: the novel PAZ domain and redefinition of the PIWI domain. Trends Biochem Sci 25:481–482

Chung H-H, Chen J-T, Chang W-C (2005) Cytokinins induce direc somatic embryogenesis of *Dendrobium chiengmai pink* and subsequent plant regeneration. In Vitro Cell Dev Biol Plant 41:765–769

Dat J, Vandenabeele S, Vranova E, Van Montagu M, Inzé D, Van Breusegem F (2000) Dual action of the active oxygen species during plant stress responses. Cell Mol Life Sci CMLS 57:779–795

Dat JF, Pellinen R, Van De Cotte B, Langebartels C, Kangasjärvi J, Inzé D, Van Breusegem F (2003) Changes in hydrogen peroxide homeostasis trigger an active cell death process in tobacco. Plant J 33:621–632

De Gara L, Tommasi F (1999) Ascorbate redox enzymes: a network of reactions involved in plant development. Recent Res Dev Phytochem 3:1–5

de Pinto M, Francis D, De Gara L (1999) The redox state of the ascorbate-dehydroascorbate pair as a specific sensor of cell division in tobacco BY-2 cells. Protoplasma 209:90–97

Doke N, Miura Y, Chai H-B, Kawakita K (1991) Involvement of active oxygen in induction of plant defense response against infection and injury. Curr Topics Plant Physiol 6:25–31

Dombrecht B, Xue GP, Sprague SJ, Kirkegaard JA, Ross JJ, Reid JB, Fitt GP, Sewelam N, Schenk PM, Manners JM (2007) MYC2 differentially modulates diverse jasmonate-dependent functions in Arabidopsis. Plant Cell 19:2225–2245

Dudits D, Györgyey J, Bögre L, Bakó L (1995) Molecular biology of somatic embryogenesis. In: Thorpe TA (ed) In vitro embryogenesis in plants. Kluwer, Dordrecht, pp 267–308

Earnshaw BA, Johnson MA (1985) The effect of glutathione on development of white carrot suspension cultures. Biochem Biophys Res Commun 133:988–993

Elhiti M, Stasolla C (2011) The use of zygotic embryos as explants for in vitro propagation: an overview. In: Thorpe TA, Yeung EC (eds) Plant embryo culture. Springer, Dordrecht, pp 229–255

Elhiti M, Tahir M, Gulden RH, Khamiss K, Stasolla C (2010) Modulation of embryo-forming capacity in culture through the expression of Brassica genes involved in the regulation of the shoot apical meristem. J Exp Bot 61:4069–4085

Elhiti M, Hebelstrup KH, Wang A, Li C, Cui Y, Hill RD, Stasolla C (2013) Function of type-2 Arabidopsis hemoglobin in the auxin-mediated formation of embryogenic cells during morphogenesis. Plant J 946:58–62

Feher A, Pasternak TP, Dudits D (2003) Transition of somatic plant cells to an embryogenic state. Plant Cell Tissue Org Cult 74:201–228

Fehér A, Ötvös K, Pasternak TP, Pettkó-Szandtner A (2008) The involvement of reactive oxygen species (ROS) in the cell cycle activation (G0-to-G1 transition) of plant cells. Plant Signal Behav 3:823–826

Filonova LH, Bozhkov PV, Brukhin VB, Daniel G, Zhivotovsky B, von Arnold S (2000) Two waves of programmed cell death occur during formation and development of somatic embryos in the gymnosperm, *Norway spruce*. J Cell Sci 113:4399–4411

Foyer CH, Harbinson J (1994) Oxygen metabolism and the regulation of photosynthetic electron transport. In: Foyer CH, Mullineaux PM (eds) Causes of photooxidative stress and amelioration of defense systems in plants. CRC Press, Boca Raton, pp 1–42

Fujita M, Fujita Y, Noutoshi Y, Takahashi F, Narusaka Y, Yamaguchi-Shinozaki K, Shinozaki K (2006) Crosstalk between abiotic and biotic stress responses: a current view from the points of convergence in the stress signaling networks. Curr Opin Plant Biol 9:436–442

Henmi K, Tsuboi S, Demura T, Fukuda H, Iwabuchi M, Ki O (2001) A possible role of glutathione and glutathione disulfide in tracheary element differentiation in the cultured mesophyll cells of *Zinnia elegans*. Plant Cell Physiol 42:673–676

Heringer AS, Vale EM, Barroso T, Santa-Catarina C, Silveira V (2013) Polyethylene glycol effects on somatic embryogenesis of papaya hybrid UENF/CALIMAN 01 seeds. Theor Exp Plant Physiol 25:116–124

Huang S, Hill R, Wally OS, Dionisio G, Ayele B, Stasolla C (2014) Hemoglobin control of cell survival/death decision regulates in vitro plant embryogenesis. Plant Physiol. doi:10.1104/pp.114.239335

Ikeda-Iwai M, Umehara M, Satoh S, Kamada H (2003) Stress induced somatic embryogenesis in vegetative tissues of *Arabidopsis thaliana*. Plant J 34:107–114

Jacobson MD (1996) Reactive oxygen species and programmed cell death. Trends Biochem Sci 21:83–86

Kairong C, Gengsheng X, Xinmin L, Gengmei X, Yafu W (1999) Effect of hydrogen peroxide on somatic embryogenesis of *Lycium barbarum* L. Plant Sci 146:9–16

Kerk NM, Feldman LJ (1995) A biochemical model for the initiation and maintenance of the quiescent center: implications for organization of root meristem. Development 121:2825–2833

Kiyosue T, Takano K, Kamada H, Harada H (1990) Induction of somatic embryogenesis in carrot by heavy metal ions. Can J Bot 68:2301–2303

Kovtun Y, Chiu W-L, Tena G, Sheen J (2000) Functional analysis of oxidative stress-activated mitogen-activated protein kinase cascade in plants. Proc Natl Acad Sci USA 97:2940–2945

Kreslavski V, Los D, Allakhverdiev S, Kuznetsov VV (2012) Signaling role of reactive oxygen species in plants under stress. Russ J Plant Physiol 59:141–154

Kuida K, Zheng TS, Na S, Kuan C-Y, Yang D, Karasuyama H, Rakic P, Flavell RA (1996) Decreased apoptosis in the brain and premature lethality in CPP32-deficient mice. Nature 384:368–372

Kuo H-L, Chen J-T, Chang W-C (2005) Efficient plant regeneration through direct somatic embryogenesis from leaf explants of Phalaenopsis 'Little Steve'. In Vitro Cell Dev Biol Plant 41:453–456

Ladyman JA, Girard B (1992) Cucumber somatic embryo development on various gelling agents and carbohydrate sources. Hortic Sci 27:164–165

Laux T, Mayer KFX, Berger J, Jurgens G (1996) The *WUSCHEL* gene is required for shoot and floral meristem integrity in Arabidopsis. Development 122:87–96

Lin Y-L, Lai Z-X (2013) Superoxide dismutase multigene family in longan somatic embryos: a comparison of CuZn-SOD, Fe-SOD, and Mn-SOD gene structure, splicing, phylogeny, and expression. Mol Breed 32:595–615

Liu X, Williams CE, Nemacheck JA, Wang H, Subramanyam S, Zheng C, Chen M-S (2010) Reactive oxygen species are involved in plant defense against a gall midge. Plant Physiol 152:985–999

Lohrmann J, Harter K (2002) Plant two-component signaling systems and the role of response regulators. Plant Physiol 128:363–369

Malabadi RB, Van Staden J (2005) Somatic embryogenesis from vegetative shoot apices of mature trees of *Pinus patula*. Tree Physiol 25:11–16

Mantiri FR, Kurdyukov S, Lohar DP, Sharopova N, Saeed NA, Wang X-D, Vanden Bosch KA, Rose RJ (2008) The transcription factor MtSERF1 of the ERF subfamily identified by transcriptional profiling is required for somatic embryogenesis induced by auxin plus cytokinin in *Medicago truncatula*. Plant Physiol 146:1622–1636

Maraschin SF, De Priester W, Spaink HP, Wang M (2005) Androgenic switch: an example of plant embryogenesis from the male gametophyte perspective. J Exp Bot 56:1711–1726

Marre E, Arrigoni O (1957) Metabolic reactions to auxin. I. The effects of auxin on glutathione and the effects of glutathione on growth of isolated plant parts. Physiol Plant 10:289–301

Miller CO (1980) Cytokinin inhibition of respiration in mitochondria from six plant species. Proc Natl Acad Sci USA 77:4731–4735

Miller G, Suzuki N, Cifci-Yilmaz S, Mittler R (2010) Reactive oxygen species homeostasis and signalling during drought and salinity stresses. Plant Cell Environ 33:453–467

Mittler R (2002) Oxidative stress, antioxidants and stress tolerance. Trends Plant Sci 7:405–410

Mittler R, Vanderauwera S, Gollery M, Van Breusegem F (2004) Reactive oxygen gene network of plants. Trends Plant Sci 9:490–498

Mizogushi T, Gotoh Y, Nishida E, Yamagushi-Shinozaki K, Hayashida N, Iwasaki T, Kamada H, Shinozaki K (1994) Characterization of two cDNAs that encode MAP kinase homologues in Arabidopsis and analysis of the possible rol of auxin in activating such kinase activities in cultured cells. Plant J 5:111–122

Mlejnek P (2013) Cytokinin-induced cell death is associated with elevated expression of alternative oxidase in tobacco BY-2 cells. Protoplasma 250:1195–1202

Mockaitis K, Howell SH (2000) Auxin induces mitogenic activated protein kinase (MAPK) activation in roots of Arabidopsis seedlings. Plant J 24:785–796

Neill S, Desikan R, Hancock J (2002) Hydrogen peroxide signalling. Curr Opin Plant Biol 5:388–395

Noctor G, Foyer CH (1998) Ascorbate and glutathione: keeping active oxygen under control. Annu Rev Plant Physiol Plant Mol Biol 49:249–279

Noctor G, Arisi A-CM, Jouanin L, Kunert KJ, Rennenberg H, Foyer CH (1998) Glutathione: biosynthesis, metabolism and relationship to stress tolerance explored in transformed plants. J Exp Bot 49:623–647

Nogler GA (1984) Genetics of apospory in apomictic *Ranunculus auricomus*. V Conclusion Bot Helv 94:411–422

Nolan KE, Saeed NA, Rose RJ (2006) The stress kinase gene MtSK1 in Medicago truncatula with particular reference to somatic embryogenesis. Plant Cell Rep 25:711–722

Orozco-Cardenas M, Ryan CA (1999) Hydrogen peroxide is generated systemically in plant leaves by wounding and systemin via the octadecanoid pathway. Proc Natl Acad Sci USA 96:6553–6557

Orozco-Cárdenas ML, Narváez-Vásquez J, Ryan CA (2001) Hydrogen peroxide acts as a second messenger for the induction of defense genes in tomato plants in response to wounding, systemin, and methyl jasmonate. Plant Cell 13:179–191

Pasternak TP, Prinsen E, Ayaydin F, Miskolczi P, Potters G, Asard H, Van Onckelen HA, Dudits D, Fehér A (2002) The role of auxin, pH, and stress in the activation of embryogenic cell division in leaf protoplast-derived cells of alfalfa. Plant Physiol 129:1807–1819

Patnaik D, Mahalakshmi A, Khurana P (2005) Effect of water stress and heavy metals on induction of somatic embryogenesis in wheat leaf base cultures. Indian J Exp Biol 43:740–743

Potters G, De Gara L, Asard H, Horemans N (2002) Ascorbate and glutathione: guardians of the cell cycle, pattern in crime? Plant Physiol Biochem 40:537–548
Quinn J, Findlay VJ, Dawson K, Millar JB, Jones N, Morgan BA, Toone WM (2002) Distinct regulatory proteins control the graded transcriptional response to increasing H_2O_2 levels in fission yeast *Schizosaccharomyces pombe*. Mol Biol Cell 13:805–816
Raghavan V (2000) Pattern formation in angiosperm embryos. Botanica 50:33–47
Raghavan C, Ong EK, Dalling MJ, Stevenson TW (2006) Regulation of genes associated with auxin, ethylene and ABA pathways by 2, 4-dichlorophenoxyacetic acid in Arabidopsis. Funct Integr Genomics 6:60–70
Rentel MC, Lecourieux D, Ouaked F, Usher SL, Petersen L, Okamoto H, Knight H, Peck SC, Grierson CS, Hirt H (2004) OXI1 kinase is necessary for oxidative burst-mediated signalling in Arabidopsis. Nature 427:858–861
Rizhsky L, Davletova S, Liang H, Mittler R (2004) The zinc finger protein Zat12 is required for cytosolic ascorbate peroxidase 1 expression during oxidative stress in Arabidopsis. J Biol Chem 279:11736–11743
Rose RJ, Nolan KE (2006) Genetic regulation of somatic embryogenesis with particular reference to *Arabidopsis thaliana* and *Medicago truncatula*. In Vitro Cell Dev Biol Plant 42:473–481
Rose RJ, Mantiri FR, Kurdyukov S, Chen S-K, Wang X-D, Nolan KE, Sheahan MB (2010) Developmental biology of somatic embryogenesis. In: Pua EC, Davey MR (eds) Plant developmental biology—biotechnological perspectives. Springer, Berlin, pp 3–26
Sagare A, Lee Y, Lin T, Chen C, Tsay H (2000) Cytokinin-induced somatic embryogenesis and plant regeneration in *Corydalis yanhusuo*: a medicinal plant. Plant Sci 160:139–147
Sakamoto M, Munemura I, Tomita R, Kobayashi K (2008) Involvement of hydrogen peroxide in leaf abscission signaling, revealed by analysis with an in vitro abscission system in Capsicum plants. Plant J 56:13–27
Shibata MA, Fukushima S, Asakawa E, Hirose M, Ito N (1992) The modifying effects of indomethacin or ascorbic acid on cell proliferation induced by different types of bladder tumor promoters in rat urinary bladder and forestomach mucosal epithelium. Jpn J Cancer Res 83:31–39
Stasolla C (2010a) Changes in the glutathione and ascorbate redox state trigger growth during embryo development and meristem reactivation at germination. In: Naser A. Anjun, Ming-Tsair Chan, Shahid Umar (eds) Ascorbate-Glutathione Pathway and Stress Tolerance in Plants. Springer, pp 231–249
Stasolla C (2010b) Glutathione redox regulation of *in vitro* embryogenesis. Plant Physiol Biochem 48:319–327
Stasolla C, Yeung EC (2001) Ascorbic acid metabolism during white spruce somatic embryo maturation and germination. Physiol Plant 111:196–205
Stasolla C, Yeung EC (2003) Recent advances in conifer somatic embryogenesis: improving somatic embryo quality. Plant Cell Tissue Org Cult 74:15–35
Stasolla C, Yeung EC (2007) Cellular ascorbic acid regulates the activity of major peroxidases in the apical poles of germinating white spruce (*Picea glauca*) somatic embryos. Plant Physiol Biochem 45:188–198
Stasolla C, Bozhkov PV, Chu T-M, van Zyl L, Egertsdotter U, Suarez MF, Craig D, Wolfinger RD, Von Arnold S, Sederoff RR (2004) Variation in transcript abundance during somatic embryogenesis in gymnosperms. Tree Physiol 24:1073–1085
Stasolla C, Belmonte MF, Tahir M, Elhiti M, Khamiss K, Joosen R, Maliepaard C, Sharpe A, Gjetvaj B, Boutilier K (2008) Buthionine sulfoximine (BSO)-mediated improvement in cultured embryo quality in vitro entails changes in ascorbate metabolism, meristem development and embryo maturation. Planta 228:255–272
Stock AM, Robinson VL, Goudreau PN (2000) Two-component signal transduction. Annu Rev Biochem 69:183–215
Tan S, Sagara Y, Liu Y, Maher P, Schubert D (1998) The regulation of reactive oxygen species production during programmed cell death. J Cell Biol 141:1423–1432
Thair M, Stasolla C (2005) Shoot apical development during in vitro embryogenesis. Can J Bot 84:1650–1659

To JP, Kieber JJ (2008) Cytokinin signaling: two-components and more. Trends Plant Sci 13:85–92

Vranová E, Atichartpongkul S, Villarroel R, Van Montagu M, Inzé D, Van Camp W (2002a) Comprehensive analysis of gene expression in *Nicotiana tabacum* leaves acclimated to oxidative stress. Proc Natl Acad Sci USA 99:10870–10875

Vranová E, Inzé D, Van Breusegem F (2002b) Signal transduction during oxidative stress. J Exp Bot 53:1227–1236

Whistler CA, Corbell NA, Sarniguet A, Ream W, Loper JE (1998) The two-component regulators GacS and GacA influence accumulation of the stationary-phase Sigma Factor and the stress response in *Pseudomonas fluorescens* Pf-5. J Bacteriol 180:6635–6641

Yeung EC (2002) The canola microspore-derived embryo as a model system to study developmental processes in plants. J Plant Biol 45:119–133

Zavattieri MA, Frederico AM, Lima M, Sabino R, Arnholdt-Schmitt B (2010) Induction of somatic embryogenesis as an example of stress-related plant reactions. Electron J Biotechnol 13:12–13

Zhang Y, Aizenman E, DeFranco DB, Rosenberg PA (2007) Intracellular zinc release, 12-lipoxygenase activation and MAPK dependent neuronal and oligodendroglial death. Mol Med 13:350–355

“Nitrosative Door” in Seed Dormancy Alleviation and Germination

Urszula Krasuska, Katarzyna Ciacka, Paulina Andryka-Dudek, Renata Bogatek, and Agnieszka Gniazdowska

1 Introduction

The life cycle of plants depends on many factors including environmental conditions. The pivotal plant strategy to survive is based on the development of dormancy, particularly seed dormancy. After loss of dormancy, seed germination is another crucial event of plant ontogeny. Development of seed dormancy (especially of perennial plants, such as trees) enables tolerance and/or avoidance of unfavorable environmental conditions and finally allows for growth of seedlings in a proper season. Seed dormancy defined as arrest of growth and development is only a temporal and of a quantitative nature. It means that during the time of dormancy this physiological state varies according to endogenous and exogenous factors. Commonly, three types of dormancy can be distinguished: ecodormancy (referring to plant response to environmental conditions), paradormancy (connected with influences of one part of the plant on another), and endodormancy (due to dormant structure by itself) (Lang 1987). Dormancy concerns various plant organs such as buds, tubers, and seeds. Dormancy protects seeds from germinating under environmental conditions not adequate for further undisturbed growth and development of seedlings (Finkelstein et al. 2008; Angelovici et al. 2010). In this meaning, dormancy is considered as temporary failure of germination of mature, viable seeds. Seeds with deep embryonic dormancy are unable to germinate even under favorable environmental conditions (Kucera et al. 2005; Bogatek and Gniazdowska 2006). Strength of seed dormancy depends on genotype and seasonal climate variation.

U. Krasuska • K. Ciacka • P. Andryka-Dudek • R. Bogatek • A. Gniazdowska (✉)
Department of Plant Physiology, Warsaw University of Life Sciences-SGGW, Nowoursynowska 159, 02-776 Warsaw, Poland
e-mail: urszula_krasuska@sggw.pl; katarzyna_budnicka@sggw.pl; paulina_andryka@sggw.pl; renata_bogatek@sggw.pl; agnieszka_gniazdowska@sggw.pl

K.J. Gupta, A.U. Igamberdiev (eds.), *Reactive Oxygen and Nitrogen Species Signaling and Communication in Plants*, Signaling and Communication in Plants 23,
DOI 10.1007/978-3-319-10079-1_11

Quantitative nature of seed dormancy leads to "opening the window" for germination process (Finch-Savage and Leubner-Metzger 2006).

If seed dormancy is due to embryonic tissues, it is embryo-based dormancy; if caused by maternal tissues, it is known as coat-based dormancy (Gubler et al. 2005). Seed dormancy can also be termed as primary dormancy (initiated during seed maturation) and secondary dormancy (initiated in mature non-dormant seed by unfavorable environmental factors) (Gubler et al. 2005).

Seed germination (germination *sensu stricto*) is commonly defined as physiological process leading to embryo activation and terminating after embryonic root protrusion throughout seed covering layers (Finch-Savage and Leubner-Metzger 2006). The process of seed germination involves three physiological phases: water imbibition, catabolic and anabolic stages. All these phases are under strict control of many agents. One of the most important factors influencing seed germination are nitrogen containing compounds such as nitrate, nitrite, or sodium azide (Taylorson and Hendricks 1977). Nowadays, it is well known that these compounds may function as potential sources of nitric oxide (NO), which belongs to the family of reactive nitrogen species (RNS) and serves as a signaling molecule in the regulation of plant growth and development.

Water uptake (imbibition) is the most important, initiating process in seed germination. To date, it was hard to believe that this physical process may be under NO control. Liu et al. (2007) showed that treatment of rice (*Oryza sativa* L.) seeds with various donors of NO led to modification of the expression of water channel genes *PIP1;1*, *PIP1;2*, *PIP1;3*, and *PIP2;8* resulting in stimulation of germination. Authors concluded that the effects of NO on rice seed germination may be partly mediated through induction of aquaporins and therefore enhanced ability of water uptake. These results suggest also the role of aquaporins as primary molecular targets of NO signaling in seed dormancy loss.

To advance the knowledge of seed dormancy and germination, especially at molecular level is a great goal for many researches, as well as for agriculture, horticulture, and forestry industry. Dormancy establishment and maintenance in seeds of crop species [such as barley (*Hordeum vulgare* L.), wheat (*Triticum aestivum* L.), rice (*Oryza sativa* L.) and corn (*Zea mays* L.)] prevents pre-harvest sprouting or vivipary (Farnsworth 2000; Gubler et al. 2005). On the other hand, seed dormancy can be noticed as a negative phenomenon in the aspect of weeds, horticulture, and forestry management. In the agriculture (crop production) seeds of weeds are the reason of loss of annual yield (Batlla and Benech-Arnold 2007). In weed management strategies special models for prediction of seed dormancy are created. The opposite complications are connected with dormant seeds of trees requiring time and money consuming treatments to promote their germination (Gubler et al. 2005). Seed dormancy breakage of some species is achieved by exposing them to low moisture conditions during different period of storage (after-ripening). Some additional (external) compounds used during after-ripening process may accelerate dormancy loss. As was shown for the hedge mustard (*Sisymbrium officinale* (L.) Scop.) seeds (after-ripened and fresh) the presence of potassium nitrate (KNO_3) was required for germination (Carillo-Barral et al. 2014). Nitrate-dependent entire dormancy lost is due to modification of phytohormone levels and metabolism. Nitrate-dependent lowering of dormancy state was observed for *Arabidopsis thaliana* seeds developed on mother plant with high nitrate supplementation (Alboresi et al. 2005). Also dormancy removal of some seeds is accelerated by stratification (seeds pregerminative treatment by low moisture content at warm (e.g. 20–25 °C) or cold (e.g. 5 °C) temperature) (Lewak 2011). During cold stratification of apple seeds (*Malus domestica* Borkh.) increased NO emission from isolated embryonic axis was observed. This enhancement of NO release was time dependent and correlated well with embryos ability to germinate (Dębska et al. 2013).

Seed dormancy is under control of exogenous (water, temperature, light) and endogenous factors (phytohormones and other plant growth regulators). Classical phytohormones determining seed dormancy or germination include abscisic acid (ABA), jasmonic acid (JA), gibberellins (GA), and ethylene (ET). Dormancy state is strictly dependent on ABA concentration and tissue sensitivity to this hormone (Finch-Savage and Leubner-Metzger 2006; Gniazdowska et al. 2010b). On the other hand, seed germination requires enhancement of GA synthesis and signaling (Finch-Savage and Leubner-Metzger 2006).

In the recent years in plant physiology the great attention was paid on beneficial role of signaling molecules such as reactive oxygen species (ROS) and RNS. Now, it is well known that NO is involved in breakage of seed dormancy of many plant species (Šírová et al. 2011 and references therein). ROS and RNS are key regulators in seed biology depending on their cellular concentration (Bailly et al. 2008 and references therein, Šírová et al. 2011). Bailly et al. (2008) proposed model of "oxidative window" which describes dual function of ROS in seed biology. Based on the model, seed dormancy alleviation and following germination are possible if ROS concentration in seed reaches level necessary to "open oxidative window." In contrast, exceeded (overloaded) ROS content leads to seed ageing and consequently to seed death (Bailly et al. 2008).

Compounds that modulate seed germination include nitrate/nitrite ions as well as commonly used NO donors such as sodium nitroprusside (SNP), *S*-nitroso-*N*-acetyl penicillamine (SNAP), or *S*-nitrosoglutathione (GSNO). Stimulation of seed germination after NO-donors treatment was observed for model plant *Arabidopsis*, but also for lettuce (*Lactuca sativa* L.), yellow lupine (*Lupinus luteus* L.), tomato (*Solanum lycopersicum* L.), apple, barley, rice, wheat, switchgrass (*Panicum virgatum* L.), troumble pigweed (*Amaranthus albus* L.), and chickweed (*Stellaria media* L.) (Taylorson and Hendricks 1977; Bogatek and Gniazdowska 2006; Bethke et al. 2007b; Piterková et al. 2012) and others. Also it was demonstrated that nontreated with NO, but germinating embryos of sorghum (*Sorghum bicolor* L.) or stratified apple seeds produced NO predominantly in early phases of germination (Simontacchi et al. 2004; Dębska et al. 2013). Moreover, NO is involved in enhancement of stress tolerance in germinating seeds. Such reaction was observed in wheat and lupin seeds exposed to heavy metals or salinity (Hu et al. 2007; Kopyra and Gwóźdź 2003; Zheng et al. 2009). NO and GA treatment of wheat seeds improved germination under chilling temperatures (Li et al. 2013). In stress tolerance, NO can be considered as a molecule that modulates ROS metabolism by stimulation of antioxidant cellular system. In cucumber (*Cucumis sativus* L.) seeds

NO-induced salt stress resistance was associated with activation of superoxide dismutase (SOD) and catalase (CAT) (Fan et al. 2013). Similar reaction was detected in mungbean (*Vigna radiata* L.) seeds under lead stress (Aly 2013). In cells of barley aleurone layer a delay of programmed cell death by NO was demonstrated, which correlated with stimulation of enzymes of antioxidant character (Beligni et al. 2002). Involvement of NO in modulation of ROS metabolism (stimulation of activity of antioxidant enzymes) was also observed during apple embryos germination under favorable conditions (Krasuska and Gniazdowska 2012).

Seed germination and the following seedling establishment are connected with cotyledon growth and greening. This process leading to the development of autotrophic metabolism requires the ability to light response. Light-dependent *Arabidopsis* seed germination is mediated by photoreceptors, mostly phytochromes (PHY), that perceive Red/Far-Red radiation (Sullivan and Deng 2003; Bai et al. 2014). It was demonstrated that NO fumigation substituted red light effect in photoblastic seeds and stimulated deetiolation of *Arabidopsis* seedlings (Giba et al. 1998; Beligni and Lamattina 2000). NO involvement in photomorphogenesis positively affected DELLA protein accumulation and negatively affected expression of one of the GA signaling component Phytochrome-Interacting Factor (*PIF*) in *Arabidopsis* seedlings (Lozano-Juste and León 2011). PIF proteins, including PIF3 is known to interact with PHY, while DELLA proteins are involved in inhibition of hypocotyls growth by decrease of PIF transcription activity (Bai et al. 2014 and citation therein). Thus, it suggests interplay of PHY, GA, and NO as NO inhibited primary root growth of *Arabidopsis* seedlings throughout a reduction of *PIF3* expression (Bai et al. 2014). Recently the release of redroot pigweed (*Amaranthus retroflexus* L.) seeds dormancy was also observed in the light in the presence of NO. Light condition of redroot pigweed seed germination enhanced tissues response to phytohormones such as GA or ethylene but not to NO (Kępczyński and Sznigir 2014).

Further studies revealed an importance of the cross-talk between NO and phytohormones for control of seed dormancy and germination. Two comprehensive reviews on the topic have been published recently (Arc et al. 2013a, b). Based on our own data (Bogatek and Gniazdowska 2006; Gniazdowska et al. 2010a, b; Krasuska et al. 2014a) and referring to direct links between hormones with NO and hormones with ROS in the regulation of seed dormancy alleviation and germination we were able to create a model of "nitrosative door," in parallel to the model of "oxidative window" (Krasuska and Gniazdowska 2012). This model describes threshold values of RNS in dormancy release and seed germination. The "nitrosative door" is "open" for seed germination when precise minimum level of RNS in tissues is reached. NO acts as signaling molecule necessary to induce metabolic changes responsible for seed transition from dormancy to germination. Seeds maintain their dormancy under conditions of low level of free RNS in cells, and in contrast an increased RNS concentration leads to seed ageing and death. The detailed description of the model, modified according to some new data on NO signaling, is presented at Fig. 1.

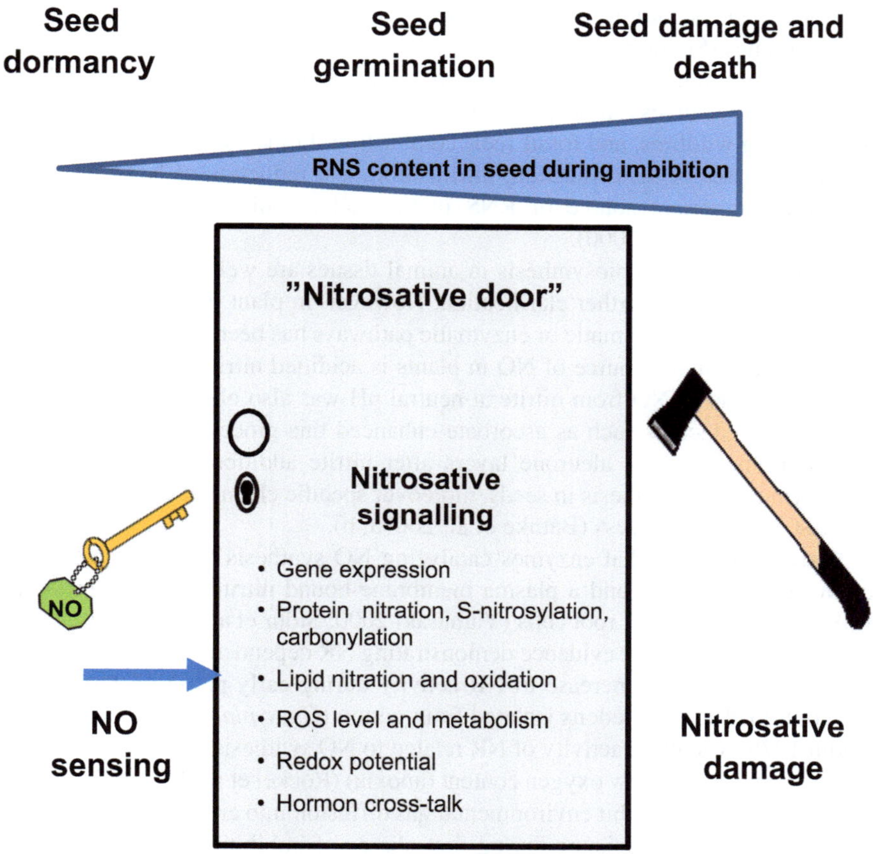

Fig. 1 The model of "nitrosative door" in seed physiology (modified according to Krasuska and Gniazdowska (2012)). RNS, mainly NO appropriate level ("nitrosative door") is required for dormancy alleviation and transition of seeds to germination. Below this level, the amount of RNS is too low to promote germination, while above, over accumulation/overproduction of RNS leads to damages of cellular components that prevent or delay germination. Thus, action of accumulated RNS may be compared with the damages caused by burglar using an ax or a crowbar to enter the building with the intention to destroy it. In the opposite, low amount of RNS induces cascade of biochemical events and acts as "golden key" that unlock the door without any noise and perturbation

2 Sources of Reactive Nitrogen Species in Seeds

Commonly, RNS are known to be potentially toxic (at high concentration and during prolonged exposition). On the other hand, at low concentrations, they act as important signaling molecules in prokaryotic and eukaryotic cells. The RNS family includes NO in its different redox forms (nitrosonium and nitroxyl anion), peroxinitrite ($ONOO^-$), nitrous anhydride (N_2O_3), and nitrogen dioxide radical (NO_2) (Stamler et al. 1992). NO in a form of a free radical dissolves in aqueous solutions and according to its hydrophobic nature easily diffuses across biomembranes (Stamler et al. 1992).

In the environment seeds are exposed to RNS originating from natural and anthropogenic sources. Atmospheric RNS come from plant emission, smoke produced during wildfires, and fossil fuels combustion. In the soil, RNS are produced during processes of denitrification, nitrification, and reduction of NO_3^- to NH_4^+. Additional important source of RNS in the soil are nitrogen fertilizers (Paul et al. 1993; Yamasaki 2000).

Mechanisms of NO biosynthesis in animal tissues are well recognized, but for plants they still need further clarification. However, in plant tissues production of NO throughout nonenzymatic or enzymatic pathways has been confirmed. The best known nonenzymatic source of NO in plants is acidified nitrite. Slow and spontaneous liberation of NO from nitrite at neutral pH was also observed, however the presence of reductans such as ascorbate enhanced this process (Yamasaki 2000). NO synthesis in barley aleurone layers after nitrite addition pointed out to the nonenzymatic NO synthesis in seeds, moreover specific chemical conditions in the apoplast favor this process (Bethke et al. 2007a, b).

There is no doubt that enzymes catalyzing NO synthesis from nitrite ions are nitrate reductase (NR) and a plasma membrane-bound nitrite:NO reductase (Ni–NOR), characteristic for root cells (Yamasaki 2000; Stöhr et al. 2001). Although at present there is no direct evidence demonstrating NR dependent evolution of NO in germinating seeds, an increase in NR activity during early phase of germination was confirmed for cotyledons isolated from cotton (*Gossypium hirsutum* L.) seeds (Radin 1976). Catalytic activity of NR related to NO synthesis was observed under specific conditions of low oxygen content (anoxia) (Rockel et al. 2002). Tight cover layers of seed coats inhibit environmental gas diffusion into embryo tissues leading to periodic oxygen deficiency during first phases of imbibition. These temporary anoxic conditions could favor NR involvement in NO enzymatic synthesis. Data obtained for sorghum embryonic axis indicated inducible NR isoform activity as an enzyme involved in NO synthesis at initiation of germination (Simontacchi et al. 2004). Moreover, a sharp increase in NR activity was also detected in germinated tomato seeds after priming with KNO_3 (Lara et al. 2014). Thus, it is quite reasonable to propose the importance of NR function in NO generation during seed germination. Additional, but indirect proof for NR engagement in NO production in plants comes from experiments on *Arabidopsis* lateral roots. During lateral root development phosphorylation of NIA2 (one of NR isoforms) by mitogen-activated protein kinase 6 (MPK6) increased NO synthesis indicating involvement of the MAPK cascade in NO metabolism and signaling (Wang et al. 2010).

Oxidative pathway of NO synthesis in plants depends on nitric oxide synthase (NOS) activity (termed NOS-like activity, as compared to mammalian NOS). This bimodal enzyme converts L-arginine (L-Arg) to L-citrulline and catalyzes NO synthesis. For catalytic activity of NOS, there is a requirement for O_2 (cosubstrate) and several cofactors including NADPH, flavin mononucleotide (FMN), FAD, and also 6R-tetra-hydrobiopterin (H_4B) (Foresi et al. 2010). NOS has a calmoduline

(CaM)-binding region which connects two domains of the protein. In human tissues three different NOS isoforms: endothelial NOS (eNOS), inducible NOS (iNOS) and neuronal NOS (nNOS), are recognized. In contrast, in Plant kingdom the only molecular proof for NOS existence was indicated in algae *Ostreococcus tauri* (Foresi et al. 2010) at both transcriptional and protein level, although NOS-like activity was detected in various plant tissues, including sorghum embryonic axis at the early phase of germination (Simontacchi et al. 2004; Mur et al. 2013 and references therein). However, strong NOS dependence on O_2 availability in the reaction medium leads to conclusion that at the early phase of seed germination, especially in seeds covered by thick seed coat, activity of this enzyme should not be suggested as crucial for NO evolution. It is possible that NOS activity could be detected mainly during the third, anabolic phase of germination and/or development of young seedlings, especially, during greening of cotyledons. The adequate point of transition from seed germination into seedling growth in some material (seeds/embryos) may be controversial. Germination of apple embryos supplied with L-Arg (NOS substrate) was stimulated and accompanied by enhanced emission of NO and $ONOO^-$ from embryonic axis and roots of growing seedlings. Addition into the imbibition medium of L-canavanine, an antimetabolite of L-Arg, strongly inhibited germination (Krasuska et al. 2014a).

Recently, NO synthesis from polyamines (PAs) was proposed in reaction catalyzed by hypothetic (still unknown) enzyme or via direct PAs conversion (nonenzymatically) (Tun et al. 2006). The existence of NO-PA cross-talk in plants at different physiological stages and conditions has been demonstrated (Wimalasekera et al. 2011). PAs belong to the group of plant growth and development regulators affecting seed dormancy and germination (Matilla 1996; Krasuska et al. 2012). As in seeds of most other plant species, putrescine or spermidine stimulated germination of dormant apple embryo, while spermine maintained dormancy. Alleviation of dormancy induced by PAs was accompanied by NO and $ONOO^-$ emission from embryonic axis or seedling roots (Krasuska et al. 2014a). Moreover, short-term pretreatment of dormant apple embryos with NO donors stimulated activity of PA oxidase (PAO), an enzyme responsible for PAs catabolism. High PAO activity was observed with spermine used as a substrate in reaction medium (Krasuska et al. 2014a). This data corresponds to alterations in PAs level during germination (Krasuska et al. 2012) and are in agreement with concept of regulatory role of ROS in seed germination, as by-product of PAO activity is hydrogen peroxide (H_2O_2), key element in seed germination "oxidative window" (Bailly et al. 2008). It is also suggested that PAO could play a role in NO synthesis (Wimalasekera et al. 2011).

Although, we are still far from precise identification of NO source in seeds, it is absolutely clear that this molecule is synthesized and released during early phase of seed germination (imbibition), while nitrosative burst occurs at the moment of termination of *sensu stricto* germination. Thus, it may be summarized that accumulation of ROS that facilitates opening of "oxidative window" and accumulation of NO at the level permitting to unlock "nitrosative door" are crucial for seed dormancy alleviation (Fig. 1).

3 Mode of Action of Reactive Nitrogen Species in Seeds

Reactivity of NO strongly depends on O_2, ROS and their initial concentration in cellular compartments. Generally, the biological half-life of NO is measured in seconds because of very unstable nature of NO. As a free radical, NO reacts rapidly with other free radicals and forms complexes with the transition metal ions (e.g., in metalloproteins). Reactivity of NO with heme-containing proteins, iron-sulfur centers in proteins, and in zinc-finger proteins have been well documented (Mur et al. 2013). RNS participate in protein posttranslational modification *via* their reaction with cysteine residue (S-nitrosylation), tyrosine residue (Tyr nitration), and tryptophan residue (Trp nitration). NO participates also in protein carbonylation as was shown for desiccated recalcitrant *Antiaris toxicaria* Lesch. seeds (Bai et al. 2011) and apple embryos (Krasuska et al. 2014b). Recently some studies were focused on NO participation in fatty acid peroxidation and nitration, which so far were considered rather to be pathobiochemical consequences of metabolic disorder than part of signaling cascade. However, products of fatty acid nitration: nitro fatty acids, nitro hydroxyl fatty acids and *trans*-fatty acids act as signaling molecules (Jain et al. 2008).

Metal-containing proteins can form complexes with NO, so they can act as NO targets. Among them are hemoglobins (Hb), which in plant tissues are present as symbiotic Hb –leghemoglobin and nonsymbiotic Hb (nsHBs). Induction of nsHb class 1 is related to hypoxic stress (Dordas et al. 2003; Matilla and Rodríguez-Gacio 2013). Early seed imbibition phase is accompanied by low O_2 diffusion, and in this condition induction of *nsHBs* expression was reported (Dordas et al. 2003). Contribution of nsHbs in the conversion of NO into nitrate (NO_3^-) modulates NO concentration/cycling at low O_2 availability (Igamberdiev et al. 2005, 2010), and plant survival is dependent on *nsHb* expression as was shown for *Arabidopsis* seedlings (Hebelstrup et al. 2006). Other metabolites acting as modulators of NO level are *S*-nitrosothiols, including nitrosoglutathione (GSNO) (Astier and Lindermayr 2012). *S*-nitrosylation is reversible process, favored by reducing environments. In seed physiology maintaining of optimal ratio of glutathione in reduced form (GSH) to glutathione in oxidized form (GSSG) is of great importance, as seed viability depends on high GSH level (Bailly et al. 2008). GSH is a key player in ROS metabolism and participates in buffering of cellular redox potential. It binds NO and GSNO is transported in long distances or acts as a reservoir of NO (Begara-Morales et al. 2014 and citation therein). Thus, in seeds GSNO probably could be also a major NO depot. Ability of GSH to scavenge NO led to *S*-nitrosylation of proteins that occurs as transnitrosylation *via* GSNO. This means that high level of GSNO stimulates formation of other *S*-nitrosothiols. Regulation of GSNO concentration is achieved by activity of GSNO reductase (GSNOR). This enzyme catalyzes oxidation of GSNO to GSSG and ammonia (Begara-Morales et al. 2014 and citation therein). Increase of GSNO as well as higher activity of GSNOR is mostly studied in plants under stress conditions (Astier and Lindermayr 2012) and so far no data referring to seeds have been presented. Another important posttranslational modification of proteins is nitration. This process involves tyrosine modification *via* covalent binding of a nitro (–NO_2) group to one of the two equivalent ortho carbons of the aromatic ring of this amino acid (Chaki et al. 2009). Nitration of tyrosine in proteins alters their structure and acts as an on/off switch of their function. Apart from being a nitrosative stress marker, the level of nitrated proteins in biological samples has been regarded as a part of cell signaling and cytoskeletal organization. This consideration has been strengthened by the analysis of nitrated proteins in the hypocotyls of sunflower (*Helianthus annuus* L.) seedlings, grown under optimal conditions (Chaki et al. 2009).

As it was mentioned previously, NO could act also as inducer of protein carbonylation (Dębska et al. 2013; Krasuska et al. 2014b). This modification is commonly known as a marker of oxidative metabolic disorder and in seeds it was linked mostly to process of aging and loss of viability (Terskikh et al. 2008; Kalemba and Pukacka 2014) characterized by over accumulation of ROS. However, carbonylation of some proteins (e.g., proteins or peptides that serve as reserved material) is a positive process required for termination of seed germination (Job et al. 2005). Proteins that are marked by *S*-nitrosylation, nitration, or carbonylation are also faster degraded by proteolysis. Activity of proteases seems to be stimulated by NO as it was reported for germinating apple embryos (Krasuska et al. 2014b). Recently, NO involvement in protein degradation by the N-end rule pathway in *Arabidopsis* seedlings was shown (Gibbs et al. 2014). The authors proposed a mechanism for NO sensing in plants *via* N-end rule pathway of the specific transcriptional regulators marked and targeted to proteolysis (Gibbs et al. 2014).

Induction of protein modifications that occur in response to NO may be a reason for the lack of specific NO receptors. Moreover, taking into account a high reactivity of NO, their existence is doubtful. Although, there are several proofs on induction of NO signal transduction cascade involving activation of guanylylcyclases (GCs) (Durner et al. 1998; Klessig et al. 2000; Pasqualini et al. 2012), activity of which leads to the biosynthesis of guanosine 3′,5″-cyclic monophosphate (cGMP). Thus, GC may be considered as a main "receptor" of NO (Pyriochou and Papapetropoulos 2005) as in plant tissues the presence of cGMP and activity of GC has been reported (Szmidt-Jaworska et al. 2008, 2009). Recently, the promotion of *Arabidopsis* seed germination by cGMP has been demonstrated, moreover, this compound was shown to be involved in phytohormonal signaling (ABA—GA) during dormancy release and germination (Teng et al. 2010). Since cGMP is involved in phytochrome-controlled physiological processes in plants, NO-mediated stimulation of germination in photoblastic seeds could be govern by a cross-talk between cGMP/GC and phytochromes.

According to the proposed model of "nitrosative door" (Fig. 1), NO seems to play a role of a "golden key" that suits to "keyhole" represented by putative NO targets (mainly proteins or lipids), modification of which "opens the door." This means activation of signal transduction cascade that allows for seed transition from dormant to nondormant "ready to germinate" state.

Since phytohormones are considered as most important endogenous agents regulating seed dormancy and germination, the next section refers to a close relationship between NO and phytohormones, their biosynthesis, degradation and signal transduction pathways.

4 Seed Dormancy Alleviation and Germination: Interplay of NO and Phytohormones

It is well documented that induction and maintenance of seed dormancy is due to ABA concentration, tissue sensitivity to ABA, and signaling pathways of this phytohormone (Bethke et al. 2004; Liu et al. 2009; Shu et al. 2013). Seed dormancy induction and reserve accumulation is due to ABA produced by maternal tissues, whereas desiccation tolerance comes from embryonic ABA (Finkelstein et al. 2008).

Inhibition of ABA biosynthesis (by fluridone) during imbibition of dormant apple embryos stimulated germination. In relation to NO, an additive effect of fluridone and NO was observed, as germination of apple embryos was very fast (Gniazdowska et al. 2010b). Strong involvement of NO in dormancy alleviation of apple embryos was indicated by the treatment with NO scavenger—2-(4-carboxyphenyl)-4,4,5,5-tetramethylimidazoline-1-oxyl-3 oxide (cPTIO), the compound which totally restricted embryo germination. In addition, a decrease in sensitivity to ABA occurred in apple embryos, dormancy of which was removed by short term pretreatment with NO donors (Gniazdowska et al. 2007). Similarly, dormancy of *Arabidopsis* and barley seeds treated with NO was alleviated due to a decline in ABA level and a lower tissue sensitivity to this hormone (Bethke et al. 2004).

Biosynthesis of ABA occurs in maturing seeds in embryonic tissues (Karrsen et al. 1983). In *Arabidopsis* five members of a gene family encoding 9-cis-epoxy-carotenoid dioxygenase (NCED), an enzyme that catalyzes a regulatory step in ABA biosynthesis have been found and *NCED6* and *NCED9* are known to play a pivotal role in seed dormancy induction (Lefebvre et al. 2006).

Treatment of the 7 months after-ripened hedge mustard seeds with 20 mM potassium nitrate (KNO_3) affected the transcripts level of *NCED5,6* which was decreasing, especially within the first 12 h of imbibition (Carillo-Barral et al. 2014). The level of *NCED5* transcripts was strongly affected (higher transcripts level) in nitrate-treated seeds compared to nontreated controls. Moreover, a prolonged imbibition was accompanied by increased *NCED5,6* transcription in nitrate-treated seeds (Carillo-Barral et al. 2014). Similarly, in apple embryos fumigated with NO an increased transcription level of *NCED9* was noticed from the beginning of imbibition, while *NCED3* declined as germination period was expanded (Andryka unpublished data).

The essential ABA signaling components (Core Signaling Elements) are three molecular units that initiate the signaling by binding to this phytohormone. These components are: ABA-binding PYR1/PYLs/RCARs (Pyrabactin Resistance 1/ PYR-like Proteins/Regulatory Components of ABA Receptor), SNF1-Related Protein Kinase 2s (SnRK2s), and a subgroup of Protein Phosphatase 2Cs (group PP2Cs) (Umezawa et al. 2010; Kim et al. 2013 and citation therein). Another part of ABA signaling pathway consists of ABA-response elements (ABREs) including ABRE-binding protein/ABRE-binding factor TFs (AREB/ABF) belonging to the positive regulators of this cascade after downstream phosphorylation by SnRK2s (Umezawa et al. 2010; Kim et al. 2013 and citation therein). In dormant seeds the expression of genes with ABREs in their promoter regions and genes coding transcription factors that bind to the ABREs is characteristic (Finch-Savage and Leubner-Metzger 2006). Recently, it was observed that short-term pretreatment of dormant apple embryos with NO donors lowered transcription level of ABA positive regulators *SnRK2s*, negative regulators *PP2Cs* and transcription of *ABF* (Andryka et al. 2013a, b). ABA-insensitive1-2 (ABI1 and ABI2, a group of PP2Cs), negative regulators of ABA signaling, bind to SnRK2s, dephosphorylate them and in consequence inhibit the positive effects of SnRK2s (Kim et al. 2013; Shu et al. 2013). This inhibition is prevented by ABA, which binds to PYR1/PYLs/ RCARs group containing ligand-binding START domains (Kim et al. 2013). ABI3 transcription factor is a positive ABA signaling element (essential for seed maturation and dormancy establishment) and negative regulator of GA biosynthesis (Shu et al. 2013). ABI3 binds to the RY *cis*-motif in DNA through its B3 domain. During seed development and dormancy release modification in ABI4 protein (ABI4 encodes an AP2/ERF type transcription factor involved in ABA signaling pathway) is observed. ABI4 is a positive regulator of ABA signaling pathway regulated by the 26S-proteasomal pathway (Shu et al. 2013 and citation herein). Seeds of *Arabidopsis* mutant *abi4* exhibited reduced primary dormancy due to lower ABA concentration in dry seeds, as compared to dry wild type seeds. The involvement of ABI4 in primary seed dormancy establishment is due to the inhibition of ABA degradation and induction of biosynthesis pathway of this phytohormone. ABI4 also plays an important role in inhibition of GA biosynthesis (Shu et al. 2013). Lately, it has been shown that *Arabidopsis* seed germination is regulated by NO-ABA cross-talk through the regulation of key repressor *ABI5* (Gibbs et al. 2014). Expression of *ABI5* is regulated by group VII ethylene response factors (ERFs). In the presence of NO these ERFs are targeted to proteolysis *via* N-end rule, and in addition it was shown that members of group VII ERFs act as homeostatic sensors of low oxygen in plants (Gibbs et al. 2014). Nitrite treatment of after-ripened hedge mustard seeds slightly stimulated *ABI5* expression during the first hours (6–12 h) of imbibition (Carillo-Barral et al. 2014).

The increase of inactive ABA metabolites such as phaseic acid and dihydrophaseic acid is aresult of ABA catabolism, the process that is involved in dormancy release. ABA can also be converted to 8′-hydroxy ABA or ABA-glucose ester (Gubler et al. 2005). Molecular analysis of *Arabidopsis* indicated the existence of a family of cytochrome P450s (CYP707A1-4) proteins that catalyze the

8′-hydroxylation of ABA to inactive form (Kushiro et al. 2004). In dry *Arabidopsis* seeds mRNAs of *CYP707A2* gene are highly accumulated. As it was mentioned above ABI4 is responsible for inhibition of *CYP707A1* and *CYP707A2* transcription. A triple mutant of *CYP707A1-3* exhibited deep dormancy (Kushiro et al. 2004; Shu et al. 2013). Moreover, lower concentration of ABA, accompanied by higher expression of *CYP707A2* was observed when *Arabidopsis* seeds were treated with nitrate. Nondormant seeds were produced by wild-type plants nourished with 50 mM nitrate, and by mutants *nia1* and *nia2*, accumulating nitrite under standard nutrition (Matakiadis et al. 2009). The negative effect of nitrate on dormancy due to the increased expression of *CYP707A2* was also proven for the after-ripened hedge mustard seeds (Carillo-Barral et al. 2014). Likewise, *Arabidopsis* seed dormancy release induced strictly by NO treatment was accompanied by an enhancement in *CYP707A2* transcription and ABA 8′-hydroxylase protein abundance (Liu et al. 2009).

The summary of NO influence on the expression of genes encoding proteins of phytohormones (GA, ABA, JA, ET) biosynthetic and degradation pathways is presented in Fig. 2. Parallel scheme related to NO impact on expression of genes of phytohormonal signal transduction pathways is shown at Fig. 3.

In contrast to inhibitory effect of ABA, active GAs induce and stimulate seed germination (Finch-Savage and Leubner-Metzger 2006; Nambara et al. 2010). Activation of embryo metabolism after water uptake requires degradation of storage materials. It is commonly accepted that GAs are involved in mobilization of storage material. On the contrary, ABA strongly inhibits metabolic changes, e.g., in *Arabidopsis* aleurone layer (Penfield et al. 2004). NO scavenger - cPTIO inhibited vacuolization (a marker of dormancy removal) of aleurone cells, and this effect was reversed by GA. *Arabidopsis* aleurone cells were also responsive to NO by increasing vacuolization. Summarizing this data, it was suggested that in GA signaling pathways NO acts as an upstream element (Bethke et al. 2007a). It has been postulated that GAs are involved in α-amylase synthesis, but later on during terminal phase of seed germination. Zhang et al. (2005) have demonstrated that not GA but NO is involved in rapid induction of β-amylase activity (and not α-amylase) at the early phase of germination. In addition, it was postulated that NO participates in direct de-polymerization of β-amylase aggregates, and as a consequence the released enzyme was quickly activated (Zhang et al. 2005). Nevertheless, both NO and GAs pretreatment led to an increased activity of β-amylase (much higher activity compared to the control) and α-amylase (only slightly higher than in control) during the first days of germination under chilling stress conditions (Li et al. 2013).

As GAs are known to govern seed germination, seed dormancy loss is accompanied by *de novo* synthesis of many of the transcripts encoding enzymes involved in metabolism of this phytohormone (Fig. 2). One of the important proteins of GAs metabolism is GA 3-oxidase (GA3ox) catalyzing the conversion of inactive forms of GAs to the active forms (GA_9 to GA_4, or GA_{20} to GA_1). It has been indicated that dormancy loss in *Arabidopsis* seeds induced by NO was accompanied by the enhanced transcription of *GA3ox1* and *GA3ox2* (Bethke et al. 2007a). High level

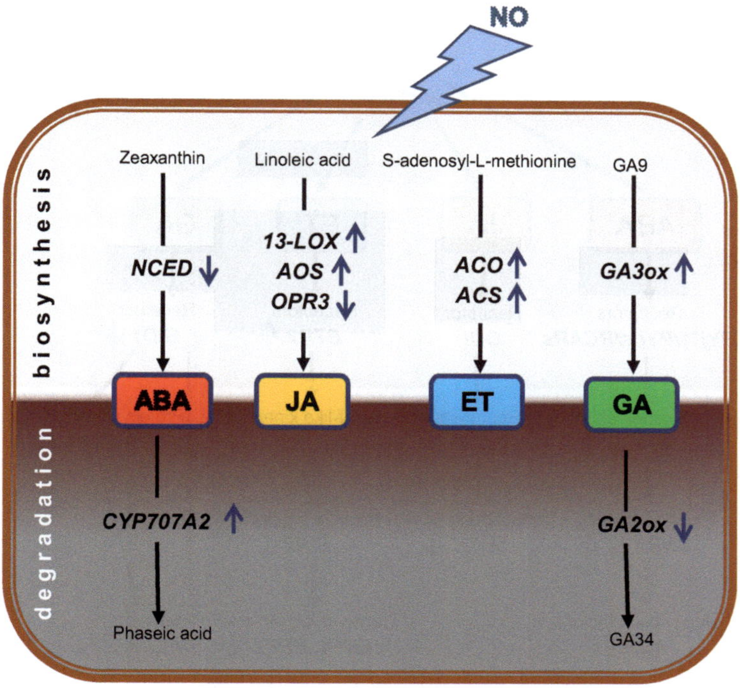

Fig. 2 Impact of NO on expression of genes encoding key enzymes of phytohormones (ABA, ET, GA, JA) metabolism (biosynthesis and degradation). Main precursors of phytohormones and products of their degradation are indicated. Data refer only to germinating seeds or embryos. Arrows accompanying gene indicate up- (↑) or downregulation (↓) of gene transcription in response to treatment with NO or chemicals that may function as NO source. Presented up- or downregulation is based on data described by: *NCED*—Carillo-Barral et al. 2014; Jacobsen et al. 2013; *CYP707A2*—Matakiadis et al. 2009; Carillo-Barral et al. 2014; Liu et al. 2009; *Ga3ox*—Bethke et al. 2007a; Carillo-Barral et al. 2014; *Ga2ox*, *13-LOX*, *AOS*, *OPR3*—Andryka unpublished data; *ACO*, *ACS*—Bogatek et al. 2011; Iglesias-Fernándes and Matilla 2009

of transcripts of *GA3ox2* was also detected in nitrate treated after-ripened hedge mustard seeds, but mostly after the prolonged imbibition (Carillo-Barral et al. 2014). GA 2-oxidase (GA2ox), the GA deactivation enzyme, converts active forms of GA to the inactive forms (GA_4 to GA_{34}, or GA_1 to GA_8) (Nonogaki et al. 2010 and references therein). In apple embryos, dormancy of which was removed by NO or SNP, a decline in transcription level of *GA2ox* was observed 24 h after treatment (Andryka unpublished results).

In *Arabidopsis* seeds key repressors of GA transduction pathway are DELLA family proteins such as: GAI, RGA, RGL1, RGL2, and RGL3 (Bai et al. 2014). In the presence of GA the degradation of RGL2 can be observed due to the presence of a conserved DELLA motif essential for its proteasome-mediated destruction (Lee et al. 2002). GA binding to GIBBERELLIN INSENSITIVE DWARF1-like receptors (GID1) enhances RGL2 interaction with the F-box protein SLEEPY1 (SLY1),

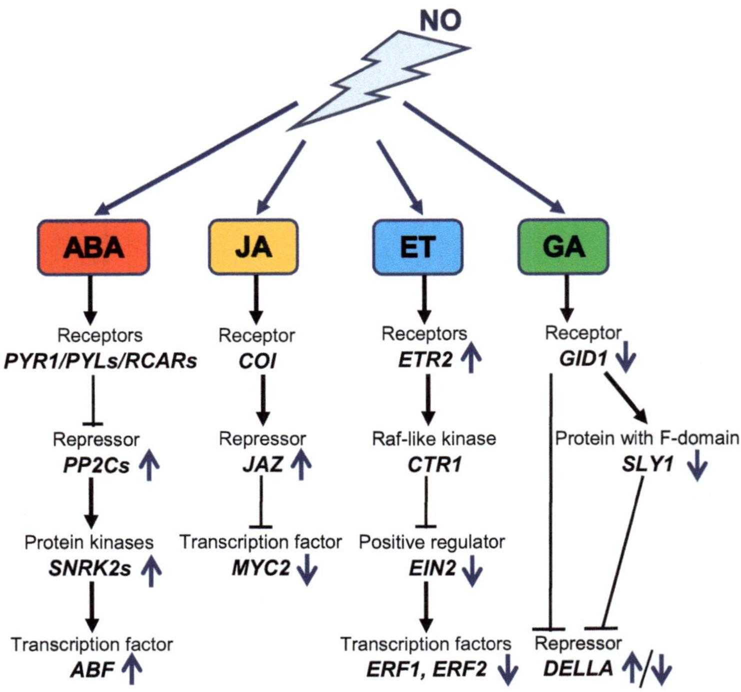

Fig. 3 Alteration of expression of genes encoding elements of the signal transduction pathway of phytohormones (ABA, ET, GA, JA) as induced in germinating seeds by treatment with NO or chemicals that may be source of NO. Arrows accompanying gene indicate their up- (↑) or downregulation (↓). Scheme refers to *SnRK2s*, *ABF*, *PP2Cs*—Andryka et al. (2013a, b); *DELLA*—Carillo-Barral et al. (2014), Andryka et al. (2013a)/Bai et al. (2014); *GID1*, *SLY1*, *JAZ*, *MYC2*—Andryka et al. (2013a); *ETR2*, *EIN2*, *ERF1*, *ERF2*—Bogatek et al. (2011)

thus facilitating RGL2 ubiquitination and stimulating subsequent degradation (Feng et al. 2008) (Fig. 3). NO evolving substances may influence GA signal transduction pathways by inhibiting biosynthesis of RGL2. Low transcript level of *RGL2* at the beginning of imbibition (until 6 h) was detected in after-ripened hedge mustard seeds germinated on nitrate (Carillo-Barral et al. 2014). In contrast, in developing *Arabidopsis* seedlings NO treatment led to the primary root growth inhibition by accumulation of RGL3, one of DELLA proteins (Bai et al. 2014).

It was confirmed that some GA-insensitive GID1 receptor mutants show reduced germination ability (Voegele et al. 2011). Voegele et al. (2011) have demonstrated using *Arabidopsis* and lettuce seeds that GID1a and GID1c are the major GA receptors involved in germination, while GID1b seems to be less important. In angiosperm seeds, the GID1-type GA receptors interact with DELLA repressor proteins, which negatively regulate GA signaling in an ontogeny stage-specific pattern (Lee et al. 2002). Recently the opposite regulation of transcript level of *DELLA* (downregulation) and *GID1, SLY1* (upregulation) was shown in dormant apple embryos shortly pretreated with NO after termination of germination *sensu stricto* (Andryka et al. 2013a, b).

In comparison to GA and ABA jasmonic acid (JA) is not so obviously identified as seed germination affecting hormone. JA promoted germination of dormant apple embryos (Ranjan and Lewak 1992) but in contrast, a decreased level of JA in stratified apple seeds was accompanied by a stimulation of germination. Biochemical action of JA in seed germination is due to the storage material mobilization (Bogatek et al. 2002). Similar to JA, influence on seed germination was detected after methyl jasmonate (Me-JA) application, a derivative of JA, although it was not so spectacular, probably due to a variation in strength of dormancy and concentration of Me-JA (Jacobsen et al. 2013 and citation herein). Jacobsen et al. (2013) demonstrated that in regulation of wheat (*Triticum aestivum* L.) seeds dormancy blue light, NO and JA play a pivotal role. Me-JA, together with NO, decreased ABA concentration by reduction of *NCED1* transcription and stimulation of expression of *ABA8'OH-1*. Moreover, it was shown, that reduction of ABA level was under interdependent control of NO and Me-JA (Jacobsen et al. 2013). Treatment of seeds with JA results in induction of genes encoding proteins of JA biosynthesis and signaling pathways. Biosynthesis of JA starts from its substrate-α-linolenic acid (α-LeA), which is liberated from galactolipids in reaction catalyzed by phospholipase1 (Wasternack and Hause 2013 and citation therein). Another enzyme involved in JA biosynthesis is lipoxygenase (LOX) that catalyzes oxygenation of α-LeA. Alleneoxide cyclase (AOC) and allene oxide synthase (AOS), partially associated with membranes of chloroplast, also take part in JA biosynthesis. This fact indicates the significance of plastid membrane metabolism in JA dependent developmental processes. Another important enzyme in JA biosynthetic pathway is *cis*-12-oxo-phytodienoic acid (OPDA) reductase3 (OPR3) (Wasternack and Hause 2013 and citation therein) (Fig. 2). OPDAs are signaling molecules (reactive electrophiles), but also of potentially cytotoxic character due to the presence of reactive α,β-unsaturated carbonyl structure (Stelmach et al. 1998). OPDAs are transferred to peroxisomes where OPRs convert them to (+)-7-iso-JA (Yan et al. 2012 and citation therein). It was recently demonstrated that short-term pretreatment of dormant apple embryos with NO or SNP stimulated the transcription level of *LOX*, *AOS*, and decreased the *OPR3* transcription during germination *sensu stricto*. Moreover, NO treatment stimulated transcripts level of JA carboxyl methyltransferase (JMT) (Andryka unpublished data), that is involved in generation of Me-JA, suggesting a positive correlation between NO and JA in regulation of seed germination.

Tissue response to JA requires receptor COI1 (Coronatine Insensitive1) (Fig. 3) belonging to the F-box proteins, a part of the E3 ubiquitin ligase SCF-complex SCFCOI1. This complex is involved in degradation of JA-signaling pathway repressors—JAZ1-12 proteins (Browse 2009; Fonseca et al. 2009). The promoters of JA-regulated genes bind to MYC2/JIN1 (transcriptional activators) released as a result of JAZ degradation (Fonseca et al. 2009). Until now, the only one presented data on the NO impact on JA-signaling cascade were done using apple embryos.

In axes of embryos fumigated by NO downregulation of *JAZ1*and *JAZ 3* and upregulation of *MYC2* transcripts level was noticed (Andryka et al. 2013a).

Involvement of ethylene (ET) in the regulation of seed dormancy and germination, although investigated for many years is yet to be clearly established (Bogatek and Gniazdowska 2012). There is never ending discussion whether ET emission in seeds is a result of the germination or ET itself initiates germination. Even without finding an answer to the question, there is no doubt that ET plays a pivotal role in seed dormancy removal (Matilla and Matilla-Vazquez 2008). ET is believed to act in concert with GA and ABA in dormancy loss and it lowers tissue sensitivity to ABA (Beaudoin et al. 2000). Blockage of ET receptors in seeds during germination resulted in total restriction in growth of root axes. Even pretreatment of embryos with NO did not overcome inhibitory effect of the ET receptors blocker. However, NO stimulated ET emission during dormancy release of apple embryos, while cPTIO decreased emission of this phytohormone (Gniazdowska et al. 2007, 2010b). ET is synthesized from its direct precursor 1-aminocyclopropane-1-carboxylic acid (ACC),with the two key enzymes involved: ACC synthase (ACS) and ACC oxidase (ACO) (Fig. 2). The activities of both of them were stimulated by NO treatment of apple embryos during dormancy alleviation (Gniazdowska et al. 2010b) and correlated with enhanced accumulation of *ACS1* and *ACO3* transcripts (Bogatek et al. 2011). Similarly, dormancy removal of non-after-ripened hedge mustard seeds treated with 20 mM KNO_3 was accompanied by increased *ACS7* and *ACO2* transcription (Iglesias-Fernándes and Matilla 2009). The same pattern of transcription of the genes encoding ACS and ACO was reported in *Arabidopsis* seeds imbibed in SNP (Parani et al. 2004). On the other hand, NO mode of action in the ET biosynthetic pathway seems to be more direct. Treatment of synthetic ACC with NO (*in vitro* experiment) resulted in ET production (Gniazdowska et al. 2010b). This direct reaction of NO with ACC could be one of the most important events in seed dormancy breakage, as it does not require high metabolic costs, and may occur at very early stages of germination.

The molecular data indicate the existence of ET receptors which act as negative regulators of ET-signaling cascade. ET receptors share structural similarity with bacterial two-component histidine kinases. In *Arabidopsis* cells five different ET receptors (ETR1, ETR2, ERS1, ERS2, and EIN4) are present. One of the first regulators of the ET-signaling pathway exhibiting negative effect is Raf-like Protein Kinase (CTR1) (Fig. 3). In the absence of ET, CTR1 keeps inactive EIN2 (Ethylene Insensitive)—positive regulator and EIN3 transcription factor, preventing ET response. Another transcription factor of ET—ERF, localized in the nucleus, initiates activation and repression of hundreds of genes involved in plant reaction to ET (Guo and Ecker 2004). Parani et al. (2004) identified ET Response Binding Protein EREBP as one of the elements induced by NO in *Arabidopsis* seeds. SNP treatment enhanced EREBP transcripts by 2–13-fold over the control expression (Parani et al. 2004). Some alteration in transcription level of other elements of ET-signaling cascade were detected also during dormancy removal of apple embryos pretreated with NO or SNP. Downregulation of *ETR2*, was accompanied by upregulation of *EIN2*, *ERF1* and mainly *ERF2* (Bogatek et al. 2011). The decrease in the expression of the gene encoding ET receptor ETR2 suggests enhanced embryo sensitivity to this phytohormone, while enhanced expression of genes encoding ERF2 can be responsible for activation of ET response genes potentially participating in the process of dormancy breaking.

The network of signals that play a role in determination of seed dormancy removal or germination involves mostly classical plant hormones (GA, ABA, JA, and ET). But to keep up links between these elements additional molecular factors are required, that can easily move from one cell compartment to another, and reacting with other molecules (mainly proteins) may induce physiological reactions. NO and other RNS as main players on the stage of seed biology may be recognized as necessary connecting agents between the members of phytohormonal orchestra (Figs. 2 and 3). A strong dependence on NO of biosynthesis of phytohormones and activation of their transduction pathways suggests that NO-mediated seed dormancy removal and stimulation of germination has to be implemented as a central statement in seed physiology.

5 Summary

Observation of the stimulatory effect of nitrate and its derivatives on seed germination comes from mid-XX century. To date, this information is complemented by understanding of beneficial effects of NO on initiation of the first phase in plant ontogeny. Numerous papers describing stimulation of germination of seeds of various plant species by treatment with NO-releasing chemicals have been published. In some cases scientists suggested that pretreatment of seeds with NO may be used as a part of the procedure leading to seed quality improvement, because NO can decrease sensitivity of plants to various abiotic stresses. All these data are rather descriptive and does not propose functional explanation of the biochemical and molecular modes of NO action in seed biology. But nowadays, due to development of omics, we have valuable tools to find the answer on most of "NO—seeds related questions." Identification of NO-modified proteins, playing crucial roles in regulation of seed dormancy and germination, and representing possible NO molecular targets is of the highest importance in the field of seed physiology.

Acknowledgments The manuscript was prepared during realization of projects financed by National Science Centre NN 303821840 given to A.G., PRELUDIUM 5 UMO-2013/09/N/NZ9/01619 given to K.C., PRELUDIUM4 UMO-2012/07/N/NZ2/01645 and the European Union funds in the frame the European Social Fund given to P.A-D.

References

Alboresi A, Gestin C, Leydecker MT, Bedu M, Meyer C, Truong HN (2005) Nitrate, a signal relieving seed dormancy in *Arabidopsis*. Plant Cell Environ 28:500–512

Aly HEM (2013) Ameliorating role of nitric oxide in germinating mung bean seeds (*Vigna radiata*) under lead stress. CATRINA 8:75–85

Andryka P, Wiśniewska A, Gniazdowska A, Krasuska U, Bogatek R (2013a) NO-induced alleviation of dormancy of apple embryos is related to modification of phytohormonal signaling pathways. BioTechnologia 94:308

Andryka P, Wiśniewska A, Krasuska U, Gniazdowska A, Bogatek R (2013b) NO modifies ABA- and GA-related gene expression in germinating apple (*Malus domestica* Borkh.) embryos. BioTechnologia 94:232

Angelovici R, Galili G, Fernie AR, Fait A (2010) Seed desiccation: a bridge between maturation and germination. Trends Plant Sci 15:1360–1385

Arc E, Galland M, Godin B, Cueff G, Rajjou L (2013a) Nitric oxide implication in the control of seed dormancy and germination. Front Plant Sci 4:346. doi:10.3389/fpls.2013.00346

Arc E, Sechet J, Corbineau F, Rajjou L, Marion-Poll A (2013b) ABA cross talk with ethylene and nitric oxide in seed dormancy and germination. Front Plant Sci 4:63. doi:10.3389/fpls.2013.00063

Astier J, Lindermayr C (2012) Nitric oxide-dependent posttranslational modification in plants: an update. Int J Mol Sci 13:15193–15208

Bai X, Yang L, Tian M, Chen J, Shi J, Yang Y, Hu X (2011) Nitric oxide enhances desiccation tolerance of recalcitrant *Antiaris toxicaria* seeds *via* protein S-nitrosylation and carbonylation. PLoS One 6:e20714

Bai S, Yao T, Li M, Guo X, Zhang Y, Zhu S, He Y (2014) PIF3 is involved in the primary root growth inhibition of *Arabidopsis* induced by nitric oxide in the light. Mol Plant 7:616–625

Bailly C, El-Maarouf-Bouteau H, Corbineau F (2008) From intracellular signaling networks to cell death: the dual role of reactive oxygen species in seed physiology. C R Biol 331:806–814

Batlla D, Benech-Arnold RL (2007) Predicting changes in dormancy level in weed seed soil banks: implication for weed management. Crop Prot 26:189–197

Beaudoin N, Serizet C, Gosti F, Giraudat J (2000) Interactions between abscisic acid and ethylene signaling cascades. Plant Cell 12:1003–1115

Begara-Morales JC, Sánchez-Calvo B, Luque F, Leyva-Pérez MO, Leterrier M, Corpas FJ, Barroso JB (2014) Differential transcriptomic analysis by RNA-Seq of GSNO-responsive genes between *Arabidopsis* roots and leaves. Plant Cell Physiol 55:1080–1095

Beligni MV, Lamattina L (2000) Nitric oxide stimulates seed germination and de-etiolation, and inhibits hypocotyls elongation, three light-inducible responses in plants. Planta 210:215–221

Beligni MV, Fath A, Bethke P, Lamattina L, Jones RL (2002) Nitric oxide acts as an antioxidant and delays programmed cell death in barley aleurone layers. Plant Physiol 129:1642–1650

Bethke PC, Gubler F, Jacobsen JV, Jones RL (2004) Dormancy of *Arabidopsis* seeds and barley grains can be broken by nitric oxide. Planta 219:847–855

Bethke PC, Libourel IGL, Aoyama N, Chung YY, Still DW, Jones RL (2007a) The *Arabidopsis* aleurone layer responds to nitric oxide, gibberellin, and abscisic acid and is sufficient and necessary for seed dormancy. Plant Physiol 143:1173–1188

Bethke PC, Libourel IGL, Jones RL (2007b) Nitric oxide in seed dormancy and germination. Annu Plant Rev 27:153–175

Bogatek R, Gniazdowska A (2006) Nitric oxide and HCN reduce deep dormancy of apple seeds. Acta Physiol Plant 28:281–287

Bogatek R, Gniazdowska A (2012) Ethylene in seed development, dormancy and germination. Annu Plant Rev 44:189–218

Bogatek R, Côme D, Corbineau F, Ranjan R, Lewak S (2002) Jasmonic acid affects dormancy and sugar catabolism in germinating apple embryos. Plant Physiol Biochem 40:167–173

Bogatek R, Szafrański K, Wiśniewska A, Krasuska U, Gniazdowska A (2011) Expression of genes related to ethylene biosynthesis and signalling pathway is modified by NO during dormancy breakage and germination of apple embryos. Informativo da ABRATES 21:168

Browse J (2009) Jasmonate passes muster: a receptor and targets for the defense hormone. Annu Rev Plant Biol 60:183–205

Carillo-Barral N, Matilla AJ, Rodríguez-Gacio MC, Iglesias-Fernández R (2014) Nitrate affects *sensu-stricto* germination of after-ripened *Sisymbrium officinale* seeds by modifying expression of *SoNCED5*, *SoCYP707A2* and *SoGA3ox2* genes. Plant Sci 217–218:99–108

Chaki M, Valderrama R, Fernández-Ocaña AM, Carreras A, López-Jaramillo J, Luque F, Palma JM, Pedrajas JR, Begara-Morales JC, Sánchez-Calvo B, Gómez-Rodríguez MV, Corpas FJ, Barroso JB (2009) Protein targets of tyrosine nitration in sunflower (*Helianthus annuus* L.) hypocotyls. J Exp Bot 60:4221–4234

Dębska K, Krasuska U, Budnicka K, Bogatek R, Gniazdowska A (2013) Dormancy removal of apple seeds by cold stratification is associated with fluctuation in H_2O_2, NO production and protein carbonylation level. J Plant Physiol 170:480–488

Dordas C, Hasinoff BB, Igamberdiev AU, Manach N, Rivoal J, Hill RC (2003) Expression of a stress-induced hemoglobin affects NO levels produced by alfalfa root cultures under hypoxic stress. Plant J 35:763–770

Durner J, Wendehenne D, Klessig DF (1998) Defense gene induction in tobacco by nitric oxide, cyclic GMP, and cyclic ADPribose. Proc Natl Acad Sci USA 95:10328–10333

Fan H-F, Du C-X, Ding L, Xu Y-L (2013) Effects of nitric oxide on the germination of cucumber seeds and antioxidant enzymes under salinity stress. Acta Physiol Plant 35:2707–2719

Farnsworth E (2000) The ecology and physiology of viviparous and recalcitrant seeds. Annu Rev Ecol Syst 31:107–138

Feng S, Martinez C, Gusmaroli G, Wang Y, Zhou J, Wang F, Chen L, Yu L, Iglesias-Pedraz JM, Kircher S, Schäfer E, Fu X, Fan LM, Deng XW (2008) Coordinated regulation of *Arabidopsis thaliana* development by light and gibberellins. Nature 451:475–479

Finch-Savage WE, Leubner-Metzger G (2006) Seed dormancy and the control of germination. New Phytol 171:501–523

Finkelstein R, Reeves W, Ariizumi T, Steber C (2008) Molecular aspects of seed dormancy. Annu Rev Plant Biol 59:387–415

Fonseca S, Chico JM, Solano R (2009) The jasmonate pathway: the ligand, the receptor and the core signalling module. Curr Opin Plant Biol 12:539–547

Foresi N, Correa-Aragunde N, Parisi G, Caló G, Salerno G (2010) Characterization of a nitric oxide from the plant kingdom: NO generation from the green alga *Ostreococcus tauri* is light irradiance and growth phase dependent. Plant Cell 22:3816–3830

Giba Z, Grubisic D, Teodorovic S, Sajc L, Stojakovic D, Konjevic R (1998) Effect of nitric oxide-releasing compounds on phytochrome-controlled germination of empress tree seeds. Plant Growth Regul 26:175–181

Gibbs DJ, Isa NM, Movahedi M, Lozano-Juste J, Mendiondo GM, Berckham S, Marín-de la Rosa N, Conde JV, Correia CS, Pearce SP, Bassel GW, Hamali B, Talloji P, Tomé DFA, Coego A, Beynon J, Alabadí D, Bachmair A, León J, Gray JE, Theodoulou FL, Holdsworth MJ (2014) Nitric oxide sensing in plants is mediated by proteolytic control of group VII ERF transcription factors. Mol Cell 53:369–379

Gniazdowska A, Dobrzyńska U, Babańczyk T, Bogatek R (2007) Breaking the apple embryo dormancy by nitric oxide involves the stimulation of ethylene production. Planta 225:1051–1057

Gniazdowska A, Krasuska U, Bogatek R (2010a) Dormancy removal in apple embryos by nitric oxide or cyanide involves modifications in ethylene biosynthetic pathway. Planta 232:1397–1407

Gniazdowska A, Krasuska U, Czajkowska K, Bogatek R (2010b) Nitric oxide, hydrogencyanide and ethylene are required in the control of germination and undisturbed development of young apple seedlings. Plant Growth Regul 61:75–84

Gubler F, Millar AA, Jacobsen JV (2005) Dormancy release, ABA and pre-harvest sprouting. Curr Opin Plant Biol 8:183–187
Guo H, Ecker JR (2004) The ethylene signaling pathway: new insights. Curr Opin Plant Biol 7:40–49
Hebelstrup KH, Hunt P, Dennis E, Jensen SB, Jensen EO (2006) Hemoglobin is essential for normal growth of *Arabidopsis* organs. Physiol Plant 127:157–166
Hu K-D, Hu L-Y, Li Y-H, Zhang F-Q, Zhang H (2007) Protective roles of nitric oxide on germination and antioxidant metabolism in wheat seeds under copper stress. Plant Growth Regul 53:173–183
Igamberdiev AU, Baron K, Manac'h-Little N, Stoimenova M, Hill RD (2005) The hemoglobin/nitric oxide cycle: involvement in flooding stress and effects on hormone signaling. Ann Bot 96:557–564
Igamberdiev AU, Bykova NV, Shah JK, Hill RD (2010) Anoxic nitric oxide cycling in plants: participating reactions and possible mechanisms. Physiol Plant 138:393–404
Iglesias-Fernándes R, Matilla A (2009) After-ripening alters the gene expression pattern of oxidases involved in the ethylene and gibberellin pathways during early imbibition of *Sisymbrium officinale* L. seeds. J Exp Bot 60:1645–1661
Jacobsen JV, Barrero JM, Hughes T, Julkowska M, Taylor JM, Xu Y, Gubler F (2013) Roles for blue light, jasmonate and nitric oxide in the regulation of dormancy and germination in wheat grain (*Triticum aestivum* L.). Planta 238:121–138
Jain K, Siddam A, Marathi A, Roy U, Falck JR, Balazy M (2008) The mechanism of oleic acid nitration by $^{\bullet}NO_2$. Free Rad Biol Med 45:269–283
Job C, Rajjou L, Lovigny Y, Belghazi M, Job D (2005) Patterns of protein oxidation in *Arabidopsis* seeds and during germination. Plant Physiol 138:790–802
Kalemba E, Pukacka S (2014) Carbonylated proteins accumulated as vitality decreases during long-term storage of beech (*Fagus sylvatica* L.) seeds. Trees 28:503–515
Karrsen CM, Brinkhorst-Van der Swan DLC, Breekland AD, Koornneef M (1983) Induction of seed dormancy during seed development by endogenous abscisic acid: studies on abscisic acid-deficient genotypes of *Arabidopsis thaliana* L. Heynh. Planta 157:158–165
Kępczyński J, Sznigir P (2014) Participation of GA3, ethylene, NO and HCN in germination of *Amaranthus retroflexus* L. seeds with various dormancy level. Acta Physiol Plant 36:1463–1472
Kim W, Lee Y, Park J, Lee N, Choi G (2013) HONSU, a protein phosphatase 2C, regulates seed dormancy by inhibiting ABA signaling in *Arabidopsis*. Plant Cell Physiol 54:555–572
Klessig DF, Durner J, Noad R, Navarre DA, Wendehenne D, Kumar D, Zhou JM, Shah J, Zhang S, Kachroo P, Trifa Y, Pontier D, Lam E, Silva H (2000) Nitric oxide and salicylic acid signaling in plant defense. Proc Natl Acad Sci USA 97:8849–8855
Kopyra M, Gwóźdź EA (2003) Nitric oxide stimulates seed germination and counteracts the inhibitory effect of heavy metals and salinity on root growth of *Lupinus luteus*. Plant Physiol Biochem 41:1011–1017
Krasuska U, Gniazdowska A (2012) Nitric oxide and hydrogen cyanide as regulating factors of enzymatic antioxidant system in germinating apple embryos. Acta Physiol Plant 34:683–692
Krasuska U, Budnicka K, Bogatek R, Gniazdowska A (2012) Polyamines in regulation of seed dormancy and germination. Adv Cell Biol 39:395–414
Krasuska U, Ciacka K, Bogatek R, Gniazdowska A (2014a) Polyamines and nitric oxide link in regulation of dormancy removal and germination of apple (*Malus domestica* Borkh.) embryos. J Plant Growth Reg 33:590–601
Krasuska U, Ciacka K, Dębska K, Bogatek R, Gniazdowska A (2014b) Dormancy alleviation by NO or HCN leading to decline of protein carbonylation levels in apple (*Malus domestica* Borkh.) embryos. J Plant Physiol 171:1132–1141
Kucera B, Cohn MA, Leubner-Matzger G (2005) Plant hormone interactions during seed dormancy release and germination. Seed Sci Res 15:281–307

Kushiro T, Okamoto M, Nakabayashi K, Yamagishi K, Kitamura S, Asami T, Hirai N, Koshiba T, Kamiya Y, Nambara E (2004) The *Arabidopsis* cytochrome P450 CYP707A encodes ABA 8'-hydroxylase: key enzymes in ABA catabolism. EMBO J 23:1647–1656

Lang GA (1987) Dormancy: a new universal terminology. Hort Sci 22:817–820

Lara TS, Lira JMS, Rodrigues AC, Rakocevic M, Alvarenga AA (2014) Potassium nitrate priming affects the activity of nitrate reductase and antioxidant enzymes in tomato germination. J Agric Sci 6:72–80

Lee S, Cheng H, King KE, Wang W, He Y, Hussain A, Lo J, Harberd NP, Peng J (2002) Gibberellin regulates *Arabidopsis* seed germination via RGL2, a GAI/RGA-like gene whose expression is upregulated following imbibition. Genes Dev 16:646–658

Lefebvre V, North H, Frey A, Sotta B, Seo M, Okamoto M, Nambara E, Marion-Poll A (2006) Functional analysis of *Arabidopsis NCED6* and *NCED9* genes indicates that ABA synthesized in the endosperm is involved in the induction of seed dormancy. Plant J 45:309–319

Lewak S (2011) Metabolic control of embryonic dormancy in apple seed: seven decades of research. Acta Physiol Plant 33:1–24

Li X, Jiang H, Liu F, Cai J, Dai T, Cao W, Jiang D (2013) Induction of chilling tolerance in wheat during germination by pre-soaking seed with nitric oxide and gibberellin. Plant Growth Regul 71:31–40

Liu H-Y, Yu X, Cui D-Y, Sun M-H, Sun W-N, Tang Z-C, Kwak S-S, Su W-A (2007) The role of water channel proteins and nitric oxide signaling in rice seed germination. Cell Res 17:638–649

Liu Y, Shi L, Ye N, Liu R, Jia W, Zhang J (2009) Nitric oxide-induced rapid decrease of abscisic acid concentration is required in breaking seed dormancy in *Arabidopsis*. New Phytol 183:1030–1042

Lozano-Juste J, León J (2011) Nitric oxide regulates DELLA content and *PIF* expression to promote photomorphogenesis in *Arabidopsis*. Plant Physiol 156:1410–1423

Matakiadis T, Alboresi A, Jikumaru Y, Tatematsu K, Pichon O, Renou J-P, Kamiya Y, Nambara E, Truong H-N (2009) The *Arabidopsis* abscisic acid catabolic gene *CYP707A2* plays a key role in nitrate control of seed dormancy. Plant Physiol 149:949–960

Matilla AJ (1996) Polyamines and seed germination. Seed Sci Res 6:81–93

Matilla AJ, Matilla-Vazquez MA (2008) Involvement of ethylene in seed physiology. Plant Sci 175:87–97

Matilla AJ, Rodríguez-Gacio MC (2013) Non-symbiotic hemoglobins in the life of seeds. Phytochemistry 87:7–15

Mur LA, Mandon J, Persijn S, Cristescu SM, Moshkov IE, Novikova GV, Hall MA, Harren FJ, Hebelstrup KH, Gupta KJ (2013) Nitric oxide in plants: an assessment of the current state of knowledge. AoB Plants 5:Pls052. doi:10.1093/aobpla/pls052

Nambara E, Okamoto M, Tatematsu K, Yano R, Seo M, Kamiya Y (2010) Abscisic acid and the control of seed dormancy and germination. Seed Sci Res 20:55–67

Nonogaki H, Bassel GW, Bewley JD (2010) Germination—still a mystery. Plant Sci 179:574–581

Parani M, Rudrabhatla S, Myers R, Weirich H, Smith B, Leaman DW, Goldman SL (2004) Microarray analysis of nitric oxide responsive transcripts in *Arabidopsis*. Plant Biotechnol J 2:359–366

Pasqualini S, Reale L, Calderini O, Pagiotti R, Ederli L (2012) Involvement of protein kinases and calcium in the NO-signalling cascade for defence-gene induction in ozonated tobacco plants. J Exp Bot 63:4485–4496

Paul JW, Beauchamp EG, Zhang X (1993) Nitrous and nitric oxide emission during nitrification and denitrification from manure-amended soil in the laboratory. Can J Soil Sci 73:539–553

Penfield S, Rylott EL, Gilday AD, Graham S, Larson TR, Graham IA (2004) Reserve mobilization in the *Arabidopsis* endosperm fuels hypocotyls elongation in the dark, is independent of abscisic acid, and requires PHOSPHOENOLOPYRUVATE CARBOXYKINASE1. Plant Cell 16:2705–2718

Piterková J, Luhová L, Hofman J, Turečková V, Novák O, Petřivalský M, Fellner M (2012) Nitric oxide is involved in light-specific responses of tomato during germination under normal and osmotic stress conditions. Ann Bot 110:767–776

Pyriochou A, Papapetropoulos A (2005) Soluble guanyl cyclase: more secrets revealed. Cell Signal 17:407–413

Radin JW (1976) Control of enzyme activities in cotton cotyledons during maturation and germination. Plant Physiol 57:902–905

Ranjan R, Lewak S (1992) Jasmonic acid promotes germination and lipase activity in non-stratified apple embryos. Physiol Plant 86:335–339

Rockel P, Strube F, Rockel A, Wildt J, Kaiser WM (2002) Regulation of nitric oxide (NO) production by plant nitrate reductase *in vivo* and *in vitro*. J Exp Bot 53:103–110

Shu K, Zhang H, Wang S, Chen M, Wu Y, Tang S, Liu C, Feng Y, Cao X, Xie Q (2013) ABI4 regulates primary seed dormancy by regulating the biogenesis of abscisic acid and gibberellins in *Arabidopsis*. PLoS Genet 9(6):e1003577

Simontacchi M, Jasid S, Puntarulo S (2004) Nitric oxide generation during early germination of sorghum seeds. Plant Sci 167:839–847

Šírová J, Sedlářová M, Piterková J, Luhová L, Petřivalský M (2011) The role of nitric oxide in the germination of plant seeds and pollen. Plant Sci 181:560–572

Stamler JS, Singel DJ, Loscalzo J (1992) Biochemistry of nitric oxide and its redox-activated forms. Science 258:1898–1902

Stelmach BA, Muller A, Henning P, Laudert D, Andert L, Weiler EW (1998) Quantitation of the octadecanoid 12-oxo-phytodienoic acid, a signaling compound in plant mechanotransduction. Phytochemistry 47:539–546

Stöhr C, Strube F, Marx G, Ullrich WR, Rockel P (2001) A plasma membrane-bound enzyme of tobacco roots catalyses the formation of nitric oxide from nitrite. Planta 212:835–841

Sullivan JA, Deng XW (2003) From seed to seed: the role of photoreceptors in *Arabidopsis* development. Dev Biol 260:289–297

Szmidt-Jaworska A, Jaworski K, Kopcewicz J (2008) Effect of light on soluble guanyl cyclase activity in *Pharbitis nil* seedlings. J Photochem Photobiol B Biol 93:9–15

Szmidt-Jaworska A, Jaworski K, Pawelek A, Kopcewicz J (2009) Molecular cloning and characterization of guanyl cyclase, *PnGC-1*, involved in light signaling in *Pharbitis nil*. J Plant Growth Regul 28:367–380

Taylorson RB, Hendricks SB (1977) Dormancy in seeds. Annu Rev Plant Physiol 28:331–354

Teng Y, Xu W, Ma M (2010) cGMP is required for seed germination in *Arabidopsis thaliana*. J Plant Physiol 167:885–889

Terskikh VV, Zeng Y, Feurtado A, Giblin M, Abrams SR, Kermode AR (2008) Deterioration of western redcedar (*Thuja plicata* Donn ex D. Don) seeds: protein oxidation and in vivo NMR monitoring of storage oils. J Exp Bot 59:765–777

Tun NN, Santa-Catarina C, Begum T, Silveira V, Handro W, Floh EIS, Scherer GFE (2006) Polyamines induce rapid biosynthesis of nitric oxide (NO) in *Arabidopsis thaliana* seedlings. Plant Cell Physiol 47:346–354

Umezawa T, Nakashima K, Miyakawa T, Kuromori T, Tanokura M, Shinozaki K, Yamaguchi-Shinozaki K (2010) Molecular basis of the core regulatory network in ABA responses: sensing, signaling and transport. Plant Cell Physiol 51:1821–1839

Voegele A, Linkies A, Müller K, Leubner-Metzger G (2011) Members of the gibberellin receptor gene family GID1 (GIBBERELLIN INSENSITIVE DWARF1) play distinct roles during *Lepidium sativum* and *Arabidopsis thaliana* seed germination. J Exp Bot 62:3151–5147

Wang P, Du Y, Li Y, Ren D, Song C-P (2010) Hydrogen peroxide–mediated activation of MAP kinase 6 modulates nitric oxide biosynthesis and signal transduction in *Arabidopsis*. Plant Cell 22:2981–2998

Wasternack C, Hause B (2013) Jasmonates: biosynthesis, perception, signal transduction and action in plant stress response, growth and development. An update to the 2007 review in *Annals of Botany*. Ann Bot 111:1021–1058

Wimalasekera R, Tebartz F, Scherer GFE (2011) Polyamines, polyamine oxidases and nitric oxide in development, abiotic and biotic stresses. Plant Sci 181:593–603

Yamasaki H (2000) Nitrite-dependent nitric oxide production pathway: implications for involvement of active nitrogen species in photoinhibition *in vivo*. Philos Trans R Soc Lond B Biol Sci 355:1477–1488

Yan Y, Christensen S, Tom Isakeit T, Engelberth J, Meeley R, Hayward A, Emery RJN, Kolomiets MV (2012) Disruption of OPR7 and OPR8 reveals the versatile functions of jasmonic acid in maize development and defense. Plant Cell 24:1420–1436

Zhang H, Shen W-B, Zhang W, Xu L-L (2005) A rapid response of β-amylase to nitric oxide but not gibberellins in wheat seeds during the early stage of germination. Planta 220:708–716

Zheng C, Jiang D, Liu F, Dai T, Liu W, Jing Q, Cao W (2009) Exogenous nitric oxide improves seed germination in wheat against mitochondrial oxidative damage induced by high salinity. Environ Exp Bot 67:222–227

Dissecting Nitric Oxide Signalling in Nucleus: Role of S-Nitrosylation in Regulating Nuclear Proteins

Ankita Sehrawat and Renu Deswal

1 Introduction

Nitric oxide (NO) is a lipophilic signalling molecule which regulates a myriad of cellular responses in plants (Kovacs and Lindermayr 2013). It is one of the smallest diatomic molecules with a high diffusivity (4.8×10^5 cm^2 s^{-1} in H_2O, Popova and Tuan 2010). It can easily migrate in the hydrophilic regions (cytoplasm) of the cell and can also freely diffuse across the lipid phase of membranes, due to its high lipid solubility. NO has a very short half-life, which makes it highly reactive. This allows it to interact directly with metal complexes and indirectly with wide range of biological targets (DNA, proteins, and lipids) in a concentration- and redox-dependent manner. NO, RNS [reactive nitrogen species, peroxynitrite ($ONOO^-$), dinitrogen trioxide (N_2O_3), *S*-nitrosoglutathione (GSNO), nitrogen dioxide ($^{\bullet}NO_2$), nitrosyl cation (NO^+), *S*-nitrosothiols (RSNOs)], and ROS [reactive oxygen species, hydroxyl (OH^-), peroxide and superoxide radicals ($O_2^{-\bullet}$)] contributes to the stress signalling. In stress, there is a change in the oxidative state of the cell which leads to oxidative stress, in which ROS are overproduced and causes cellular damage. At low concentration, NO itself acts as antioxidant and reacts with harmful $O_2^{\bullet-}$ to produce $ONOO^-$. It also scavenges ROS by increasing the activities of antioxidant enzymes, leading to accumulation of antioxidants, which in turn, decrease ROS levels (Groß et al. 2013). Additionally, NO prevents ROS-induced chlorophyll degradation (Beligni and Lamattina 1999). High level of NO enhances $O_2^{\bullet-}$ production in mitochondria by inhibiting electron flow through cytochrome c oxidase (Millar and Day 1996), and therefore causes higher ROS production.

A. Sehrawat • R. Deswal (✉)
Molecular Plant Physiology and Proteomics Laboratory, Department of Botany, University of Delhi, Delhi 110007, India
e-mail: rdeswal@botany.du.ac.in

K.J. Gupta, A.U. Igamberdiev (eds.), *Reactive Oxygen and Nitrogen Species Signaling and Communication in Plants*, Signaling and Communication in Plants 23,
DOI 10.1007/978-3-319-10079-1_12

Since the discovery of NO as a key signalling molecule in providing disease resistance in plants (Delledonne et al. 1998), researchers are trying to examine the mechanisms underlying the NO production in plants. Eight major pathways (including non-enzymatic nitrite reduction pathway) are known for NO production, belonging to two broad categories, namely oxidative pathway and reductive pathways (Mur et al. 2013a; Gupta et al. 2011; Bethke et al. 2004). Various biosynthetic pathways known to be involved in the NO production in different cellular compartments are diagrammatically represented in Fig. 1. Involvement of L-arginine- [by Nitric-oxide synthase (NOS)-like enzyme and polyamine oxidation], hydroxylamine-, and nitrite-dependent [by nitrate reductase (NR)] NO production was observed in the cytosol, chloroplast, and peroxisomes. In contrast, in mitochondria, only nitrite reduction-dependent NO production was observed (Modolo et al. 2005; Gupta et al. 2010). Similarly, in the apoplast, non-enzymatic nitrite reduction (Bethke et al. 2004) and in the plasma membrane, nitrite:NO reductase (Ni–NOR) mediated nitrite reduction (Stöhr and Stremlau 2006), indicated the involvement of specific pathways of NO production. However, metabolic pathways involved in the NO production in vacuoles, Golgi apparatus, endoplasmic reticulum, and nucleus need investigation. Nevertheless, identification of NR in the dehydration responsive nuclear proteome of *Cicer arietinum* (Subba et al. 2013), indicated possibility of NR-mediated NO production in the nucleus. Additionally, sodium nitroprusside (SNP, a NO donor) treatment caused chromatin condensation and altered the shape of the nuclei of the *Sechium edule* nucellus, from spherical to half-moon shaped (Lombardi et al. 2010), therefore suggesting probable role of NO signalling in the nucleus.

NO controls physiological processes by either activating/inhibiting gene transcription (Sha and Marshall 2012) or directly altering protein functions by post-translational modifications (PTMs). NO binds to cysteine or tyrosine residue of the protein by S-nitrosylation (Romero-Puertas et al. 2013) and tyrosine nitration respectively (Corpas et al. 2013) and thereby regulates the activity (Begara-Morales et al. 2014) and trans-localization (Tada et al. 2008) of the proteins. Additionally, NO interact with transition metals of metalloproteins to form metal–nitrosyl complexes (Astier and Lindermayr 2012). S-Nitrosylation is the most studied and only NO-based PTM which has been analysed in isolated sub-organelle. Mitochondria (Camejo et al. 2012) and peroxisomes (Ortega-Galisteo et al. 2012) showed majority of S-nitrosylated targets associated with energy metabolism and photorespiration. Similarly, apoplastic S-nitrosylation analysis showed more targets with stress, redox and metabolism related functions (Sehrawat and Deswal 2014). Nucleus, "the regulatory hub" or "brain of the cell" modulates cell signalling by controlling gene expression or by regulating activity of proteins (Narula et al. 2013). In the past decade many studies have focused on the identification of stress (cold and drought) responsive nuclear proteins from *A. thaliana*, *Oryza sativa*, *Cicer arietinum*, *Xerophyta viscosa*, *Medicago truncatula*, and *Zea mays* (extensively reviewed in Erhardt et al. 2010; Casati 2012). These studies without doubt have shown the modulation of signal transduction, translation, DNA replication, protein folding, and metabolism-related nuclear proteins in stress. Mechanism of NO signalling of

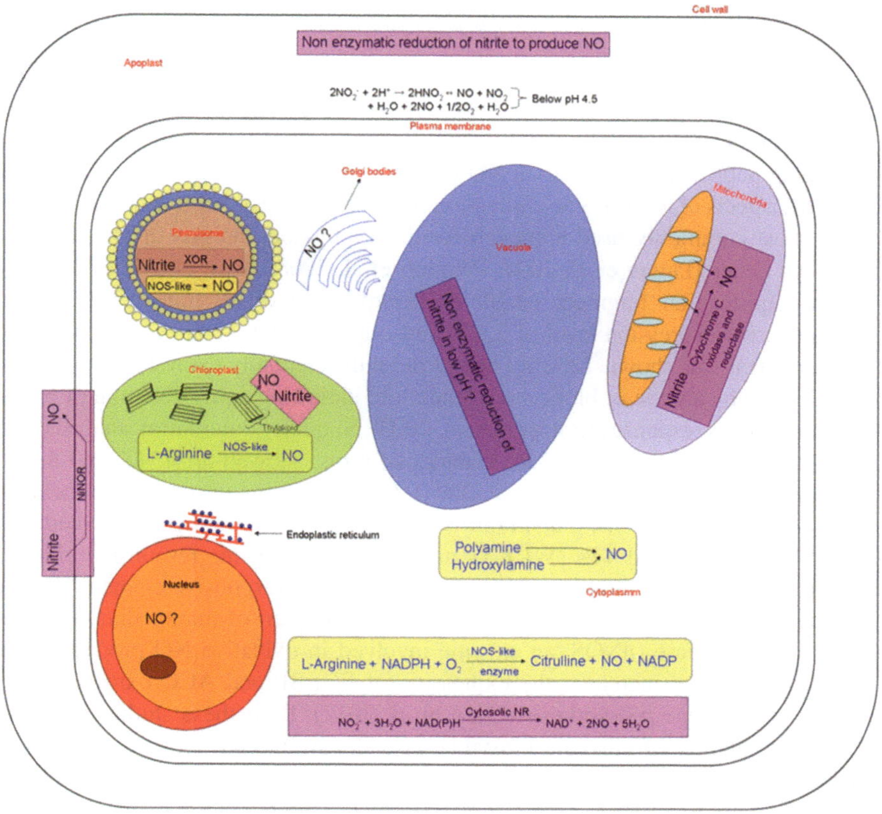

NO, nitric oxide; NiNOR, nitrite:NO reductase; NOS-like, nitric oxide synthase-like; XOR, xanthine oxidoreductase; Nitrite, NO_2^-; HNO_2, nitrous acid; NO_2, nitrogen dioxide

Fig. 1 Pictorial representation of a plant cell showing routes of nitric oxide production in subcellular compartments

these regulatory and stress responsive proteins is not known. Nevertheless, S-nitrosylation analysis has identified various nuclear proteins and transcription factors (TFs, as shown in detail in the Sect. 3). *S*-Nitrosoproteome analysis in the nucleus would give information about NO-modulated targets involved in controlling gene expression. Tyrosine nitration analysed using confocal laser scanning microscopy and anti-nitrotyrosine antibody showed the abundance of tyrosine nitrated proteins around the nuclei in the actively dividing cells of the root primordium in the *Helianthus annuus* seedlings, therefore giving hints of tyrosine nitration in the nucleus (Yadav et al. 2013).

NO-mediated regulation of genes and proteins (via S-nitrosylation) will be summarized in more detail in the following sections.

2 Genes Regulated by Nitric Oxide

Pubmed literature survey (As of February 2014) showed 1,435 published manuscripts with NO signalling in either title or abstract. This explains the amount of research focusing on deciphering the role(s) of NO. However, in spite of the huge amount of data being published every year on NO signalling, not much information is available about the effects of NO on gene expression (Grun et al. 2006). Microarray technology is the most commonly used approach for the expression analysis of NO induced/ repressed genes (Parani et al. 2004; Palmieri et al. 2008; Badri et al. 2008). Using cDNA-amplification fragment length polymorphism (AFLP) transcript profiling, changes in the expression level of 2,500 transcripts on SNP (1 mM, 10 min, 1, and 3 h) treatment was shown. Seventy one of these differentially expressed genes were also affected by abiotic or biotic stress, therefore indicating a probable role of NO in regulating stress responses (Polverari et al. 2003). Similarly, microarray analysis with SNP (0.1 and 1 mM) treatment to Arabidopsis roots, showed up- (342) and down-regulation (80) of genes (Parani et al. 2004). Genes showing up-regulation included disease-resistance proteins, glutathione *S*-transferase (GST), mitogen-activated protein (MAP) kinase and TFs [WRKYs and EREBPs (ethylene responsive element binding proteins], zinc finger proteins, and dehydration responsive element binding proteins (DREB1 and DREB2). Recently, overexpression of mammalian neuronal nitric oxide synthase (nNOS, an enzyme involved in oxidative NO production) in transgenic tobacco, increased the expression of EREBP and ACC oxidase (a key enzyme of ethylene biosynthesis, Chun et al. 2012). Both these studies indicated positive regulation of ethylene biosynthesis pathway by NO. However, this is in contrast to the earlier observation that NO inhibits ethylene biosynthesis and acts antagonistically to ethylene during fruit ripening (Zaharah and Singh 2011).

NOR3 {(E)-ethyl-2-[(E)-hydroxyimino]-5-nitro-3-hexene-amide, a NO donor} and gaseous NO treatment to Arabidopsis cell suspension cultures (7 days old) and plants (4–5 weeks old) showed enrichment of GBOX, OCSE (ocs element-like sequences) and L1BX (L1 box) elements in the genes up-regulated in both these tissues (Palmieri et al. 2008). OCSE elements are essential for the expression of GST6 and pathogenesis-related genes (PR1), indicating positive regulation of these by NO. Down-regulated gene in cell cultures showed increased occurrence of TBPF (TATA-binding protein factors) and MIIG (a MYB TF) elements. Interestingly, MYB2 and MYB 30 (members of MYR family) are negatively regulated by S-nitrosylation (explained in the Sect. 3). 12-oxophytodienoate reductases (OPR1, OPR2, and OPR3) and two lipoxygenases (LOX3 and a putative lipoxygenase protein) genes, involved in jasmonic acid biosynthesis pathway were also up-regulated after NO treatment. Similarly, NO increased the expression of LOX3, OPR1, OPR2, and OPR3, while it decreased the expression of allene oxide cyclase (AOC), therefore indicating the role of NO in regulating jasmonic acid pathway (Mur et al. 2013b).

These studies provided information about the role of NO in regulating genes associated with diverse functions such as defense-related (PR1), redox regulation (GST), ethylene (ACC oxidase) and jasmonic acid (OPR1, OPR2, OPR3, LOX3 and AOC) biosynthesis pathway. However, most of these studies used SNP (a NO donor) for experiments, which might not reflect the actual physiological state in a cell during stress. Therefore, analysis of NO regulated genes in stress/in vivo conditions is important. One of the mechanism by which NO regulates these genes and TFs is S-nitrosylation (Grun et al. 2006). S-Nitrosylation involves the reaction of a cysteine thiol in a protein with NO in the presence of an electron acceptor to form an *S*-nitrosothiol (SNOs, Kovacs and Lindermayr 2013). S-Nitrosylation can either affect the binding of TFs to the DNA or could control the localization of the nuclear proteins and thereby regulating gene transcription or nuclear trafficking (Mengel et al. 2013).

3 S-Nitrosylation of the Nuclear Proteins

S-Nitrosylation has not been reported in the nucleus until recently and only few targets are now reported to be S-nitrosylated (Table 1). To know the putative S-nitrosylated proteins in the nucleus, an in silico approach was followed, wherein the presence of S-nitrosylation site(s) was predicted in the cold-induced nuclear proteins of *A. thaliana* (Bae et al. 2003) using two softwares (dbSNO and GPS-SNO 1.0) as shown in Table 2. The dbSNO database is a freely accessible database

Table 1 Overview of S-nitrosylated targets in nucleus

Protein	System	Effect of S-nitrosylation	References
AtMYB2	*Arabidopsis thaliana*	SNP and GSNO treatment inhibited M2D (a fully active minimal DNA-binding domain of AtMYB2 spanning residues) DNA binding	Serpa et al. (2007)
AtMYB30	*Arabidopsis thaliana*	SNP treatment inhibited DNA-binding domain (DBD) of AtMYB30 (a fully active minimal DNA-binding domain of AtMYB30 spanning residues)	Tavares et al. (2014)
NPR1	*Arabidopsis thaliana*	GSNO promoted oligomerization of NPR1 through S-nitrosylation at Cys 156 and blocks its nuclear translocation	Lindermayr et al. (2010), Tada et al. (2008)
TGA1	*Arabidopsis thaliana*	GSNO enhanced DNA binding of TGA1 to the activation sequence-1 (as-1) element of promoter region of defense proteins	Lindermayr et al. (2010)
GAPDH	*Nicotiana tabacum*	S-nitrosylation mediates nuclear translocation of GAPDH	Wawer et al. (2010)
Probable WRKY transcription factor	*Arabidopsis thaliana*	NA	Maldonado-Alconada et al. (2010)
Zinc finger protein	*Arabidopsis thaliana*	NA	Maldonado-Alconada et al. (2010)

NA—Effect of S-nitrosylation was not analysed

Table 2 Prediction of S-nitrosylation sites in the cold responsive nuclear protein from *Arabidopsis thaliana* using SNOSite (http://csb.cse.yzu.edu.tw/SNOSite/index.html) and GPS-SNO (http://sno.biocuckoo.org)

Protein name	Acc. no.	Cysteine location	S-nitrosylation sites	The matched MDD-clustered Motifs	Functional categorization
HSP90-like protein	15233740	670	–	–	Protein folding
HSP70-like protein	15233779	170 380*	–	–	Protein folding
dnaK-type molecular chaperone hsc70.1	15241849	20 326*** 366 486 609*** 319*** 273**	IGIDLGTTYSCVGVWQHDRVE FRKCMEPVEKCLRDAKMDKST LQDFFNGKELCKSINPDEAVA APRGVPQITVCFDIDANGILN EDKMKELESICNPIIAKMYQG	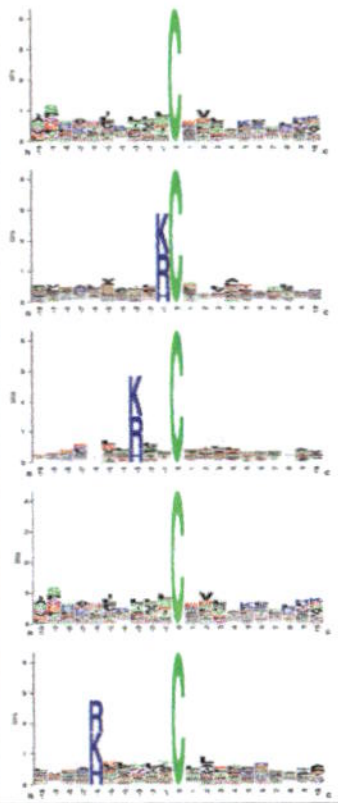	Protein folding
Myrosinase binding like protein	12230212	339	–	–	Structure
Myrosinase-binding protein homolog	2997767	339	–	–	Structure
(AY088397) Putative lectin	21594017	–	–	–	Structure
Putative lectin	15228216	–	–	–	Structure

Myrosinase-associated protein	15231805	215 229 293 294 314	FVVQLLAPLGCLPIVRQEYKT VRQEYKTGNECYELLNDLAKQ NYRFFVTNTSCCGVGTHNAYG YRFFVTNTSCCGVGTHNAYGC CGKGNVHSKLCEYQRSYFFFD	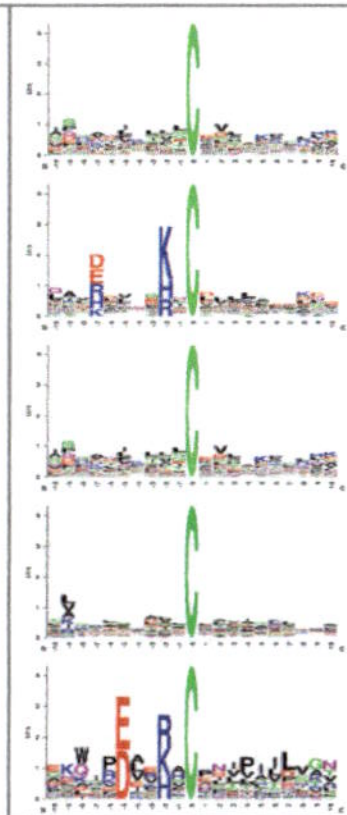	Structure
Putative Helicase C contained in DEAD/ DEAH box	15215678	94* 165 212 266 281	GKTSMIALSVCQVVDTSSREV HVVSGTPGRVCDMIKRRSLRT YRYLPPDLQVCLVSATLPHEI EKEEWKFDTLCDLYDTLTITQ TLTITQAVIFCNTKRKVDYLS	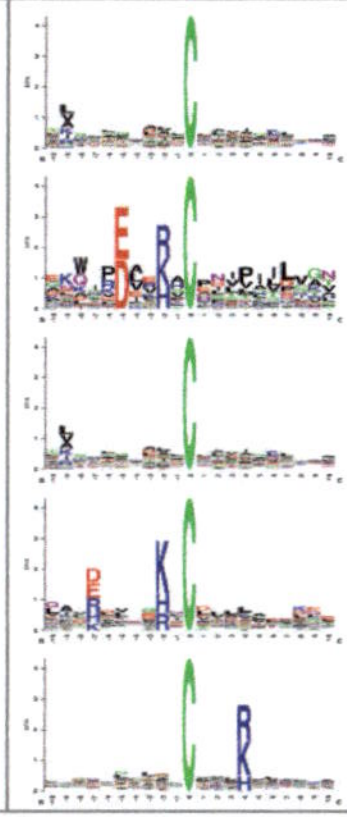	RNA metabolism

(continued)

Table 2 (continued)

Protein name	Acc. no.	Cysteine location	S-nitrosylation sites	The matched MDD-clustered Motifs	Functional categorization
RNA-binding protein homolog F18A5.250	7488314	14	FSIPPKLLILCLLKPNFDIHL LFDQQSTFLSCPLRLLPSSSS SSSSNRNSLVCIVKSFSSSAT PSDWKTAKAYCKSGDTFEGEV	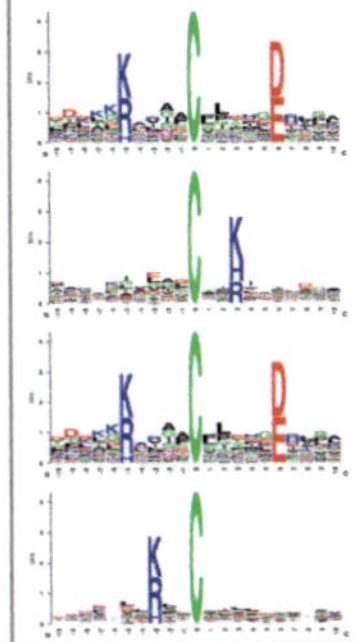	RNA metabolism
Expressed protein (serine-rich region profile, S1 domain)	18403854	29*** 49*** 104*	–	–	RNA metabolism
U2 small nuclear ribonucleoprotein A’	15218274	–	–	–	RNA metabolism
60S acidic ribosomal protein P2	15226231	–	–	–	Translation
60S acidic ribosomal protein P2	21593095	–	–	–	Translation
(AF085279) hypothetical EIF-2-Alpha	4588003	102 226 251** 300 10**	RRVSEEDIQTCEERYNKSKLV MKIRADIELKCFQFDGVVHIK KNAEAAGNEDCPVKIKLVAPP IEILNKAIAACTETIETHKGK	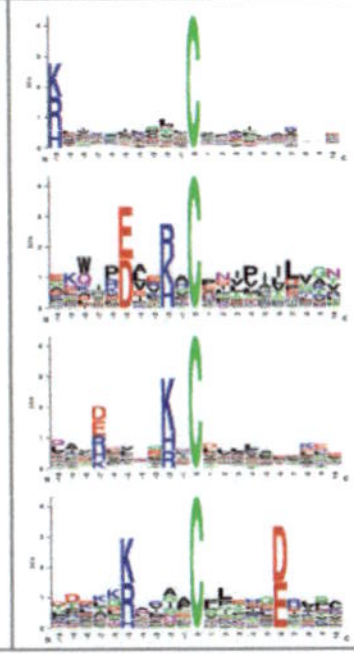	Translation

Putative 60S acidic ribosomal protein P0	15232603	18	EKKIAYDTKLCQLIDEYTQIL	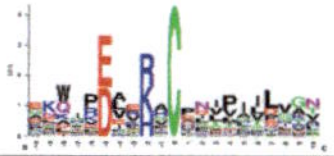	Translation
Ribosomal protein L11 family protein	15228098	137* 142	SGTVREILGTCVSVGCTVDGK EILGTCVSVGCTVDGKDPKDI	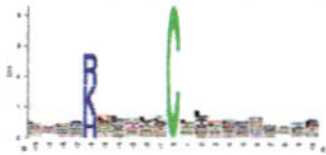 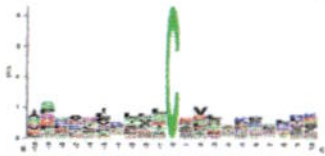	Translation
60S ribosomal protein-like protein	15229631	–	–	–	Translation
60S ribosomal protein L12	15228098	137* 142	SGTVREILGTCVSVGCTVDGK EILGTCVSVGCTVDGKDPKDI	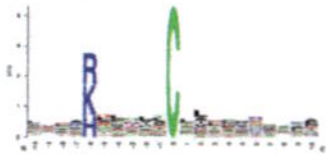 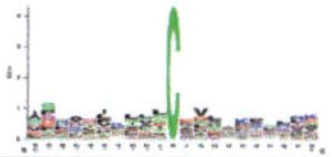	Translation
DNA-damage-repair/toleration protein DRT102	18397039	72 73*** 94 109 186 189** 190	SSSSEVRGLVCCGTGVGVAMF SSSEVRGLVCCGTGVGVAMFA NKFPGVYAATCLSVEDAVNAR DAVNARSISNCNVLAFSGIKT TKKIDETTASCVICCLAKNRE IDETTASCVICCLAKNREFTP DETTASCVICCLAKNREFTPV	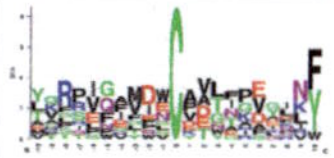 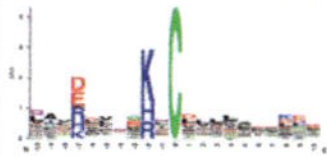 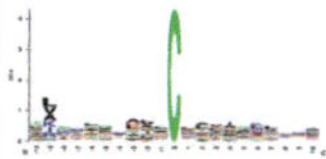 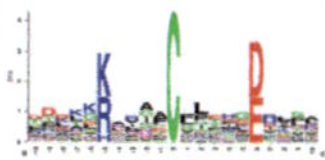 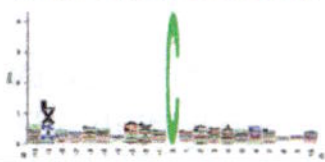	DNA replication/repair/ modification

(continued)

Table 2 (continued)

Protein name	Acc. no.	Cysteine location	S-nitrosylation sites	The matched MDD-clustered Motifs	Functional categorization
				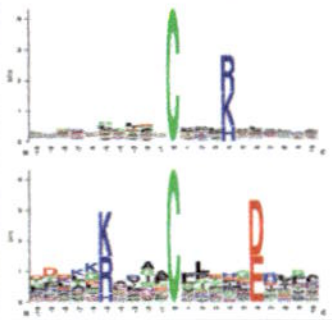	
Glyceraldehyde-3-phosphate dehydrogenase	15222848	156*** 160***	SDLDIVSNASCTTNCLAPLAK IVSNASCTTNCLAPLAKVIND	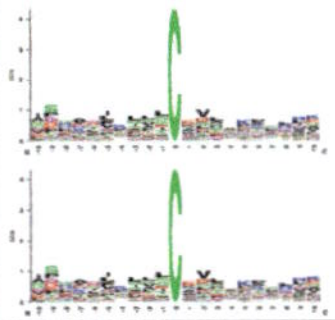	DNA replication repair/ modification
(BT000613) T5I7-ATPase families	23308421	175 224 282 316 451***	GGKGQGKSFQCELVMAKMGIN ADLIKKGKMCCLLINDLDAGA KEENARVPIICTGNDFSTLYA APTREDRIGVCKGIFRTDKIK	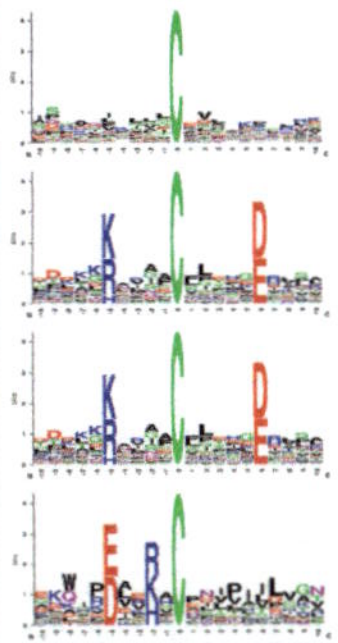	Signaling and gene regulation
Phosphoglycerate kinase	15230595	173 292	–	–	Signaling and gene regulation

(AY088437) lipase/hydrolase, putative	21594055	32 160 225*** 240*** 299 300 311	FKVKAEPQVPCYFIFGDSLVD EYTAADYLKRCIYSVGMGSND FALVGIGAIGCSPNALAQGSE LAQGSEDGTTCVERINSANRI AYGFTNTNTACCGIGRNGGQL YGFTNTNTACCGIGRNGGQLT GIGRNGGQLTCLPGEPPCLNR	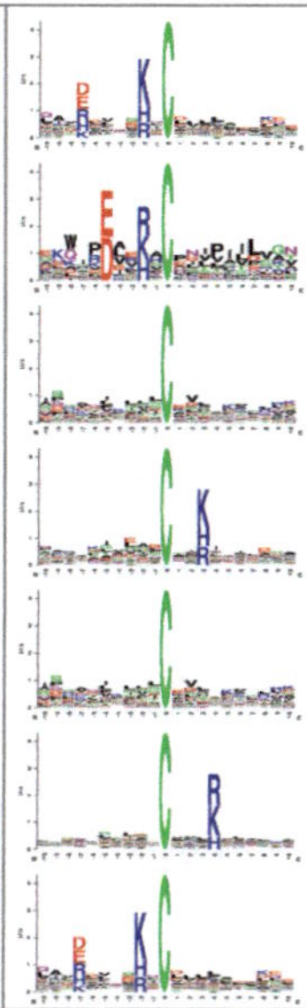	Signaling and gene regulation
(NM_127384) syntaxin SYP112	15224143	131*** 196	ANRTSFKEGSCVDRTRTSITN MEKMISGSGSCSDLVKTFEVK	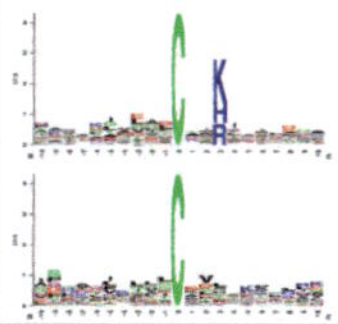	Signaling and gene regulation

(continued)

Table 2 (continued)

Protein name	Acc. no.	Cysteine location	S-nitrosylation sites	The matched MDD-clustered Motifs	Functional categorization
Nuclear receptor binding factor-like protein	11358586	33 76 158 201 257* 318	SIRRGETPTLCIKSFSTIMSP PPVEVKENDVCVKMIAAPINP ESVWHKIDKECPMEYAATITV QNGATSIVGQCVIQLARLRGI NLPEPALGFNCVGGNAASLVL SWLSMGKVKECREMIDYLLGL	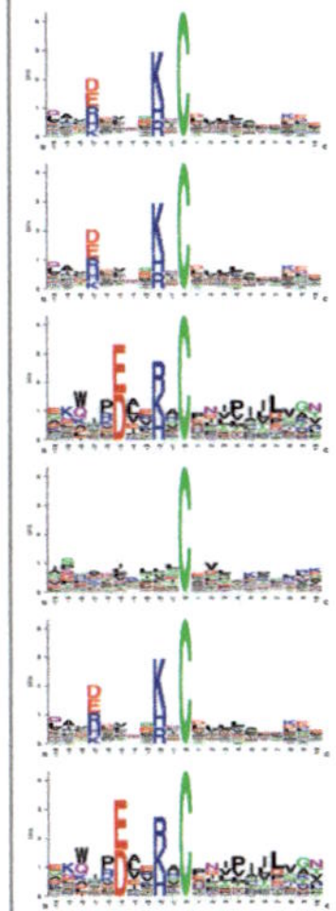	Signaling and gene regulation
MYB transcription factor AtMYB2	14970972	61 129** 161* 243	SSGVKRTGKSCRLRWLNYLRP RVQKQAKHLRCDVNSNLFKET RINAQSLPTTCEQVESMITDP TGFGEFNDWGCVGGDNMWTDE	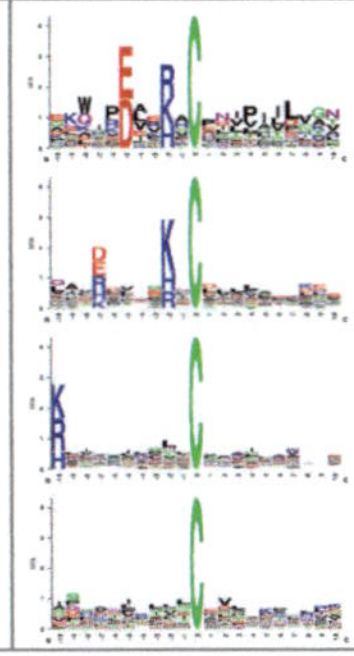	Signaling and gene regulation

Serine acetyltransferase	15222799	21** 22 134 246 297 4*** 8***	GNTQDDDSRFCCIKNFFRPGF NTQDDDSRFCCIKNFFRPGFS LIAVKERDPACISYVHCFLGF GDGVLIGAGSCILGNITIGEG ENPRKHDKIPCLTMDQTSYLT	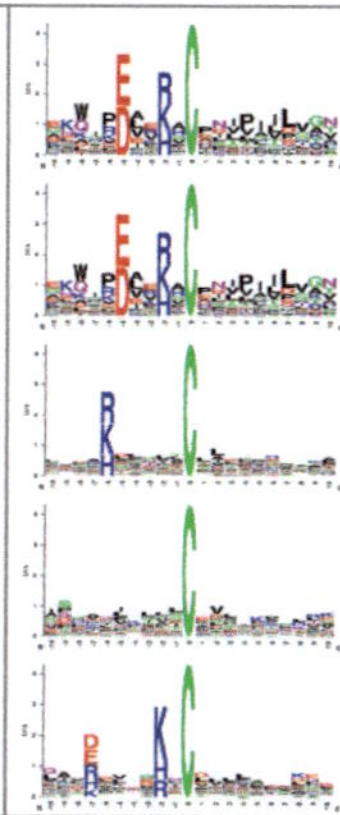	Signaling and gene regulation
Hypothetical calcium-binding protein	11135591	**55** **63** **117***** **172***** **219** **6*****	VSHRNHRGNICGVDFIWLCNP NICGVDFIWLCNPSDSRKPLR HSDSLTYSDFCLALQKVNLTG NMTMVNQPGNCKESVMESKKE PENYNISDHACLTVQFSPVKM	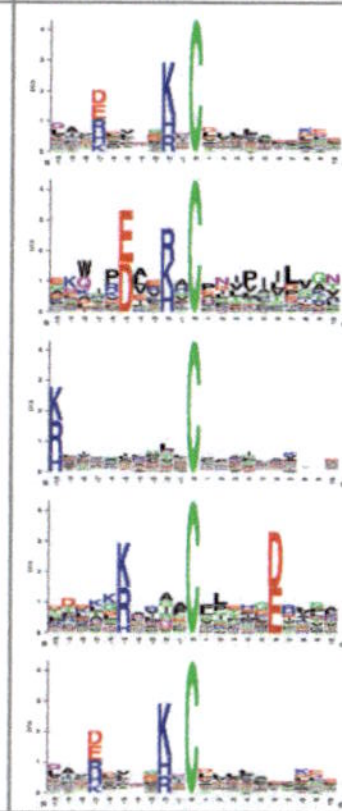	Signaling and gene regulation

(continued)

Table 2 (continued)

Protein name	Acc. no.	Cysteine location	S-nitrosylation sites	The matched MDD-clustered Motifs	Functional categorization
TATA-binding protein-associated phosphoprotein Dr1	2129748	46 47	ARDAQDLLIECCVEFINLVSS RDAQDLLIECCVEFINLVSSE	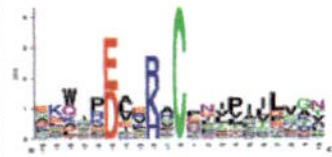 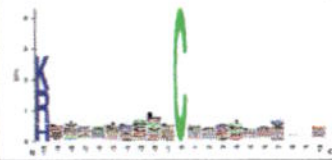	Signaling and gene regulation
Putative protein	7267238	50 53 68 96 99 121	VNFDKEGVYGCVGCNAPLYKS DKEGVYGCVGCNAPLYKSTTK YKSTTKFNAGCGWPAFFEGIP DPDGRRIEINCATCGGHLGHV GRRIEINCATCGGHLGHVFKG GFATPTDERHCVNSVSLKFTP	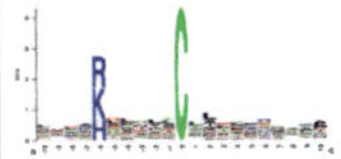 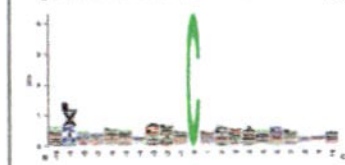 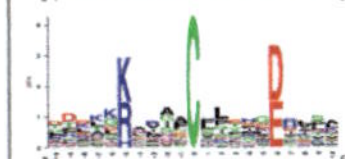 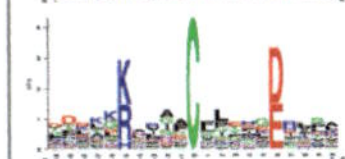 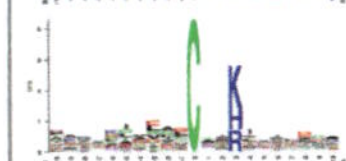 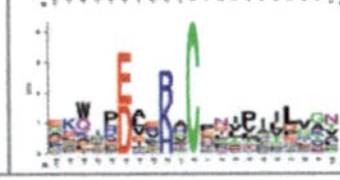	Signaling and gene regulation

Transcription factor TGA4	15238046	168** 256 262* 300	HWVEEQNRQICELRTVLHGQV PLTDQQLLDVCNLRQSCQQAE LLDVCNLRQSCQQAEDALSQG GEGSYIPQMTCAMERLEALVS	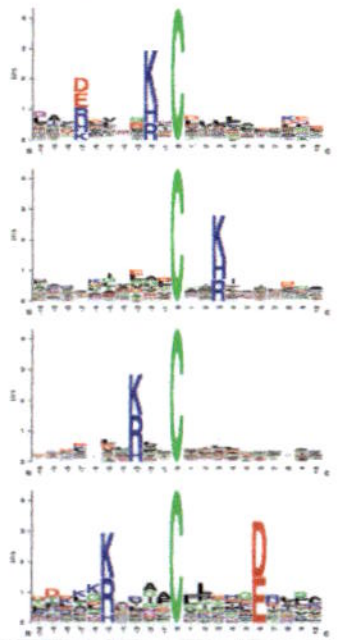	Signaling and gene regulation
CHP-rich zinc finger protein, putative	15225540	21 32 33 36 48 51 59 91 113 143 146 158 161 169 198 201 214 217 226 283	CKLPVVPLFWCNNKEPNGADF NNKEPNGADFCCSACKLTNLG NKEPNGADFCCSACKLTNLGT PNGADFCCSACKLTNLGTAYY LTNLGTAYYFCATCRKKFHKE LGTAYYFCATCRKKFHKECVE ATCRKKFHKECVESPLEIKHP SPFALVNCICCDISFQDIIYH SNHKLSMHPVCAMKPVPIFID TFSPREAFLPCNVCGLIKESI PREAFLPCNVCGLIKESILTY LIKESILTYVCVPCDFVVHQD ESILTYVCVPCDFVVHQDCLY VPCDFVVHQDCLYFTHVIKIS TSSLPYGKWSCGVCRREVDNH LPYGKWSCGVCRREVDNHYGA EVDNHYGAYSCIKCGDYFVHT NHYGAYSCIKCGDYFVHTRCA KCGDYFVHTRCALRKDVWNGE INRVYDENKFCQACILPIYEG	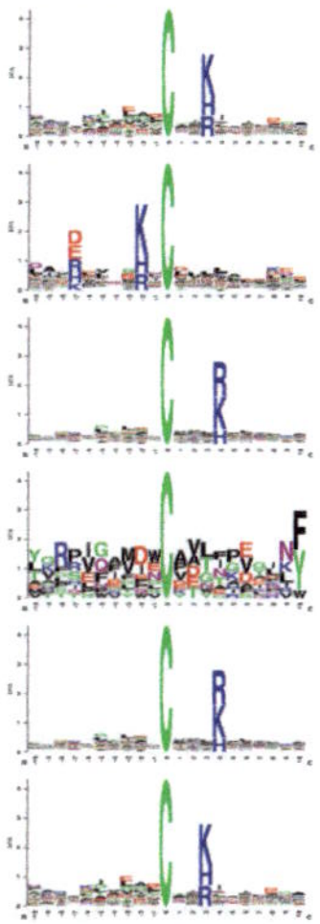	Signaling and gene regulation

(continued)

Table 2 (continued)

Protein name	Acc. no.	Cysteine location	S-nitrosylation sites	The matched MDD-clustered Motifs	Functional categorization
		286	VYDENKFCQACILPIYEGTYY	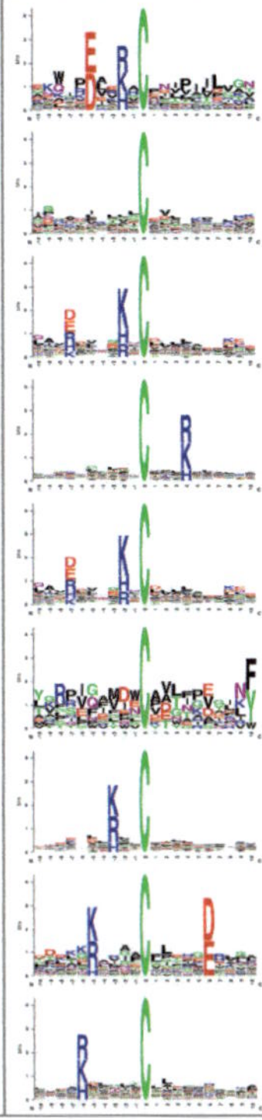	
		298	LPIYEGTYYSCLDKCEFILHE		
		302	EGTYYSCLDKCEFILHEACVN		
		310	DKCEFILHEACVNAPRKKHHP		
		347	HYRRRGNYFDCGACKRASCGF		
		350	FDCGACKRASCGFVYDSQDFS		
		370	DSQDFSLDLRCASISEPFQYQ		
		400	LEPDEEKAAICQICQEDGDGN		
		403	DEEKAAICQICQEDGDGNNYI		
		429	MECKYIICFRCATLPYKAKYK		
		460	GEEGSDQLDWCEVCEKKLIYS		
		478	IYS RKGGFYACDDCCTTLHVD		
		481	RKGGFYACDDCCTTLHVDCLL		
		482	KGGFYACDDCCTTLHVDCLLG		
		463***			

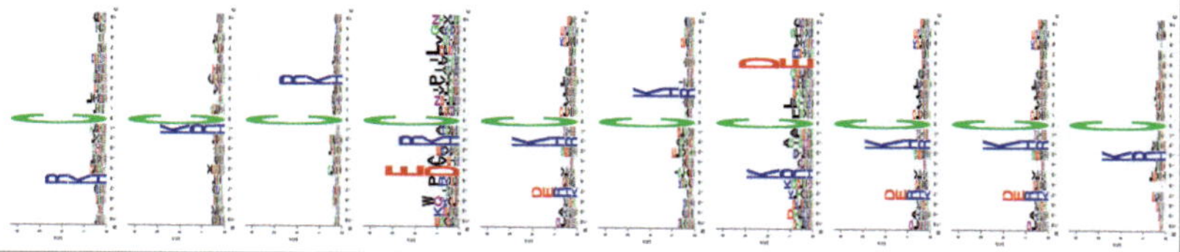

(continued)

Table 2 (continued)

Protein name	Acc. no.	Cysteine location	S-nitrosylation sites	The matched MDD-clustered Motifs	Functional categorization
				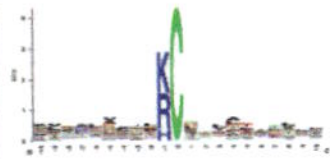	
				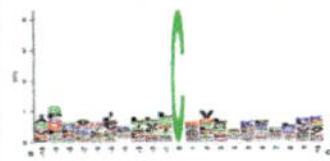	
				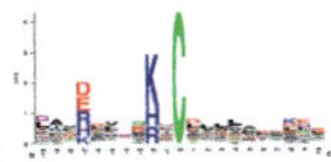	
				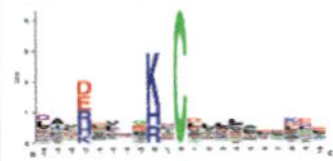	
				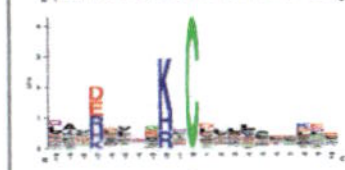	
				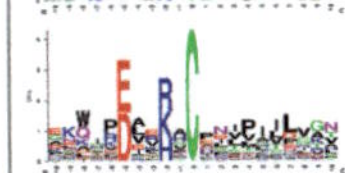	
				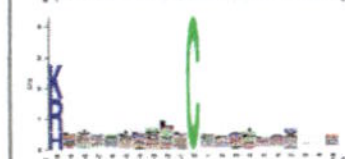	
				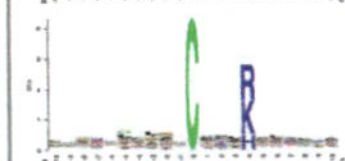	
				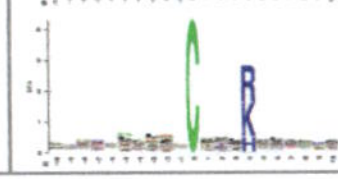	

Putative bHLH transcription factor	20127049	228** 243**	IQEKLSTEEVCVVPREMVQVL EMVQVLKAEECILTNPKISRD	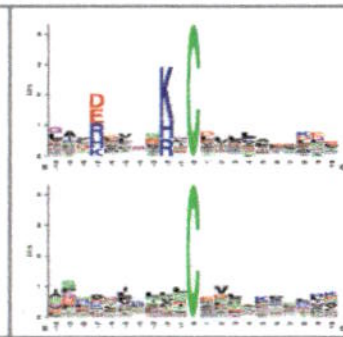	Signaling and gene regulation
Probable transcription factor MYB34	11358685	–	–	–	Signaling and gene regulation
Putative bHLH transcription factor similar to the MYC	22329272	146 172 278*** 288 302 489 567 572*** 596*	ALSNGEPIWLCNAETADSKVF AKSASLQTVVCFPFLGGVLEI WQFVGEEISNCIHQSLNSSDC CIHQSLNSSDCVSQTFVGTTG TFVGTTGRLACDPRKSRIQRL DDEEERASANCMNSKRKGSDV VQSSTGDGLLCLTVNCKHKGT GDGLLCLTVNCKHKGTKIATT	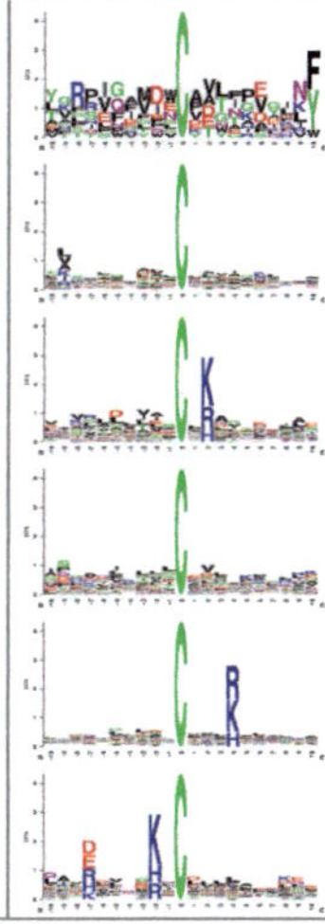	Signaling and gene regulation

(continued)

Table 2 (continued)

Protein name	Acc. no.	Cysteine location	S-nitrosylation sites	The matched MDD-clustered Motifs	Functional categorization
				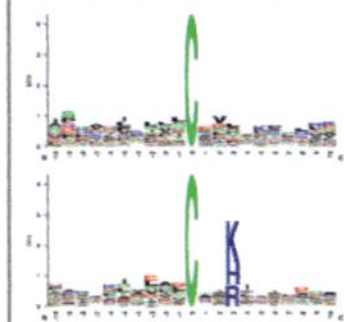	
Hypothetical protein	15217502	18** 61 122*	KTQDGEENLKCSLEKLAKKII FGRERPGRVRCVGLGPTPSSF	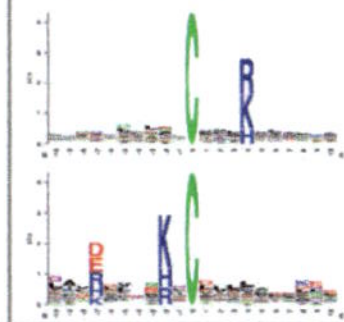	Uncharacterized
Unknown protein	19423977	109 216 240** 248***	THIPQHLDKRCFDFSEVDEGY LEDVKIAYRACALKWHPDRHH KNEAEEKFKLCTVAYQSLCEK	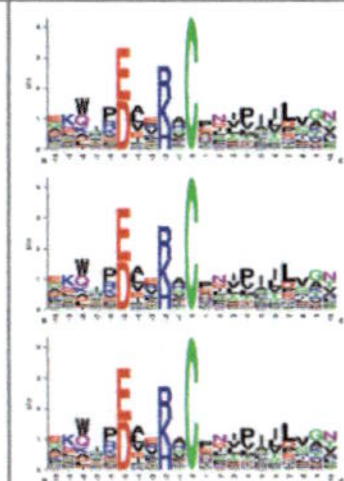	Uncharacterized
Unknown protein	17529354	96 114 127*	FAAFKISFLKCSDIMILSTKL TKLFDIYLPRCFFVIFLDHIS	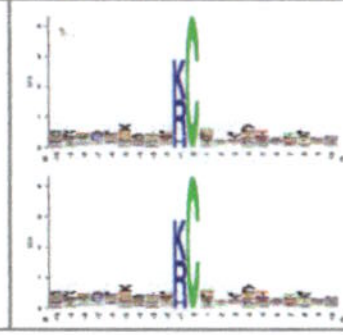	Uncharacterized

Hypothetical protein	15220664	178 213 247**	LDKSSKSEKSCQPFPQGNLEN HKLLFPLLISCDVPVFGKEKF PSTSKFVEALCVVQRLSDKNI	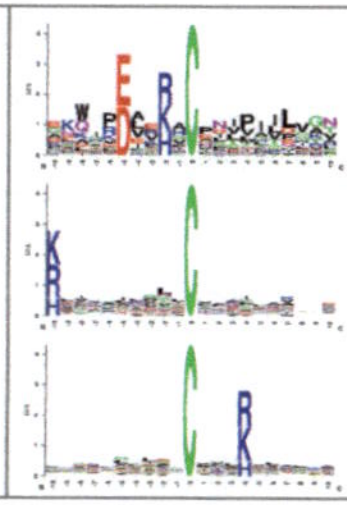	Uncharacterized

MDD maximal dependence decomposition. Target cysteine with predicted S-nitrosylation sites is shown in the *blue colour* and the amino acids that match with the respective MDD-clustered motif are shown in *red*. Target cysteine of the MDD-clustered motif is shown in the *green colour*, while the maximal dependence with the occurrence of basic or acid amino acids in each position is marked with *blue* and *red colour*, respectively. Cysteine marked with *** are showing high threshold for the S-nitrosylation, while cysteine with ** and * indicates medium and low threshold, respectively, by GPS-SNO

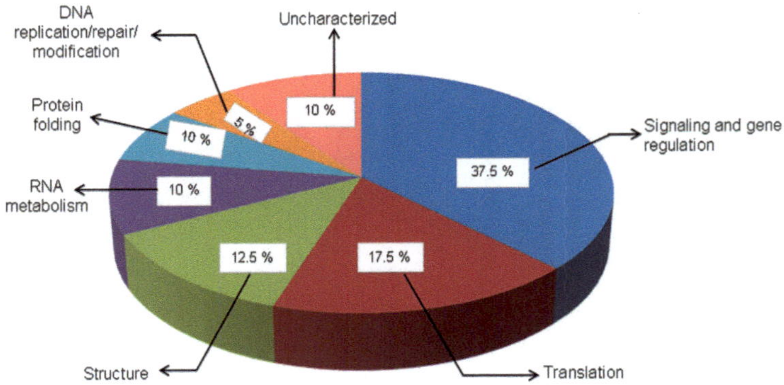

Fig. 2 Functional categorization of cold-induced nuclear proteins predicted as S-nitrosylated by dbSNO and GPS-SNO

(http://dbSNO.mbc.nctu.edu.tw) which contains manually accumulated >3,000 experimentally verified S-nitrosylated peptides from 219 research articles. Moreover, maximal dependence decomposition (MDD) is also used to cluster all the sequences of S-nitrosylation sites into subgroups, with statistically significant motifs. Additionally, freely available online GPS-SNO 1.0 (group-based prediction system) was used, which it is based on the information obtained from literature studies and from public databases with 504 experimentally verified S-nitrosylation sites in 327 proteins (http://sno.biocuckoo.org/). The targets showing maximum S-nitrosylated sites in Arabidopsis nuclear proteome were signalling and gene regulation related (37.5 %, Fig. 2). This reflects the role of S-nitrosylation in regulating mainly signalling and gene regulation pathways in the nucleus. The next largest class is of translation associated targets (17.5 %). The other categories included proteins associated with maintenance of cellular structure (12.5 %), RNA metabolism (10 %), unknown function (10 %), protein folding (7.5 %) and DNA repair/replication/modification (5 %). In putative helicase C (gi|15215678), 5 Cys matched with a MDD-motif and one Cys (Cys 94) was selected by GPS algorithm as a potential site for the S-nitrosylation. MYB34 (gi|11358685) did not show any S-nitrosylation site by these two softwares. In contrast, in MYB2 (gi|14970972), 4 cys matched with MDD-motif, while Cys 129 and Cys 161 were predicted as probable S-nitrosylated sites by GPS algorithm. Additionally, literature studies showed S-nitrosylation of AtMYB2 and AtMYB30 (details in Sect. 3.1), therefore indicating the specificity of S-nitrosylation in regulating MYB family members. In some targets like T5I7-ATPase families (gi|23308421), serine acetyltransferase (gi|15222799), dnaK-type molecular chaperone hsc70.1 (gi|15241849), and unknown protein (gi|19423977), a difference in the S-nitrosylation prediction site was observed by dbSNO and GPS-SNO. Although Cys in the TATA-binding protein-associated phosphoprotein Dr1 (gi|2129748), putative protein (gi|7267238), and myrosinase-associated protein (gi|15231805) matched with a MDD-motif, no S-nitrosylation site was predicted by the GPS algorithm. Similarly,

34 Cys in CHP-rich zinc finger protein matched with a MDD-motif, while only one Cys was predicted by GPS-SNO. Cys 380 was predicted as S-nitrosylation site in HSP70-like protein (gi|15233779) by GPS algorithm, while no S-nitrosylation site was predicted by dbSNO. Nonetheless, prediction of 72.5 % of the cold induced nuclear proteins for S-nitrosylation by dbSNO and GPS-SNO, suggested greater involvement of NO signalling in regulating nuclear proteins. Nuclear proteins validated for S-nitrosylation in plants are described below.

3.1 MYBs

MYB TF consists of a conserved DNA-binding domain (DBD) and R2R3-MYBs constitute the largest family of plant MYB TFs (Du et al. 2012). R2R3-MYB DBD contains two adjacent MYB repeats (R2 and R3) with a very conserved Cys (Cys 53) in the DNA-recognition helix of R2. Reduction of this Cys is required for its DNA binding. AtMYB2 is a R2R3-MYB from *Arabidopsis thaliana* involved in regulating hormones, developmental process, and metabolic pathways. Inhibition of AtMYB2 binding to DNA by S-nitrosylation at Cys 53 was shown (Serpa et al. 2007). M2D (a fully active minimal DNA-binding domain of AtMYB2 spanning residues) was cloned and NO donor (SNP and GSNO) treatment inhibited M2D DNA binding in EMSA assays, while DTT reversed the results. Similarly, S-nitrosylation of AtMYB30 (at Cys 49 and Cys 53) was recently shown to inhibit its DNA binding (Tavares et al. 2014). As these studies were performed with overexpressed (in *Escherichia coli*) and purified MYB protein, it's in planta S-nitrosylation needs validation.

3.2 NPR1–TGA1 System

Non-expressor of Pathogenesis Related-1 (NPR1) is an important component of the salicylic acid (SA)–mediated signal transduction pathway. In control condition, NPR1 remains in multimeric form in the cytosol, while during stress it is reduced to the monomeric form (active form) and is translocated to the nucleus. In nucleus, NPR1 interact with TGA1 (a bZIP TF) and enhances the DNA binding of TGA1 to the promoter of defense related genes and stimulates their expression. GSNO promoted the oligomerization of Arabidopsis NPR1 through S-nitrosylation of Cys 156 and blocks its nuclear translocation (Tada et al. 2008). Both NPR1 and TGA1 are S-nitrosylated by GSNO treatment (Lindermayr et al. 2010). However, contrary to Tada et al. (2008), NPR1 translocation to the nucleus was shown to be induced by GSNO. S-Nitrosylation-mediated oligomerization was suggested as a primary step before monomer accumulation. Cytosolic hydrogen peroxide (H_2O_2) prevented the SA-induced NPR1 translocation to the nucleus and inhibited induction of PR genes in Arabidopsis (Grossman et al. 2010), therefore indicating negative regulation of NPR1 by H_2O_2. As GSNO enhanced the DNA binding of TGA1 on the promoter region of defense proteins in the presence of NPR1, it could be assumed that NPR1 is S-nitrosylated again in the nucleus to promote the DNA binding activity of TGA1. These results showed NO mediated regulation of both NPR1 and TGA1 systems.

3.3 Glyceraldehyde-3-Phosphate Dehydrogenase

Glyceraldehyde-3-phosphate dehydrogenase (GAPDH) is a glycolytic enzyme that catalyses the conversion of glyceraldehyde-3-phosphate to 1,3-bisphosphoglyceric acid. GAPDH is detected in the cytoplasm as well as in the nucleus and therefore, it is considered to have multifunctional properties. NO produced in the HEK293 cells in apoptotic conditions caused S-nitrosylation of GAPDH (Hara et al. 2005). S-Nitrosylated GAPDH binds to Siah1 (seven in absentia homologue 1, an E3-ubiquitin ligase) and forms SNO-GAPDH/Siah1 complex. This complex translocates to the nucleus and mediates cell death. Additionally, S-nitrosylated GAPDH mediates transnitrosylation of the nuclear proteins, [deacetylating enzyme (SIRT1), histone deacetylase-2 (HDAC2), and DNA-activated protein kinase (DNA-PK)] in HEK293 and 293T cells (Kornberg et al. 2010).

In contrast to animal cells, where only one isoform is found, in plants, several isoforms of GAPDH are reported (Zaffagnini et al. 2013). In plants, the function of nuclear GAPDH is not completely known. In *Nicotiana tabacum* cell suspensions, NtGAPCa (GAPDH isoform) and NtGAPCb were found to interact with NtOSAK (*Nicotiana tabacum* osmotic stress-activated protein kinase), both in the cytoplasm and in the nucleus (Wawer et al. 2010). Interaction of GAPDH with NtOSAK along with its increased S-nitrosylation during salt stress, suggested its alternate function in plant stress signalling cascades, besides its regular glycolytic function (Wawer et al. 2010). GAPC mutants with no cysteines in the active site (mutated to serines) interacted with NtOSAK only in the cytoplasm, while no complex was detected in the nucleus, suggesting that cysteine is required for the nuclear translocation of GAPC. Translocation of GAPC1 from cytosol to nucleus was shown in the root tip of cadmium stress-treated Arabidopsis seedlings (Vescovi et al. 2013). However, stress-induced nuclear relocalization was stimulated by mutating the catalytic cysteine with serine. This showed that S-nitrosylation is not involved in nuclear transport of Arabidopsis GAPDH in cadmium stress and indicates that probably some other modification is regulating stress-induced relocalization. These results are in contrast with the S-nitrosylation-induced nuclear translocation of GAPDH during apoptosis in animals (Hara et al. 2005) and osmotic stress in *Nicotiana tabacum* (Wawer et al. 2010). Therefore, to have a clear view of the regulation of GAPDH by S-nitrosylation, more detailed analysis is required.

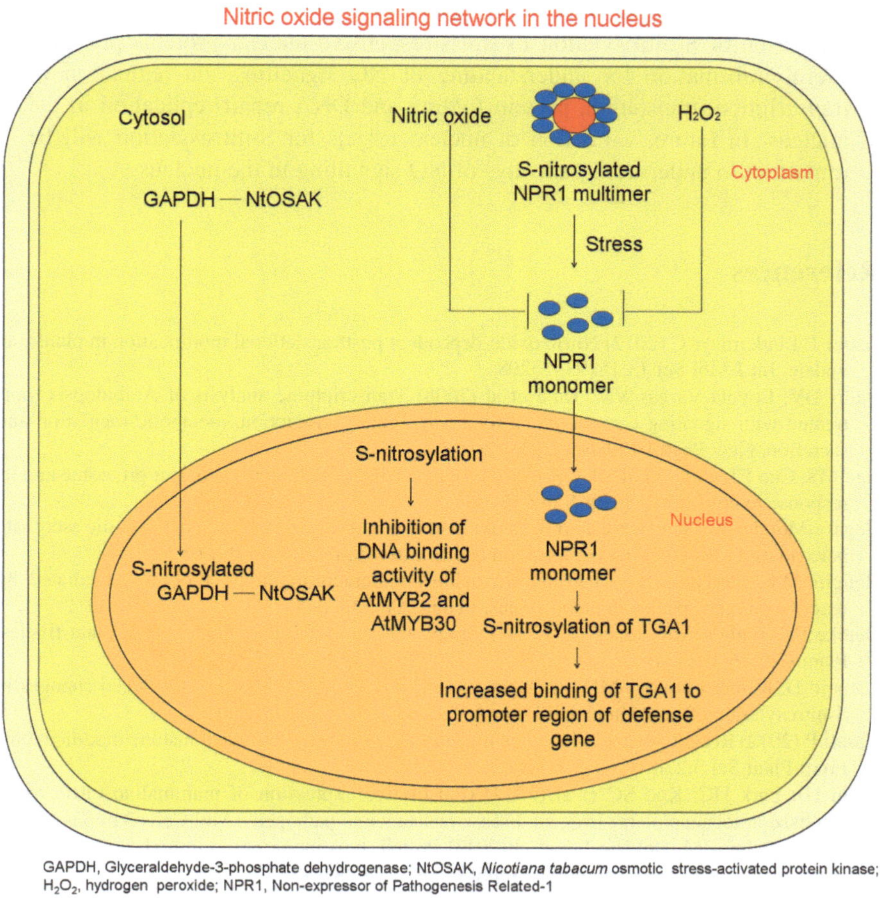

Fig. 3 Putative nitric oxide signalling network in the nucleus

Conclusion

In the past decade, a number of articles have addressed the role of NO signalling in regulating many physiological processes. However, the current understanding of NO signalling in the nucleus is still incomplete. Deciphering the route of NO synthesis in the nucleus would be the primary step to understand the enzymes involved in NO production. Additionally, analysis of modulation of NO levels in stress would be important. In the recent past, S-nitrosylation has emerged as very important NO-based PTM and a comprehensive analysis of S-nitrosylation in the nucleus is needed. On the basis of current information, a putative NO signalling network in the nucleus is presented in Fig. 3, which suggests S-nitrosylation-mediated regulation of MYB2, MYB 30, NPR1–TGA1 system, and GAPDH. Furthermore, in silico

(continued)

prediction of S-nitrosylation in stress-responsive nuclear proteins provided useful information for understanding of NO signalling via regulation of transcription, translation, protein folding and DNA repair/replication in the nucleus. In future, validation of nuclear targets for S-nitrosylation will be important to understand pathways of NO signalling in the nucleus.

References

Astier J, Lindermayr C (2012) Nitric oxide-dependent posttranslational modification in plants: an update. Int J Mol Sci 13:15193–15208

Badri DV, Loyola-Vargas VM, Du J et al (2008) Transcriptome analysis of Arabidopsis roots treated with signaling compounds: a focus on signal transduction, metabolic regulation and secretion. New Phytol 179:209–223

Bae MS, Cho EJ, Choi EY, Park OK (2003) Analysis of the Arabidopsis nuclear proteome and its response to cold stress. Plant J 36:652–663

Begara-Morales JC, Sánchez-Calvo B, Chaki M et al (2014) Dual regulation of cytosolic ascorbate peroxidase (APX) by tyrosine nitration and S-nitrosylation. J Exp Bot 65:527–538

Beligni MV, Lamattina L (1999) Nitric oxide counteracts cytotoxic processes mediated by reactive oxygen species in plant tissues. Planta 208:337–344

Bethke PC, Badger MR, Jones RL (2004) Apoplastic synthesis of nitric oxide by plant tissues. Plant Cell 16:332–341

Camejo D, Romero-Puertas MD, Rodriguez-Serrano M et al (2012) Salinity-induced changes in S-nitrosylation of pea mitochondrial proteins. J Proteomics 79:87–99

Casati P (2012) Recent advances in maize nuclear proteomic studies reveal histone modifications. Front Plant Sci 3:278

Chun HJ, Park HC, Koo SC et al (2012) Constitutive expression of mammalian nitric oxide synthase in tobacco plants triggers disease resistance to pathogens. Mol Cells 34:463–471

Corpas FJ, Palma JM, Del Río LA et al (2013) Protein tyrosine nitration in higher plants grown under natural and stress conditions. Front Plant Sci 4:29

Delledonne M, Xia Y, Dixon RA, Lamb C (1998) Nitric oxide functions as a signal in plant disease resistance. Nature 394:585–588

Du H, Feng BR, Yang SS et al (2012) The R2R3-MYB transcription factor gene family in maize. PLoS One 7:e37463

Erhardt M, Adamska I, Franco OL (2010) Plant nuclear proteomics—inside the cell maestro. FEBS J 277:3295–3307

Groß F, Durner J, Gaupels F (2013) Nitric oxide, antioxidants and prooxidants in plant defence responses. Front Plant Sci 4:419

Grossman S, Melamed-Book N, Cohen G et al (2010) Cytoplasmic H_2O_2 prevents translocation of NPR1 to the nucleus and inhibits the induction of PR genes in Arabidopsis. Plant Signal Behav 5:1401–1406

Grun S, Lindermayr C, Sell S, Durner J (2006) Nitric oxide and gene regulation in plants. J Exp Bot 57:507–516

Gupta KJ, Igamberdiev AB, Kaiser WB (2010) New insights into the mitochondrial nitric oxide production pathways. Plant Signal Behav 5:999–1001

Gupta KJ, Fernie AR, Kaiser WM et al (2011) On the origins of nitric oxide. Trends Plant Sci 16:160–168

Hara MR, Agrawal N, Kim SF et al (2005) S-nitrosylated GAPDH initiates apoptotic cell death by nuclear translocation following Siah1 binding. Nat Cell Biol 7:665–674

Kornberg MD, Sen N, Hara MR et al (2010) GAPDH mediates nitrosylation of nuclear proteins. Nat Cell Biol 12:1094–1100

Kovacs I, Lindermayr C (2013) Nitric oxide-based protein modification: formation and site-specificity of protein S-nitrosylation. Front Plant Sci 4:137

Lindermayr C, Sell S, Müller B et al (2010) Redox regulation of the NPR1-TGA1 system of Arabidopsis thaliana by nitric oxide. Plant Cell 22:2894–2907

Lombardi L, Ceccarelli N, Picciarelli P et al (2010) Nitric oxide and hydrogen peroxide involvement during programmed cell death of *Sechium edule* nucellus. Physiol Plant 140:89–102

Maldonado-Alconada AM, Zomeno SE, Lindermayr C et al (2010) Proteomic analysis of Arabidopsis protein S-nitrosylation in response to inoculation with *Pseudomonas syringae*. Acta Physiol Plan 33:1493–1514

Mengel A, Chaki M, Shekariesfahlan A et al (2013) Effect of nitric oxide on gene transcription—S-nitrosylation of nuclear proteins. Front Plant Sci 4:293

Millar AH, Day DA (1996) Nitric oxide inhibits the cytochrome oxidase but not the alternative oxidase of plant mitochondria. FEBS Lett 398:155–158

Modolo LV, Augusto O, Almeida IM, Magalhaes JR, Salgado I (2005) Nitrite as the major source of nitric oxide production by *Arabidopsis thaliana* in response to *Pseudomonas syringae*. FEBS Lett 579:3814–3820

Mur LA, Mandon J, Persijn S et al (2013a) Nitric oxide in plants: an assessment of the current state of knowledge. AoB Plants 5:1–17

Mur LA, Prats E, Pierre S et al (2013b) Integrating nitric oxide into salicylic acid and jasmonic acid/ethylene plant defense pathways. Front Plant Sci 4:215

Narula K, Datta A, Chakraborty N, Chakraborty S (2013) Comparative analyses of nuclear proteome: extending its function. Front Plant Sci 4:100

Ortega-Galisteo AP, Rodríguez-Serrano M, Pazmiño DM et al (2012) S-Nitrosylated proteins in pea (*Pisum sativum* L.) leaf peroxisomes:changes under abiotic stress. J Exp Bot 63:2089–2103

Palmieri MC, Sell S, Huang X et al (2008) Nitric oxide-responsive genes and promoters in *Arabidopsis thaliana*: a bioinformatics approach. J Exp Bot 59:177–186

Parani M, Rudrabhatla S, Myers R et al (2004) Microarray analysis of nitric oxide responsive transcripts in Arabidopsis. Plant Biotechnol J 2:359–366

Polverari A, Molesini B, Pezzotti M et al (2003) Nitric oxide-mediated transcriptional changes in *Arabidopsis thaliana*. Mol Plant Microbe Interact 16:1094–1105

Popova L, Tuan T (2010) Nitric oxide in plants: properties, biosynthesis and physiological functions. Iranian J Sci Technol 34:A3

Romero-Puertas MC, Rodríguez-Serrano M, Sandalio LM (2013) Protein S-nitrosylation in plants under abiotic stress: an overview. Front Plant Sci 4:373

Sehrawat A, Deswal R (2014) S-nitrosylation analysis in *Brassica juncea* apoplast highlights the importance of nitric oxide in cold stress signaling. J Proteome Res. doi:10.1021/pr500082u

Serpa V, Vernal J, Lamattina L et al (2007) Inhibition of AtMYB2 DNA-binding by nitric oxide involves cysteine S-nitrosylation. Biochem Biophys Res Commun 361:1048–1053

Sha Y, Marshall HE (2012) S-nitrosylation in the regulation of gene transcription. Biochim Biophys Acta 1820:701–711

Stöhr C, Stremlau S (2006) Formation and possible roles of nitric oxide in plant roots. J Exp Bot 57:463–470

Subba P, Kumar R, Gayali S et al (2013) Characterisation of the nuclear proteome of a dehydration-sensitive cultivar of chickpea and comparative proteomic analysis with a tolerant cultivar. Proteomics 13:1973–1992

Tada Y, Spoel SH, Pajerowska-Mukhtar K et al (2008) Plant immunity requires conformational changes of NPR1 via S-nitrosylation and thioredoxins. Science 321:952–956

Tavares CP, Vernal J, Delena RA et al (2014) S-nitrosylation influences the structure and DNA binding activity of AtMYB30 transcription factor from *Arabidopsis thaliana*. Biochim Biophys Acta 1844:810–817

Vescovi M, Zaffagnini M, Festa M et al (2013) Nuclear accumulation of cytosolic glyceraldehyde-3-phosphate dehydrogenase in cadmium-stressed Arabidopsis roots. Plant Physiol 162:333–334

Wawer I, Bucholc M, Astier J et al (2010) Regulation of *Nicotiana tabacum* osmotic stress-activated protein kinase and its cellular partner GAPDH by nitric oxide in response to salinity. Biochem J 429:73–83

Yadav S, David A, Baluška F et al (2013) Rapid auxin-induced nitric oxide accumulation and subsequent tyrosine nitration of proteins during adventitious root formation in sunflower hypocotyls. Plant Signal Behav 8:e23196

Zaffagnini M, Morisse S, Bedhomme M et al (2013) Mechanisms of nitrosylation and denitrosylation of cytoplasmic glyceraldehyde-3-phosphate dehydrogenase from *Arabidopsis thaliana*. J Biol Chem 288:22777–22789

Zaharah SS, Singh Z (2011) Mode of action of nitric oxide in inhibiting ethylene biosynthesis and fruit softening during ripening and cool storage of 'Kensington Pride' mango. Postharvest Biol Technol 62:258–266

Nitration and S-Nitrosylation: Two Post-translational Modifications (PTMs) Mediated by Reactive Nitrogen Species (RNS) and Their Role in Signalling Processes of Plant Cells

Francisco J. Corpas, Juan C Begara-Morales, Beatriz Sánchez-Calvo, Mounira Chaki, and Juan B. Barroso

1 Introduction

Nitric oxide (NO) is widely recognized as an outstanding signal molecule because in animal and plant systems it has been demonstrated to have a wide range of direct as well as indirect pleiotropic functions. In higher plants, the NO generation was reported earlier than in animal systems (Fewson and Nicholas 1960; Keppler 1979). However, its biochemical and physiological relevance in higher plants started to be recognized in the 1990s, when several reports showed the implication of NO in different processes in plants (Cueto et al. 1996; Leshem and Haramaty 1996; Ninnemann and Maier 1996; Delledonne et al. 1998; Durner et al. 1998; Barroso et al. 1999). A significant number of reviewers papers have updated the relevance of NO as a signal molecule or as a molecule involved in the response mechanism against (a)biotic stresses (Lamattina et al. 2003; Shapiro 2005; Arasimowicz and Floryszak-Wieczorek 2007; Besson-Bard et al. 2008; Gupta et al. 2011; Baudhouin 2011; Corpas et al. 2011). This chapter offers a general view of current state of art knowledge on the mechanism of NO signalling through post-translational modifications (PTMs), mainly nitration and S-nitrosylation in plant cells. Figure 1 shows a schematic model of PTMs mediated by nitric oxide (NO) in plant cells where all these processes, together or separately, could participate to cell signalling.

F.J. Corpas (✉)
Group of Antioxidants, Free Radicals and Nitric Oxide in Biotechnology, Food and Agriculture, Department of Biochemistry, Cell and Molecular Biology of Plants, Estación Experimental del Zaidín, CSIC, Apartado 419, 18080 Granada, Spain
e-mail: javier.corpas@eez.csic.es

J.C. Begara-Morales • B. Sánchez-Calvo • M. Chaki • J.B. Barroso
Group of Biochemistry and Cell Signaling in Nitric Oxide, Department of Biochemistry and Molecular Biology, University of Jaén, Campus "Las Lagunillas", 23071 Jaén, Spain

K.J. Gupta, A.U. Igamberdiev (eds.), *Reactive Oxygen and Nitrogen Species Signaling and Communication in Plants*, Signaling and Communication in Plants 23,
DOI 10.1007/978-3-319-10079-1_13

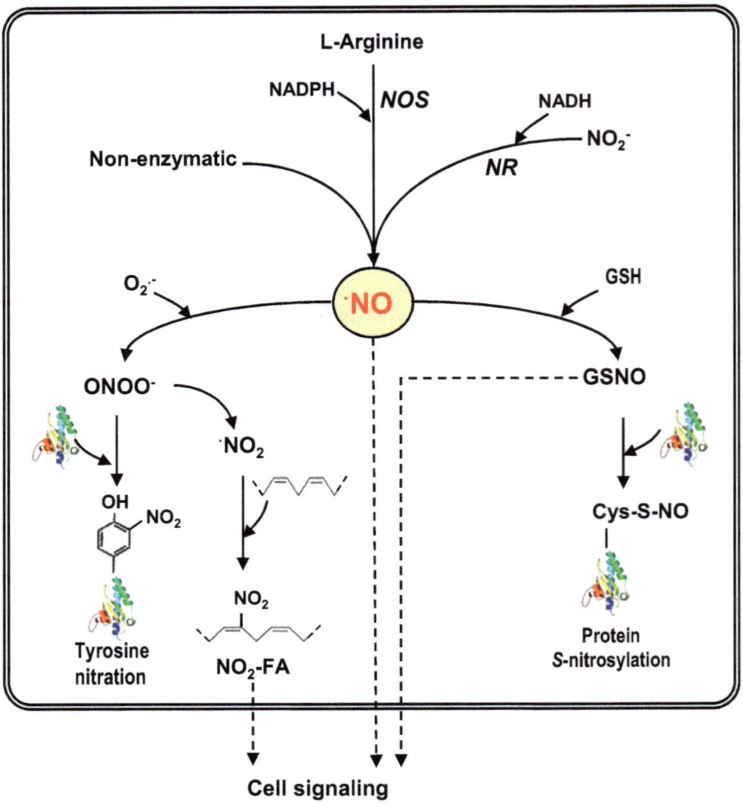

Fig. 1 Schematic model of postranslational modifications (PTMs) mediated by nitric oxide (NO) in plant cells. L-Arginine-dependent nitric oxide syntase (NOS), nitrate reductase (NR) or other non-enzymatic processes can generate NO which can react with reduced glutathione (GSH) in the presence of O_2 to form *S*-nitrosoglutathione (GSNO) which could mediate protein S-nitrosylation. On the other hand, nitric oxide can react with superoxide radicals ($O_2^{\bullet -}$) to generate peroxynitrite ($ONOO^-$) a powerful oxidant molecule that can mediate the tyrosine nitration of proteins and fatty acids (FA) where nitrogen dioxide ($^\bullet NO_2$) seems to be the intermediary. Nitric oxide and related molecules could be part of cell signaling processes

2 Nitration

Nitration involves the addition of a nitro group ($–NO_2$) to a specific molecule. This process can affect proteins, nucleic acids, and lipids. In higher plants most of the studies focus on protein nitration. More recently some preliminary data has started to indicate that the nitration of fatty acids could have an important physiological relevance such as that described in animal systems (Sánchez-Calvo et al. 2013).

2.1 *Protein Nitration*

Several amino acids are preferentially nitrated, such as tyrosine (Tyr), tryptophan (Trp), cysteine (Cys), and methionine (Met). However, most plant studies concern tyrosine nitration (Tyr-NO_2), which consists of adding a nitro group to one of the two equivalent ortho-carbons of the aromatic ring of tyrosine residues. This converts tyrosine into a negatively charged hydrophilic nitrotyrosine moiety and causes a marked shift in the local pK_a of the hydroxyl group from 10.1 in tyrosine to 7.5 in nitrotyrosine (Abello et al. 2009; Turko and Murad 2002). To date, several mechanisms are known to mediate protein tyrosine nitration in vivo involving either peroxynitrite ($ONOO^-$), nitrite/H_2O_2/haemperoxidase, or transition metals (Radi 2004; Corpas et al. 2013c). Moreover, this PTM is regarded as a process that depends also on several factors such as protein structure, the nitration mechanism, and the environmental compartments where the protein is located. The nitration of the tyrosine residues is an irreversible process and can alter the normal protein's function such as loss of protein function, gain-of-function, or no change in function (Souza et al. 2008; Radi 2013). In higher plants, research has focused on identifying potential protein targets under natural and stress conditions mainly by proteomic approaches.

During the last decade, research in higher plants concerning protein nitration has increased gradually and thus far it has been accepted that there is a physiological level of protein nitration which must have a regulatory relevance, although under stress conditions its increase could be considered a reliable marker of nitrosative stress (Corpas et al. 2007; Astier and Lindermayr 2012).

2.1.1 Protein Nitration and Plant Development

Currently, information on protein nitration under physiological conditions is meagre in comparison with the data available on stress situations. However, available data show physiological protein nitration in the different plant organs. In 9-day-old sunflower (*Helianthus annuus* L.) seedlings grown under optimal conditions, nitroproteome analysis of hypocotyls has demonstrated the presence of 21 nitrotyrosine-immunopositive proteins involved in different functional categories including photosynthesis and antioxidative, ATP, carbohydrate, and nitrogen metabolisms (Chaki et al. 2009). In 2-week-old Arabidopsis (*Arabidopsis thaliana*) seedlings, by an alternative technical approach based on the purification of nitrated proteins by immunoprecipitation with well-characterized anti-3-nitroTyr antibody and further identification by LC-MS/MS, a total of 127 potentially nitrated proteins have been identified (Lozano-Juste et al. 2011). However, the nitroproteome analysis of bitter orange *Citrus aurantium* roots revealed 26 potential candidate proteins to nitration (Tanou et al. 2012). In the case of green pepper (*Capsicum annuum* L.) fruits, the analysis of the nitroproteoma has resulted in the identification of 21 immunoreactive spots, and this number increases to 31 during fruit ripening when fruits turn red (Chaki et al. unpublished results). This suggests that increased protein nitration could be a potential indicator of the ripening process in fruits. In rice seedlings and in tobacco, BY-2 suspension cells grown under normal conditions, tyrosine nitration of α-tubulin seems to inhibit cell division and consequently cell growth (Jovanović et al. 2010). Recently, the nitration analysis during development and senescence of different organs from 8, 12, 14, and 16-day-old (young), and 71-day-old (senescent) pea (*Pisum sativum* L.) plants has indicated that each organ has a specific protein-nitration pattern (Begara-Morales et al. 2013a, b). Therefore, independently of the general view of the specific pattern of protein nitration in the different plant species and organs during plant development, a significant advance has been achieved in the characterization and identification of the residues affected in a specific protein and how this nitration influences the function of the target protein. Table 1 summaries the nitrated proteins where the nitration site and the effect of protein function have been identified.

Table 1 List of nitrated proteins in higher plants where the nitrated Tyr residue has been identified together with the effect of this PTM on their function

Protein	Plant species	Effects	Identified nitrated Tyr	References
S-adenosyl homocysteine hydrolase (SAHH)	*Helianthus annuus*	Decreased activity	Tyr448[a]	Chaki et al. (2009a)
Carbonic anhydrase (β-CA)	*Helianthus annuus*	Decreased activity	Tyr205[a]	Chaki et al. (2013)
Peroxisomal hydroxypyruvate reductase (HPR1)	*Pisum sativum*, *A. thaliana*	Decreased activity	Tyr198[b,c]	Corpas et al. (2013a, b, c)
PSBA(D1) of Photosystem II complex	*A. thaliana*	Disassembly of PSII dimers	Tyr262[b]	Galetskiy et al. (2011)
Methionine synthase	*A. thaliana*	Decreased activity	Tyr 287[b]	Lozano-Juste et al. (2011)
O-acetylserine(thiol) lyase A1	*A. thaliana*	Decreased activity	Tyr302[b]	Álvarez et al. (2011)
NADP-isocitrate dehydrogenase	*Pisum sativum*, *A. thaliana*	Decreased activity	Tyr392[b]	Begara-Morales et al. (2013a)
Glutathione reductase (GR)	*Helianthus annuus*, *Pisum sativum*	No effect	Tyr23[a]	Chaki et al. (2009a) Begara-Morales (2012)
Cytosolic ascorbate peroxidase (APX)	*Pisum sativum*	Decreased activity	Tyr235[b]	Begara-Morales et al. (2013a, b)

[a]In silico identification
[b]Mass spectrometric techniques (LC-MS/MS)
[c]Site-Directed Mutagenesis

2.1.2 Protein Nitration Under Adverse Environmental Conditions

When plants are exposed to adverse environmental conditions, a rise in the level of nitrated tyrosine both free or in proteins could be considered a footprint of nitro-oxidative stress (Corpas et al 2007; Cecconi et al. 2009; Berton et al. 2012; Corpas and Barroso 2013). Thus, the number of studies on protein nitration have significantly increased under different stress conditions and plant species, work which can provide a broad view of the relevance of nitration in the physiology of plants (Corpas et al. 2013b, c). In this context, several sets of reports have studied tyrosine nitration under different perspectives. In the first group of studies, the nitration is analysed as a reliable marker of nitro-oxidative stress; the second group focuses on the identification/classification of nitrated proteins under specific stress conditions, invariably using proteomic approaches. Finally, the third group of studies analyses how nitration affects the function of the protein under a specific stress.

Examples of the first set of studies are a group of pioneer works in plant-protein nitration. Thus, in nitrite reductase, antisense tobacco leaves showed induction of several tyrosine-nitrated polypeptides with molecular masses of between 10- and 50-kDa (Morot-Gaudry-Talarmain et al. 2002). In suspension cultures of *Taxus cuspidata*, an increase of 31 % was found in the free 3-nitrotyrosine content during shear stress (Gong and Yuan 2005). Similarly, in olive (*Olea europaea*) trees grown under in vitro conditions and exposed to 200 mM salinity it was observed that among the different parameters studied, such as NO and *S*-nitrosothiols, protein nitration was induced, indicating that salinity provoked nitrosative stress (Valderrama et al. 2007). The analysis of RNS metabolism in pea plants subjected to six different abiotic stress conditions using protein tyrosine nitration as a marker of nitrosative stress revealed that low and high temperature, continuous light, and high light intensity induced nitrosative stress in pea plants (Corpas et al. 2008). Likewise, other studies have shown that nitration is a reliable marker of nitrosative stress in different plant and stress conditions such as in sunflower hypocotyls infected by the pathogen *Plasmopara halstedii* (Chaki et al. 2009), *Prunus* genotypes under high bicarbonate and high pH (Cellini et al. 2011), in pepper leaves under low temperature (Airaki et al. 2012), in arabidopsis seedlings under arsenic stress (Leterrier et al. 2012), and in *Lotus japonicus* roots and leaves exposed to water stress (Signorelli et al. 2013).

A good example of the second set of studies is the proteomic analysis of sunflower hypocotyls exposed to high temperature (38 °C for 4 h), which showed that this stress induced the protein expression of 13 tyrosine-nitrated proteins of different functional categories such as photosynthesis, carbohydrate, and antioxidant metabolism, among others (Chaki et al. 2011). A similar study was done in bitter orange trees exposed to 150 mM NaCl stress. Thus mass-spectrometry analysis both of roots and of leaves identified candidate nitrated proteins in roots and leaves that were then classified into functional categories (Tanou et al. 2012).

Finally, the last set of papers focused on analysing specific proteins that are targets of nitration and how their function could be modulated by this PTM under stress conditions. Thus, several reports of this type have been mentioned previously. In sunflower seedlings exposed to high temperature (38 °C for 4 h) where the content of tyrosine nitration (NO_2-Tyr) was induced, the nitroproteome analysis under this condition confirmed that this stress induced the protein expression of 13 tyrosine-nitrated proteins. Among the induced proteins, ferredoxin–NADP reductase (FNR) was evaluated and nitration inhibited its activity (Chaki et al. 2011). Similar behaviour has been described for the carbonic anhydrase activity in the same plant species (Chaki et al 2013).

2.2 Nitro-Fatty Acids

Although protein nitration is the most studied process, the nitration of unsaturated fatty acids is a new area of plant research which could open novel lines of research. The potential relevance of Nitro-Fatty Acids (NO_2-FAs) as powerful signal molecules in higher plants is based on the information available in animal systems where it has been demonstrated that these molecules have anti-inflammatory properties and can also modulate a significant number of cell-signalling pathways by their capacity to interact with specific cellular targets (Baker et al. 2009). The most significant NO_2-FAs studied in animal systems are nitro-oleic acid (NO_2-OA), nitro-linoleic acid (NO_2-L), nitro-linolenic acid (NO_2-Ln), and nitro-arachidonic acid (NO_2-AA). Thus, a characteristic of these NO_2-FAs is that they have different bioactivities from their corresponding precursor. The NO_2-FAs result from the non-enzymatic interaction of unsaturated fatty acids with NO-derived species such as nitrogen dioxide (NO_2), nitrite (NO_2^-), or peroxynitrite ($ONOO^-$); however, the close mechanism of fatty acid nitration in vivo remains unknown (Rubbo 2013). On the other hand, the nitrated unsaturated fatty acids have chemically electrophilic properties by which these molecules mediate reversible nitroalkylation reactions (Michael reaction) with deprotonated thiolate anions such as thiol group of cysteine present in proteins or peptides such as glutathione. Moreover, the generation of NO_2-FAs in a lipophilic environment such as the bilayer of cellular membranes provides the appropriate mechanism for signal transduction. Nitrated lipids also function as signalling molecules, since small amounts of these molecules function as effective mediators for signal-transduction cascades.

In higher plants the presence and physiological function of NO_2-FAs can be considered an unexplored area (Sánchez-Calvo et al. 2013). Very recently, the presence of endogenous nitro-conjugated linoleic acid (NO_2-cLA) has been reported in extra-virgin olive oil and NO_2-OA-cysteine adducts in fresh olives (Fazzari et al. 2014). Moreover, preliminary data indicate the presence of endogenous nitro-linolenic acid in Arabidopsis. Moreover in vitro and in vivo analyses reveal that NO_2-LnA is available to release nitric oxide (NO) and modulate *S*-nitrosothiol (SNO) and *S*-nitrosogluthatione (GSNO) levels (Sánchez-Calvo et al. unpublished results).

3 Protein S-Nitrosylation

S-Nitrosylation is a redox-mediated PTM that regulates protein function via covalent reaction of nitric oxide (NO) with a cysteine thiol group on the target protein. This process is considered the most widespread and functionally important form of physiological NO-dependent PTM. S-Nitrosylated plant proteins have been identified on a proteome-wide scale from different plant systems usually treated with an exogenous agent (NO-donor) to increase the possibility of detecting S-nitrosylated proteins, but this has also been achieved without the addition of the NO donor (Lindermayr et al. 2005; Romero-Puertas et al. 2008; Fares et al. 2011, 2014). Thus, over 60 endogenous protein targets have been identified, mainly in Arabidopsis as model plant (Puyaubert et al. 2014). However, the analysis of others plant species such as *Kalanchoe pinnata* (Abat et al. 2008), *Brassica juncea* (Abat and Deswal 2009), *Citrus aurantium* (Tanou et al. 2012), *Nicotiana tabacum* (Astier et al. 2012b), *Solanum tuberosum* (Kato et al. 2013), *Pisum sativum*, and *Helianthus annuus* (Begara-Morales et al. 2013a, b) has also extended the identification of new S-nitrosylated proteins. Table 2 summaries the S-nitrosylated proteins where the S-nitrosylated Cys residues have been identified, and the effect of this PTM on their function.

3.1 *Protein S-Nitrosylation and Plant Development*

Little information is available on the specific modification of the content of proteins S-nitrosylation during specific process of development such as germination, root growth, or senescence, being limited basically to the identification of endogenous S-nitrosylated proteins in the different plant organs analysed (Tanou et al. 2012; Begara-Morales et al. 2013a, b).

3.2 *Protein S-Nitrosylation Under Adverse Environmental Conditions*

Under normal physiological conditions S-nitrosylation can act as a modulator of signal-transduction pathways but, under adverse environmental conditions, NO generation could be affected in both directions (over- or under-production), provoking anomalous S-nitrosylated proteins which could be specifically affected and consequently so could the signal pathways where this process is involved. Thus, proteomic studies have been conducted under biotic stress (Romero-Puertas et al. 2008; Holzmeister et al. 2011; Astier et al. 2012a, b), low temperature (Abat and Deswal 2009), and salinity stress (Tanou et al. 2009). More recently, using 14-day-old Arabidopsis seedlings exposed to cold treatment (4 °C for 4 h) 62 endogenously nitrosylated proteins were identified, but after cold stress, 20 of these proteins were over-nitrosylated (Puyaubert et al. 2014).

Table 2 List of S-nitrosylated proteins in higher plants where the S-nitrosylated Cys residues have been identified, and the effect of this PTM on their function

Protein	Plant species	Effects	Identified S-nitrosylated Cys	References
Methionine adenosyltransferase	*A. thaliana*	Inhibition	114[a]	Lindermayr et al. (2006)
Cytosolic NAD-dependent glyceraldehyde 3-P dehydrogena (GAPDH)	*A. thaliana*	Inhibition	155,159[a]	Holtgrefe et al. (2008) Wawer et al. (2010) Zaffagnini et al. (2013)
Arabidopsis histidine-containing phosphotransfer protein 1 AHP1	*A. thaliana*	Inhibition	115[a]	Feng et al. (2013)
Rubisco large subunit	*A. thaliana*, *Kalanchoe pinnata*, *B. juncea*	Inhibition carboxylase activity	175[b]	Fares et al. (2011) Abat et al. (2008), Abat and Deswal (2009)
Cytosolic ascorbate peroxidase (APX)	*A. thaliana*, *Pisum sativum*	Increased activity	32[b]	Fares et al. (2011), Begara-Morales et al. (2014)
Photosystem I apoprotein A2	*A. thaliana*	Unknown	559[b]	Fares et al. (2011)
Tubulin β	*A. thaliana*	Unknown	56[b]	Lindermayr et al. (2005), Fares et al. (2011)
Auxin receptor (TIR1)	*A. thaliana*	Increased activity	140[c]	Terrile et al. (2012)
Transcription factor MYB2	*A. thaliana*	Inhibition	53[d]	Serpa et al. (2007)
Transcription factor NPR1	*A. thaliana*	Inhibition	156[c]	Tada et al. (2008)
Transcription factor TGA1	*A. thaliana*	Activation	172,287,260[a]	Lindermayr et al. (2010)
Calnexin (CNX1)	*A. thaliana*	Unknown	108[b]	Fares et al. (2011)
Elongation factorEF1-a	*A. thaliana*, *B. juncea*	Unknown	87[b]	Lindermayr et al. (2005), Tanou et al. (2009), Fares et al. (2011)
Vacuolar ATPase subunit		Unknown	201[b]	Fares et al. (2011)
Phosphate transporter (PHT3;1)	*A. thaliana*	Unknown	104[b]	Fares et al. (2011)

(continued)

Table 2 (continued)

Protein	Plant species	Effects	Identified S-nitrosylated Cys	References
Adenylate translocater	*A. thaliana*	Unknown	130[b]	Fares et al. (2011)
NADPH oxidase	*A. thaliana*	Inhibition	890[a,c]	Yun et al. (2011)
Peroxiredoxin II E	*A. thaliana*	Inhibition	121[a]	Romero-Puertas et al. (2007)
Metacaspase AtMC9	*A. thaliana*	Inhibition	147[c]	Belenghi et al. (2007)
Phytochelatin (PC2, PC3 and PC4)	*A. thaliana*	No effect	2[a]	Elviri et al. (2010)
Cysteine protease RD21	*A. thaliana*	Unknown	233,342[b]	Fares et al. (2011)
NADP-dependent isocitrate dehydrogenase	*A. thaliana*	Unknown	75,269,363[b]	Fares et al. (2011)
Ferredoxin-NADP reductase	*A. thaliana*, *Pisum sativum*	Unknown	108[e]	Holzmeister et al. (2011), Begara-Morales et al. (2013a, b)
CDC48 (cell division cycle 48)	*Nicotiana tabacum*	Inhibition	526[a]	Astier et al. (2012a, b)

[a]Mass spectrometric techniques
[b]Combination of Biotin Switch (BS) and labeling with isotope-coded affinity tags (ICAT)
[c]Combination of Biotin Switch (BS) and Site-Directed Mutagenesis
[d]Biotin Switch (BS)
[e]In silico identification

It is well known that plants under heavy-metal stress rapidly respond by producing phytochelatins (PCs), which have a high capacity to bind toxic metals (Pal and Rai 2010; Gupta et al. 2013). PCs are synthesised by PC synthase from GSH and related thiol tripeptides and consequently they are also overproduced under heavy-metal stress. Recently, it has been established that, under cadmium stress, nitric oxide modulates the PC contents by a S-nitrosylation process (De Michele et al. 2009). Thus, S-nitrosylated PCs expressed in cadmium-stressed Arabidopsis cells have been detected by mass spectrometry analysis and three phytochelatins (PC2, PC3 and PC4) exhibited the presence of the nitro group on the first Cys residue, starting from the N-terminal (De Michele et al. 2009; Elviri et al. 2010).

4 Ascorbate Peroxidase (APX) as an Example of Dual Regulation by Tyrosine Nitration and S-Nitrosylation

APX is a key enzyme in the regulation mechanism of the cellular content of H_2O_2 during plant development and under different types of environmental stress (Shigeoka et al. 2002). It catalyses the electron transfer from ascorbate to H_2O_2, thereby giving rise to dehydroascorbate and water as products. Different proteomic studies have identified APX as potential targets both of nitration (Lozano-Juste et al. 2011; Tanou et al. 2012) and of S-nitrosylation processes (Fares et al. 2011). In relation to S-nitrosylation, two recent reports have shown that APX was modulated by S-nitrosylation but contradictory effects were described. Thus, whereas in tobacco cell culture the S-nitrosylation of cytosolic APX was responsible for the rapid decline in its activity (de Pinto et al. 2013), in Arabidopsis roots S-nitrosylation reportedly stimulated cytosolic APX activity (Correa-Aragunde et al. 2013). To discern how APX is affected by NO-derived molecules, an additional study using cytosolic APX from pea demonstrated that this antioxidant enzyme is modulated by both processes, i.e. irreversible tyrosine nitration and reversible S-nitrosylation, which had antagonistic effects. Thus, the nitration of Tyr235 inhibited APX activity, whereas S-nitrosylation of Cys32 boosted APX activity (Begara-Morales et al. 2014), indicating interplay between NO metabolism and this relevant antioxidant enzyme involved in ROS metabolism under normal and stress situations involving special situations of nitro-oxidative stress (Corpas and Barroso 2013).

Conclusions

In plants, knowledge of the metabolism of endogenous NO and other RNS has increased during the last decade. However, there are still several key challenges that remain to be addressed or need a fuller knowledge. These questions include the identification and characterization of the enzymatic source(s) and subcellular localization of endogenous NO. The identification and functional characterization of endogenous targets of both NO-mediated PTMs are also needed. On the other hand, as mentioned above, unexplored areas of research remain, such as nitration of fatty acids which reveal new signalling pathways in plants. Similarly, the regulation mechanism of gene expression by NO-derived molecules is also an area to be explored. Consequently, further research will be necessary to identify the pathways interconnecting reactive oxygen species (ROS) and reactive nitrogen species (RNS) in (a)biotic stresses, and establish biotechnological strategies against these stresses, which are responsible for major annual losses in plant yield and world crop production.

Acknowledgements Work in our laboratories is supported by ERDF-cofinanced grants from the Ministry of Science and Innovation (BIO2009-12003-C02-01, BIO2009-12003-C02-02BIO2012-33904 and RECUPERA2020), Junta de Andalucía (groups BIO286 and BIO192).

References

Abat JK, Deswal R (2009) Differential modulation of *S*-nitrosoproteome of *Brassica juncea* by low temperature: change in *S*-nitrosylation of Rubisco is responsible for the inactivation of its carboxylase activity. Proteomics 9:4368–4380

Abat JK, Mattoo AK, Deswal R (2008) S-nitrosylated proteins of a medicinal CAM plant *Kalanchoe pinnata*—ribulose-1,5-bisphosphate carboxylase/oxygenase activity targeted for inhibition. FEBS J 275:2862–2872

Abello N, Kerstjens HA, Postma DS, Bischoff R (2009) Protein tyrosine nitration: selectivity, physicochemical and biological consequences, denitration, and proteomics methods for the identification of tyrosine-nitrated proteins. J Proteome Res 8:3222–3238

Airaki M, Leterrier M, Mateos RM, Valderrama R, Chaki M, Barroso JB, Del Río LA, Palma JM, Corpas FJ (2012) Metabolism of reactive oxygen species and reactive nitrogen species in pepper (Capsicum annuum L.) plants under low temperature stress. Plant Cell Environ 35:281–295

Álvarez C, Lozano-Juste J, Romero LC, García I, Gotor C, León J (2011) Inhibition of Arabidopsis O-acetylserine(thiol)lyase A1 by tyrosine nitration. J Biol Chem 86:578–586

Arasimowicz M, Floryszak-Wieczorek J (2007) Nitric oxide as a bioactive signalling molecule in plant stress responses. Plant Sci 172:876–887

Astier J, Lindermayr C (2012) Nitric oxide-dependent posttranslational modification in plants: an update. Int J Mol Sci 13:15193–15208

Astier J, Kulik A, Koen E, Besson-Bard A, Bourque S, Jeandroz S, Lamotte O, Wendehenne D (2012a) Protein S-nitrosylation: what's going on in plants? Free Radic Biol Med 53:1101–1110

Astier J, Besson-Bard A, Lamotte O, Bertoldo J, Bourque S, Terenzi H, Wendehenne D (2012b) Nitric oxide inhibits the ATPase activity of the chaperone-like AAA^+ ATPase CDC48, a target for *S*-nitrosylation in cryptogein signalling in tobacco cells. Biochem J 447:249–260

Baker PR, Schopfer FJ, O'Donnell VB, Freeman BA (2009) Convergence of nitric oxide and lipid signaling: anti-inflammatory nitro-fatty acids. Free Radic Biol Med 46(8):989–1003

Barroso JB, Corpas FJ, Carreras A, Sandalio LM, Valderrama R, Palma JM, Lupiáñez JA, del Río LA (1999) Localization of nitric-oxide synthase in plant peroxisomes. J Biol Chem 274:36729–36733

Baudhouin E (2011) The language of nitric oxide signaling. Plant Biol (Stuttg) 13:233–242

Begara-Morales JC (2012) Functional analysis and characterization of cellular targets of reactive nitrogen species (RNS) during the plant development and under abiotic stress. Ph.D. Thesis, University of Jaén

Begara-Morales JC, Chaki M, Sánchez-Calvo B, Mata-Pérez C, Leterrier M, Palma JM, Barroso JB, Corpas FJ (2013a) Protein tyrosine nitration in pea roots during development and senescence. J Exp Bot 64:1121–1134

Begara-Morales JC, López-Jaramillo FJ, Sánchez-Calvo B, Carreras A, Ortega-Muñoz M, Santoyo-González F, Corpas FJ, Barroso JB (2013b) Vinyl sulfone silica: application of an open preactivated support to the study of transnitrosylation of plant proteins by *S*-nitrosoglutathione. BMC Plant Biol 13:61

Begara-Morales JC, Sánchez-Calvo B, Chaki M, Valderrama R, Mata-Pérez C, López-Jaramillo J, Padilla MN, Carreras A, Corpas FJ, Barroso JB (2014) Dual regulation of cytosolic ascorbate peroxidase (APX) by tyrosine nitration and *S*-nitrosylation. J Exp Bot 65:527–538

Belenghi B, Romero-Puertas MC, Vercammen D, Brackenier A, Inzé D, Delledonne M, Van Breusegem F (2007) Metacaspase activity of *Arabidopsis thaliana* is regulated by S-nitrosylation of a critical cysteine residue. J Biol Chem 282:1352–1358

Berton P, Domínguez-Romero JC, Wuilloud RG, Sánchez-Calvo B, Chaki M, Carreras A, Valderrama R, Begara-Morales JC, Corpas FJ, Barroso JB, Gilbert-López B, García-Reyes JF, Molina-Díaz A (2012) Determination of nitrotyrosine in Arabidopsis thaliana cell cultures with a mixed-mode solid-phase extraction cleanup followed by liquid chromatography time-of-flight mass spectrometry. Anal Bioanal Chem 404:1495–1503

Besson-Bard A, Pugin A, Wendehenne D (2008) New insights into nitric oxide signaling in plants. Annu Rev Plant Biol 59:21–39

Cecconi D, Orzetti S, Vandelle E, Rinalducci S, Zolla L, Delledonne M (2009) Protein nitration during defense response in *Arabidopsis thaliana*. Electrophoresis 30:2460–2468

Cellini A, Corpas FJ, Barroso JB, Masia A (2011) Nitric oxide content is associated with tolerance to bicarbonate-induced chlorosis in micropropagated Prunus explants. J Plant Physiol 168 (13):1543–1549

Chaki M, Valderrama R, Fernández-Ocaña AM, Carreras A, López-Jaramillo J, Luque F et al (2009) Protein targets of tyrosine nitration in sunflower (Helianthus annuus L.) hypocotyls. J Exp Bot 60:4221–4234

Chaki M, Valderrama R, Fernández-Ocaña AM, Carreras A, Gómez-Rodríguez MV, López-Jaramillo J, Begara-Morales JC, Sánchez-Calvo B, Luque F, Leterrier M, Corpas FJ, Barroso JB (2011) High temperature triggers the metabolism of S-nitrosothiols in sunflower mediating a process of nitrosative stress which provokes the inhibition of ferredoxin-NADP reductase by tyrosine nitration. Plant Cell Environ 34:1803–1818

Chaki M, Carreras A, López-Jaramillo J, Begara-Morales JC, Sánchez-Calvo B, Valderrama R, Corpas FJ, Barroso JB (2013) Tyrosine nitration provokes inhibition of sunflower carbonic anhydrase (β-CA) activity under high temperature stress. Nitric Oxide 29:30–33

Corpas FJ, Barroso JB (2013) Nitro-oxidative stress vs oxidative or nitrosative stress in higher plants. New Phytol 199:633–635

Corpas FJ, del Río LA, Barroso JB (2007) Need of biomarkers of nitrosative stress in plants. Trends Plant Sci 12:436–438

Corpas FJ, Chaki M, Fernández-Ocaña A, Valderrama R, Palma JM, Carreras A, Begara-Morales JC, Airaki M, del Río LA, Barroso JB (2008) Metabolism of reactive nitrogen species in pea plants under abiotic stress conditions. Plant Cell Physiol 49:1711–1722

Corpas FJ, Leterrier M, Valderrama R, Airaki M, Chaki M, Palma JM, Barroso JB (2011) Nitric oxide imbalance provokes a nitrosative response in plants under abiotic stress. Plant Sci 181:604–611

Corpas FJ, Alché JD, Barroso JB (2013a) Current overview of *S*-nitrosoglutathione (GSNO) in higher plants. Front Plant Sci 4:126

Corpas FJ, Leterrier M, Begara-Morales JC, Valderrama R, Chaki M, López-Jaramillo J, Luque F, Palma JM, Padilla MN, Sánchez-Calvo B, Mata-Pérez C, Barroso JB (2013b) Inhibition of peroxisomal hydroxypyruvate reductase (HPR1) by tyrosine nitration. Biochim Biophys Acta 1830:4981–4989

Corpas FJ, Palma JM, del Río LA, Barroso JB (2013c) Protein tyrosine nitration in higher plants grown under natural and stress conditions. Front Plant Sci 4:29

Correa-Aragunde N, Foresi N, Delledonne M, Lamattina L (2013) Auxin induces redox regulation of ascorbate peroxidase 1 activity by *S*-nitrosylation/denitrosylation balance resulting in changes of root growth pattern in Arabidopsis. J Exp Bot 64:3339–3349

Cueto M, Hernández-Perea O, Martín R, Ventura ML, Rodrigo J, Lamas S, Golvano MP (1996) Presence of nitric oxide synthase activity in roots and nodules of Lupinus albus. FEBS Lett 398:159–164

De Michele R, Vurro E, Rigo C, Costa A, Elviri L, Di Valentin M, Careri M, Zottini M, Sanità di Toppi L, Lo Schiavo F (2009) Nitric oxide is involved in cadmium-induced programmed cell death in Arabidopsis suspension cultures. Plant Physiol 150(1):217–228

de Pinto MC, Locato V, Sgobba A, Romero-Puertas Mdel C, Gadaleta C, Delledonne M, De Gara L (2013) S-nitrosylation of ascorbate peroxidase is part of programmed cell death signaling in tobacco Bright Yellow-2 cells. Plant Physiol 163:1766–1775

Delledonne M, Xia Y, Dixon RA, Lamb C (1998) Nitric oxide functions as a signal in plant disease resistance. Nature 394:585–588

Durner J, Wendehenne D, Klessig DF (1998) Defense gene induction in tobacco by nitric oxide, cyclic GMP, and cyclic ADP-ribose. Proc Natl Acad Sci USA 95:10328–10333

Elviri L, Speroni F, Careri M, Mangia A, di Toppi LS, Zottini M (2010) Identification of in vivo nitrosylated phytochelatins in *Arabidopsis thaliana* cells by liquid chromatography-direct electrospray-linear ion trap-mass spectrometry. J Chromatogr A 1217:4120–4126

Fares A, Rossignol M, Peltier JB (2011) Proteomics investigation of endogenous S-nitrosylation in Arabidopsis. Biochem Biophys Res Commun 416:331–336

Fares A, Nespoulous C, Rossignol M, Peltier JB (2014) Simultaneous identification and quantification of nitrosylation sites by combination of biotin switch and ICAT labeling. Methods Mol Biol 1072:609–620

Fazzari M, Trostchansky A, Schopfer FJ, Salvatore SR, Sánchez-Calvo B, Vitturi D, Valderrama R, Barroso JB, Radi R, Freeman BA, Rubbo H (2014) Olives and olive oil are sources of electrophilic fatty acid nitroalkenes. PLoS One 9(1):e84884

Feng J, Wang C, Chen Q, Chen H, Ren B, Li X, Zuo J (2013) S-nitrosylation of phosphotransfer proteins represses cytokinin signaling. Nat Commun 4:1529

Fewson CA, Nicholas DJD (1960) Utilization of nitric oxide by microorganisms and higher plants. Nature 188:794–796

Galetskiy D, Lohscheider JN, Kononikhin AS, Popov IA, Nikolaev EN, Adamska I (2011) Phosphorylation and nitration levels of photosynthetic proteins are conversely regulated by light stress. Plant Mol Biol 77(4–5):461–473

Gong YW, Yuan YJ (2005) Nitric oxide mediates inactivation of glutathione S-transferase in suspension culture of *Taxus cuspidata* during shear stress. J Biotechnol 23:185–192

Gupta KJ, Fernie AR, Kaiser WM, van Dongen JT (2011) On the origins of nitric oxide. Trends Plant Sci 16:160–168

Gupta DK, Huang HG, Corpas FJ (2013) Lead tolerance in plants: strategies for phytoremediation. Environ Sci Pollut Res Int 20:2150–2161

Holtgrefe S, Gohlke J, Starmann J, Druce S, Klocke S, Altmann B, Wojtera J, Lindermayr C, Scheibe R (2008) Regulation of plant cytosolic glyceraldehyde 3-phosphate dehydrogenase isoforms by thiol modifications. Physiol Plant 133:211–228

Holzmeister C, Fröhlich A, Sarioglu H, Bauer N, Durner J, Lindermayr C (2011) Proteomic analysis of defense response of wildtype *Arabidopsis thaliana* and plants with impaired NO-homeostasis. Proteomics 11:1664–1683

Jovanović AM, Durst S, Nick P (2010) Plant cell division is specifically affected by nitrotyrosine. J Exp Bot 61:901–909

Kato H, Takemoto D, Kawakita K (2013) Proteomic analysis of *S*-nitrosylated proteins in potato plant. Physiol Plant 148:371–386

Keppler LA (1979) Nitric oxide (NO) and nitrogen dioxide (NO_2) emissions from herbicidetreated soybean plants. Atmos Environ 13:537–542

Lamattina L, García-Mata C, Graziano M, Pagnussat G (2003) Nitric oxide: the versatility of an extensive signal molecule. Annu Rev Plant Biol 54:109–136

Leshem YY, Haramaty E (1996) The characterization and contrasting effects of the nitric oxide free radical in vegetative stress and senescence of Pisum sativum Linn. foliage. J Plant Physiol 148:258–263

Leterrier M, Airaki M, Palma JM, Chaki M, Barroso JB, Corpas FJ (2012) Arsenic triggers the nitric oxide (NO) and S-nitrosoglutathione (GSNO) metabolism in Arabidopsis. Environ Pollut 166:136–143

Lindermayr C, Saalbach G, Durner J (2005) Proteomic identification of *S*-nitrosylated proteins in Arabidopsis. Plant Physiol 137:921–930

Lindermayr C, Saalbach G, Bahnweg G, Durner J (2006) Differential inhibition of Arabidopsis methionine adenosyl transferases by protein *S*-nitrosylation. J Biol Chem 281:4285–4291

Lindermayr C, Sell S, Müller B, Leister D, Durner J (2010) Redox regulation of the NPR1-TGA1 system of *Arabidopsis thaliana* by nitric oxide. Plant Cell 22:2894–2907

Lozano-Juste J, Colom-Moreno R, León J (2011) *In vivo* protein tyrosine nitration in *Arabidopsis thaliana*. J Exp Bot 62:3501–3517

Morot-Gaudry-Talarmain Y, Rockel P, Moureaux T, Quillere I, Leydecker MT, Kaiser WM, Morot-Gaudry JF (2002) Nitrite accumulation and nitric oxide emission in relation to cellular signaling in nitrite reductase antisense tobacco. Planta 215:708–715

Ninnemann H, Maier J (1996) Indications for the occurrence of nitric oxide synthases in fungi and plants and the involvement in photoconidiation of *Neurospora crassa*. Photochem Photobiol 64:393–398

Pal R, Rai JP (2010) Phytochelatins: peptides involved in heavy metal detoxification. Appl Biochem Biotechnol 160:945–963

Puyaubert J, Fares A, Rézé N, Peltier JB, Baudouin E (2014) Identification of endogenously S-nitrosylated proteins in Arabidopsis plantlets: effect of cold stress on cysteine nitrosylation level. Plant Sci 215–216:150–156

Radi R (2004) Nitric oxide, oxidants, and protein tyrosine nitration. Proc Natl Acad Sci USA 101:4003–4008

Radi R (2013) Protein tyrosine nitration: biochemical mechanisms and structural basis of functional effects. Acc Chem Res 46:550–559

Romero-Puertas MC, Laxa M, Mattè A, Zaninotto F, Finkemeier I, Jones AM, Perazzolli M, Vandelle E, Dietz KJ, Delledonne M (2007) *S*-nitrosylation of peroxiredoxin II E promotes peroxynitrite-mediated tyrosine nitration. Plant Cell 19:4120–4130

Romero-Puertas MC, Campostrini N, Mattè A, Righetti PG, Perazzolli M, Zolla L, Roepstorff P, Delledonne M (2008) Proteomic analysis of *S*-nitrosylated proteins in *Arabidopsis thaliana* undergoing hypersensitive response. Proteomics 8:1459–1469

Rubbo H (2013) Nitro-fatty acids: novel anti-inflammatory lipid mediators. Braz J Med Biol Res 46(9):728–734

Sánchez-Calvo B, Barroso JB, Corpas FJ (2013) Hypothesis: nitro-fatty acids play a role in plant metabolism. Plant Sci 199–200:1–6

Serpa V, Vernal J, Lamattina L, Grotewold E, Cassia R, Terenzi H (2007) Inhibition of AtMYB2 DNA-binding by nitric oxide involves cysteine S-nitrosylation. Biochem Biophys Res Commun 361:1048–1053

Signorelli S, Corpas FJ, Borsani O, Barroso JB, Monza J (2013) Water stress induces a differential and spatially distributed nitro-oxidative stress response in roots and leaves of Lotus japonicus. Plant Sci 201–202:137–146

Shapiro AD (2005) Nitric oxide signaling in plants. Vitam Horm 72:339–398

Shigeoka S, Ishikawa T, Tamoi M, Miyagawa Y, Takeda T, Yabuta Y, Yoshimura K (2002) Regulation and function of ascorbate peroxidase isoenzymes. J Exp Bot 53:1305–1319

Souza JM, Peluffo G, Radi R (2008) Protein tyrosine nitration–functional alteration or just a biomarker? Free Radic Biol Med 45:357–366

Tada Y, Spoel SH, Pajerowska-Mukhtar K, Mou Z, Song J, Wang C, Zuo J, Dong X (2008) Plant immunity requires conformational changes of NPR1 via *S*-nitrosylation and thioredoxins. Science 321:952–956

Tanou G, Job C, Rajjou L, Arc E, Belghazi M, Diamantidis G, Molassiotis A, Job D (2009) Proteomics reveals the overlapping roles of hydrogen peroxide and nitric oxide in the acclimation of citrus plants to salinity. Plant J 60:795–804

Tanou G, Filippou P, Belghazi M, Job D, Diamantidis G, Fotopoulos V, Molassiotis A (2012) Oxidative and nitrosative-based signaling and associated post-translational modifications orchestrate the acclimation of citrus plants to salinity stress. Plant J 72:585–599

Terrile MC, París R, Calderón-Villalobos LI, Iglesias MJ, Lamattina L, Estelle M, Casalongué CA (2012) Nitric oxide influences auxin signaling through S-nitrosylation of the Arabidopsis TRANSPORT INHIBITOR RESPONSE 1 auxin receptor. Plant J 70:492–500

Turko IV, Murad F (2002) Protein nitration in cardiovascular diseases. Pharmacol Rev 54:619–634

Valderrama R, Corpas FJ, Carreras A, Fernández-Ocaña A, Chaki M, Luque F, Gómez-Rodríguez MV, Colmenero-Varea P, del Río LA, Barroso JB (2007) Nitrosative stress in plants. FEBS Lett 581:453–461

Wawer I, Bucholc M, Astier J, Anielska-Mazur A, Dahan J, Kulik A, Wyslouch-Cieszynska A, Zareba-Koziol M, Krzywinska E, Dadlez M, Dobrowolska G, Wendehenne D (2010) Regulation of *Nicotiana tabacum* osmotic stress-activated protein kinase and its cellular partner GAPDH by nitric oxide inresponsetosalinity. Biochem J 429:73–83

Yun BW, Feechan A, Yin M, Saidi NB, Le Bihan T, Yu M, Moore JW, Kang JG, Kwon E, Spoel SH, Pallas JA, Loake GJ (2011) *S*-nitrosylation of NADPH oxidase regulates cell death in plant immunity. Nature 478:264–268

Zaffagnini M, Fermani S, Costa A, Lemaire SD, Trost P (2013) Plant cytoplasmic GAPDH: redox post-translational modifications and moonlighting properties. Front Plant Sci 4:450

S-Nitrosoglutathione Reductase: A Key Regulator of S-Nitrosylation in Plant Development and Stress Responses

Marek Petřivalský, Lucie Kubienová, Tereza Tichá, and Lenka Luhová

1 Introduction

Nitric oxide (NO), originally identified as endothelium-relaxation factor, has been considered as an important signalling molecule in a plethora of biological processes in a wide range of organisms. In animals it is widely documented that NO participates in various physiological processes like vasorelaxation, neurotransmission, and immunity regulation and that disturbances of cellular levels of NO and related reactive nitrogen species (RNS) underlie multiple pathological disorders (Luiking et al. 2010). NO is involved in the regulation of plant developmental and growth processes and signalling cascades of plant responses to stress conditions (Groß et al. 2013; Yu et al. 2014)

The biosynthesis of NO can proceed by various enzymatic or non-enzymatic reactions depending on the site and the nature of stimulus for NO production. Constitutive and inducible isoforms of nitric oxide synthase (NOS, EC 1.14.13.39) represent major enzyme source in mammals and bacteria, whereas their occurence in plants has not been unequivocally demonstrated in plants where nitrate reductase (NR, EC 1.6.6.1) was identified as possible main enzymatic NO source, together with NO production by nitrite reduction in electron transport chains of mitochondria or chloroplasts (Mur et al. 2013).

S-nitrosothiols (SNO) are formed by covalent modification of cysteine thiol with nitroso (NO–) group both in low-molecular weight thiols and proteins. According to actual knowledge, SNO represents relatively stable reserve and transport form of NO in vivo (Gaston 1999; Handy and Loscalzo 2006). S-Nitrosylation is currently considered as one of the most important post-translational protein modifications

M. Petřivalský (✉) • L. Kubienová • T. Tichá • L. Luhová
Faculty of Science, Department of Biochemistry, Palacky University, Šlechtitelů 11, 78371 Olomouc, Czech Republic
e-mail: marek.petrivalsky@upol.cz

K.J. Gupta, A.U. Igamberdiev (eds.), *Reactive Oxygen and Nitrogen Species Signaling and Communication in Plants*, Signaling and Communication in Plants 23,
DOI 10.1007/978-3-319-10079-1_14

(Hess and Stamler 2011) and increasing number of S-nitrosylation targets has been identified among proteins and enzymes involved in crucial cellular processes like apoptosis, gene transcription, or iron homeostasis (Seth and Stamler 2011), therefore it is not surprising that disturbances in the control of protein S-nitrosylation formation and catabolism are tightly related with series of pathological states (Foster et al. 2009).

S-Nitrosoglutathione (GSNO) represents the major low molecular weight *S*-nitrosothiol, which is capable under specific conditions to release free NO or to participate in NO transport and in transnitrosation reactions, when nitroso group is transferred from SNO to thiol group of another molecule (Martínez-Ruiz and Lamas 2004; Marino and Gladyshev 2010; Derakhshan et al. 2007). In an aerobic environment, GSNO can be formed by a reaction of glutathione tripeptide (GSH, γ-Glu–Cys–Gly) with reactive nitrogen species mediated by dinitrosyl–iron complexes and it has a key role in the cellular homeostasis of physiological levels of protein *S*-nitrosothiols (Bosworth et al. 2009).

Similarly to animal cells, GSNO is supposedly the most abundant low-weight molecular *S*-nitrosothiol in plant cells, with high significance to the signalling pathways and metabolism of NO in plants. However, our knowledge on the distribution, intracellular levels, and modulation of GSNO under natural and stress conditions is still very limited (Corpas et al. 2013).

2 *S*-Nitrosoglutathione Reductase: Key Enzyme of the Regulation of S-Nitrosylation and Formaldehyde Detoxification

Published results in previous years have indicated that *S*-nitrosoglutathione reductase (GSNOR) plays the crucial role in GSNO catabolism, in the regulation of protein *S*-nitrosothiol levels, and in the metabolism of NO and RNS in general (reviewed by Benhar et al. 2009). GSNOR activity was found to be associated with another previously known enzyme, in older literature referred to as glutathione-dependent formaldehyde dehydrogenase (FALDH; EC 1.2.1.1) (Uotila and Koivusalo 1979). The elucidation of exact reaction mechanism, where *S*-(hydroxymethyl)glutathione (HMGSH) as a spontaneous adduct of glutathione and formaldehyde was identified as the proper substrate, led to the enzyme later reclassified as *S*-(hydroxymethyl)glutathione dehydrogenase (EC 1.1.1.284), which catalyses the oxidation of HMGSH to *S*-formylglutathione using NAD^+ as a coenzyme (Koivusalo et al. 1989) (Fig. 1A).

However, later the NADH-dependent reduction of GSNO, leading to the formation of oxidised glutathione (GSSG) and ammonium, has been uncovered as a more physiologically relevant reaction (Liu et al. 2001; Jensen et al. 1998; Staab et al. 2008b). For this reason, although this designation has not been yet accepted by IUBMB nomenclature commission, the denomination of this enzyme as GSNOR

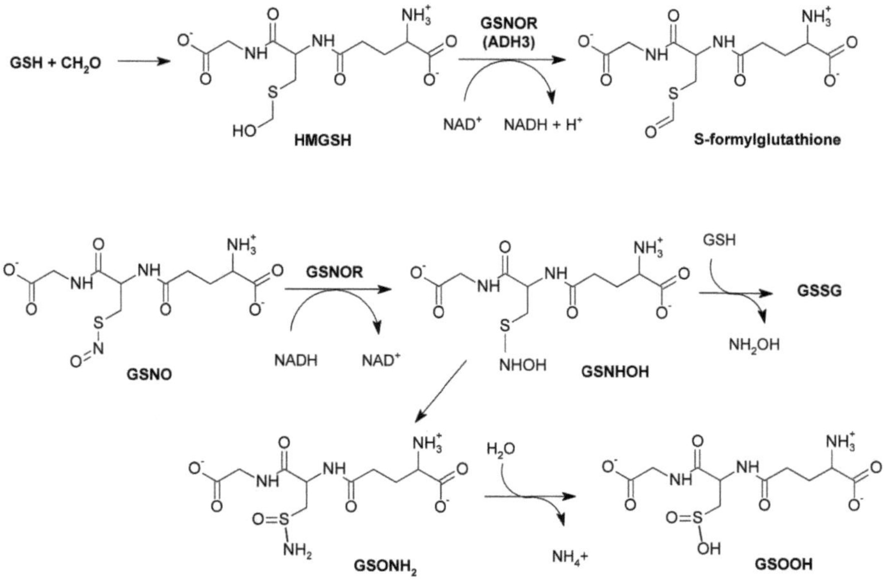

Fig. 1 Biochemical reactions catalysed by S-nitrosoglutathione reductase (class III alcohol dehydrogenase, ADH3). (**a**) In dehydrogenase mode GSNOR/ADH3 catalyses the oxidation of formaldehyde-glutathione adduct, S-(hydroxymethyl)glutathione (HMGSH), to S-formylglutathione. (**b**) In reductase mode GSNOR catalyses NADH-dependent reduction of S-nitrosoglutathione (GSNO) to N-hydroxysulfinamide (GSNHOH), an unstable intermediate, which in reaction with another molecule of reduced glutathione (GSH) yield the final products—oxidised glutathione (GSSG) and ammonium ions. Alternatively, under decreased intracellular levels of reduced glutathione, N-hydroxysulfinamide undergoes spontaneous rearrangement to glutathione sulphinamide ($GSONH_2$), which can be converted further to glutathione sulfinic acid (GSOOH)

is currently highly extended within scientific literature. According to formal enzyme classification, GSNOR belongs to a family of Zn-dependent class III alcohol dehydrogenases (ADH3; EC 1.1.1.1), which is distinct from members of class I ADH family by their high affinity towards long-chain alcohols (Koivusalo and Uotila 1991; Staab et al. 2008b).

Cellular redox potential is likely an important factor in the regulation of GSNOR-catalysed GSNO reduction (Fig. 1B). The reaction mechanism of GSNO reduction includes the formation of unstable intermediate *N*-hydroxysulfinamide (GSNHOH). The next reaction step to the final product is apparently dependent on the local concentration of GSH: in high GSH levels a nucleophilic attack on intermediate results in glutathione disulfide (GSSG) with concomitant release of hydroxylamine, whereas at low GSH levels, the reaction intermediate GSNHOH can spontaneously convert to glutathione sulfinamide ($GSONH_2$) which is further hydrolysed to glutathione sulfinic or even sulfonic acid (Jensen et al. 1998).

Physiological GSH concentrations in the cells are found in milimolar range which favours the reaction directed towards GSSG and hydroxylamine (Staab et al. 2009). However, the intracellular level of GSH is known to widely fluctuate under stress conditions, e.g. under exposure to toxic xenobiotic compounds or under oxidative stress associated with pathological states. Studies in vitro demonstrated decreased GSH levels resulted in a reaction shift from GSNHOH intermediate to the formation of $GSONH_2$ and GSO_2H, which can be under severe oxidative stress further oxidised even to glutathione sulfonic acid (GSO_3H) (Staab et al. 2008a). These metabolites were shown to exert inhibitory effect towards glutathione transferases, the enzymes crucial in the detoxification of a wide range of xenobiotics by their conjugation with GSH. Inhibition of GSH-dependent enzymes can results in further increase of nitrosative or oxidative injury of stressed cells (Staab et al. 2009).

The availability of NADH as GSNOR coenzyme is considered another factor regulating GSNO turnover. Under normal conditions, the value of cytoplasmic ratio of oxidised and reduced cofactor (NAD^+/NADH) is about 700, whereas the similar ratio for NADPH coenzymes is much lower, what enables NADPH to function as reductant in biosynthetic pathways (Veech et al. 1972). All GSNOR described so far show cytosolic localisation and high preference for NADH coenzyme so they cannot be supposed to use NADPH in GSNO reduction. GSNOR enzymes, due to their dehydrogenase activity, are capable to produce reduced NADH in the catabolism of formaldehyde and this reaction was proposed to provide NADH required for GSNO reduction (Staab et al. 2008a). Crotty(2009) suggested that the reductive function of Arabidopsis GSNOR in cytoplasmic compartment is possible due to its high affinity to NADH compared to NAD^+, as determined by isothermal titration calorimetry.

The toxicity of formaldehyde as a product of carbon compounds metabolism is connected with high reactivity of carbonyl group, which can react with DNA and proteins to stable carboxylated products. In plants, formaldehyde originates mainly by dissociation of 5,10-methylentetrahydrofolate and by oxidation of methanol formed in pectin demethylation (Hanson et al. 2000). Formaldehyde can be formed also in other oxidative and demethylation reactions, like decarboxylation of glyoxylate and cytochrome P450-dependent oxidation of pesticides (Espunya et al. 2006). ADH3 can be considered the most effective formaldehyde-metabolising enzyme in vitro (Staab et al. 2008b). NAD^+-dependent oxidation of *S*-(hydroxymethyl)glutathione (HMGSH), spontaneous adduct of formaldehyde with major intracellular thiol glutathione, catalysed by ADH3 thus represents a key step in the detoxification of exo- and endogenous formaldehyde. In certain bacteria, the formation of ADH3 substrate HMGSH is accelerated by the action of specific enzyme *S*-(hydroxymethyl)glutathione synthase (EC 4.4.1.22). GSNOR/ADH3 therefore participates in all organisms in metabolic pathways involved in the detoxification of intracellular formaldehyde. The product of formaldehyde oxidation *S*-formylglutathione is than decomposed by *S*-formylglutathione hydrolase (EC 3.1.2.12) (Espunya et al. 2006).

Besides GSNOR, it has been demonstrated that GSNO can be metabolised by other enzymes, and that GSNO-reducing activity is often associated with enzyme activities of previously known enzymes, such as thioredoxins, glutathione peroxidase, glutamyl peptidase, Cu/Zn-superoxide dismutase, and nitroreductase (Benhar et al. 2009; Sengupta and Holmgren 2013), but description of mentioned enzymes is out of scope of this chapter.

3 Molecular Properties of *S*-Nitrosoglutathione Reductase

3.1 GSNOR Structure

Similarly to other enzymes from the alcohol dehydrogenase family, GSNOR has a homodimeric structure containing two zinc atoms. Each monomer is composed of a big catalytic domain and a smaller non-catalytic domain with a binding site for NAD^+ coenzyme (Engeland et al. 1993; Sanghani et al. 2002a, b). Both Zn ions are attached to the catalytic domain, but only one ion is involved in the catalytic mechanism, the other zinc having purely a structural role. Active sites of GSNOR monomers are localised in a cleft between catalytic and cofactor domains,

In human GSNOR, the catalytic zinc atom is bonded to His69, Cys177 and Cys47, while the structural zinc is coordinated to four cysteine residues. The formation of protein dimer is based on the interaction of 12 pseudo-continuous beta-sheets, which comprise the major part of the coenzyme domain. Determination of the crystal structure of an enzyme ternary complex with HMGSH substrate and NADH coenzyme in 1.6 Å resolution enabled to elucidate the role of specific residues in the active site during the binding of physiologically relevant substrate with concomitant structural changes within the catalytic cycle of GSNOR dehydrogenase activity (Sanghani et al. 2002b, 2003). The formation of the ternary complex includes a movement of catalytic domain towards the coenzyme-binding domain. Close localisations of one water molecule to ribose 2′-hydroxyl suggest that the proton from substrate is transferred to the solvent directly from the coenzyme, and not through the terminal histidine like in case of proton transfer in ADH1. HMGSH substrate is coordinated directly to active site zinc and interacts with highly conserved residues Arg114, Asp55, Glu57, and Thr46. The binding site of catalytic Zn has tetrahedral coordination which includes Cys44, His66, Cys173, and HMGSH. This is in contrast to the Zn coordination in GSNOR binary complex with the coenzyme, where all ligands originate from the protein and include the forth amino acid Glu67. Observed changes in Zn coordination in the ternary complex are achieved by a movement of catalytic domain within a distance range of 2.3 Å (Sanghani et al. 2002b, 2003).

Recent advances in detailed biochemical and structural characterization of plant GSNOR enzymes confirmed their similarities to mammalian homologues (Crotty 2009; Kubienová et al. 2013; Xu et al. 2013). In comparison to human GSNOR, plant GSNORs exhibit a difference in the composition of the anion-binding pocket, which negatively influences the affinity for the carboxyl group of ω-hydroxyfatty acids (Kubienová et al. 2013). Plant GSNORs are also unusually cysteine-rich proteins, which suggested their possible regulation by post-translational modifications of several conserved cysteine residues exposed at the protein dimer surface (Xu et al. 2013).

3.2 GSNOR Substrate Specificity and Inhibition

Active site structure of GSNOR as a representative of ADH class III is significantly different compared to classical class I ADH. The residues 53–59 and 113–120 are situated further from the active site, which is thus enlarged and forms a widened entrance for substrates to binding domain. These structural changes result in different substrate specificity of ADH3, which enlarged active site cannot be saturated by ethanol like in ADH1, but is arranged to bind bigger substrates like GSNO, HMGSH, long-chain alcohols, and ω-hydroxyfatty acids. Namely Arg114 is known to play a key role in the binding of HSGSH and GSNO due to its positive charge which enables correct orientation of negatively charged substrates in the binding site (Staab et al. 2009; Sanghani et al. 2003).

GSNOR specifically catalyses the oxidation of long-chain alcohols like cinnamyl alcohol, farnesol, and geraniol to corresponding aldehydes with participation of NAD^+. Alcohols with chains longer than four carbons are oxidised much more effectively than ethanol, which is a very weak substrate (Staab et al. 2009). However, for other plausible substrates such as ω-hydroxyfatty acids (e.g. 12-hydroxydodecanoic acid) GSNOR shows much lower catalytic efficiency (Achkor et al. 2003). Thus according to current knowledge, *S*-nitrosoglutathione is considered as the physiologically most relevant substrate, as the K_m value of human ADH3 for HMGSH is approx. 100-times higher as compared to GSNO. The catalytic constant is about 10-times higher for GSNO reduction, but the catalytic efficiency is about 2 times higher for HMGSH (Hedberg et al. 2003; Sanghani et al. 2000; Green et al. 2012).

4-Pentylpyrazol and 1,10-phenanthrolin are known inhibitors of alcohols oxidation in various ADH classes, whereas dodecanoic acid behaves as a specific GSNOR inhibitor due to the interaction with Arg114. Various substrate analogues, including medium-chain fatty acids and glutathione derivatives have been tested as potential GSNOR inhibitors and in majority of cases a non-competitive type of inhibition was observed (Staab et al. 2009). Inhibition constants for ω-fatty acids fall within micromolar range and showed dependence on the chain length, when dodecanoic and dodecandioic acid were the most potent inhibitors. Similarly, all inhibitors derived from glutathione (*S*-methylglutathione, glutathione sulfonic acid, *S*-acetamidoglutathione, and glutathione disulfide) had inhibition constants in millimolar range, which is three orders of magnitude higher than K_m values of GSNOR substrates HMGSH and GSNO. This study also concluded that the active site Zn moiety had a crucial role in the correct orientation and binding of hydrophilic glutathione-derived substrates within predominantly hydrophobic active site.

S-Nitrosoglutathione reductase activity of the plant enzyme shows similar kinetic and molecular properties as mammalian GSNOR (Sakamoto et al. 2002; Crotty 2009; Kubienová et al. 2013). Arabidopsis and tomato GSNOR exhibit 10-fold higher velocity of NADH-dependent GSNO reduction compared to HMGSH oxidation (Crotty 2009; Kubienová et al. 2013). Similarly to human GSNOR, fatty acids as substrate analogues show weak non-competitive inhibition of plant GSNOR, whereas GSH and *S*-methylglutathione were reported to non-competitively inhibit GSNO reduction in the millimolar range (Kubienová et al. 2013). Several new potent pyrrole-based GSNOR inhibitors including the compound N6022 have been described for the human enzyme (Sun et al. 2011). The binding of N6022 includes an interaction of the propionic acid side chain of the inhibitor with all three residues forming the GSNOR anionic pocket (Green et al. 2012). N6022 was found to inhibit tomato GSNOR at concentrations of 10^{-7} M and thus represents an attractive tool for GSNOR inhibition studies in plants (Kubienová et al. 2013).

4 GSNOR Role in Plants

4.1 GSNOR in Plant Growth and Development

The plant enzyme has been widely characterized in biochemical and genetic aspects in several reports either previously as glutathione-dependent formaldehyde dehydrogenase (Uotila and Koivusalo 1979) or as ADH class III (Martínez et al. 1996; Dolferus et al. 1997; Achkor et al. 2003). GSNOR as the key enzyme catalysing GSNO reduction and involved in the regulation of intracellular levels of protein *S*-nitrosothiols (Fig. 2), was described in *Arabidopsis thaliana* (Sakamoto et al. 2002).

The corresponding gene *Atgsnor1* (originally termed as *adh5*) is found in a single copy in Arabidopsis and most plant genomes, although new bioinformatics analysis reported the presence of multiple GSNOR gene copies in several plant species, whose phylogenetic analysis suggested recent and sporadic occurrence of GSNOR-encoding genes duplication among plant species (Xu et al. 2013). Although the accumulated experimental evidence suggests cytosolic localization (Xu et al. 2013), proteomic

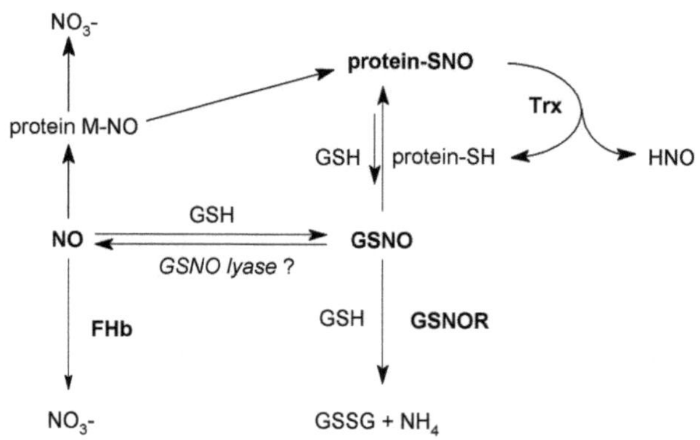

Fig. 2 GSNOR role within S-nitrosothiol catabolism. GSNOR is considered as main pathway of irreversible degradation of low molecular S-nitrosothiols, namely GSNO, to oxidised glutathione. GSNOR is also believed to control the level of protein S-nitrosothiols, which are formed in trans-nitrosylation reactions of protein and GSNO. S-nitrosylated proteins can also result from rearrangements of metal-nitrosyl complexes formed on Fe atoms of hemoglobins and other hem-containing proteins. Thioredoxin (Trx) is involved in the NADPH-dependent reduction of protein S-nitrothiols, which do not belong among GSNOR substrates. Flavohemoglobin (FHb) with NO dioxygenase activity represent another important enzymatic route of NO conversion to nitrate

analyses of Arabidopsis peroxisomes have suggested that this protein could be located also in this organelle (Reumann et al. 2007). Using electron microscopy and immunogold labelling, GSNOR from pea leaf cells was observed to be localized in different subcellular compartments including cytosol, chloroplasts, mitochondria, and peroxisomes (Barroso et al. 2013). In contrast, in Arabidopsis plants expressing GFP-tagged GSNOR, the enzyme localization at cellular level was observed only in cytosol and nucleus, excluding nucleolus. This is in agreement with bioinformatics analysis, which predicted mitochondrial targeting sequence only for *Physcomitrella* GSNOR paralog (Xu et al. 2013). Interestingly, studies in *Arabidopsis thaliana* cell suspension cultures demonstrated GSNOR importance in the regulation of mitochondrial functionality under normal and stress conditions (Frungillo et al. 2013). Activities of mitochondrial complex I, external NADH dehydrogenase, alternative oxidase, and uncoupling protein were responsive to modulations in GSNOR levels. GSNOR activity is involved in modulating the activity of the mitochondrial respiratory chain in Arabidopsis cells via controlling SNO homeostasis under normal growth conditions and under nutritional stress. Taken together, these results suggest that GSNOR activity may control the redox state of plant cells by affecting mitochondrial bioenergetics under stress conditions (Salgado et al. 2013).

GSNOR/ADH3 has been supposed to be constitutively expressed throughout the plant; however, using histochemical activity staining and immunolocalization, GSNOR was found to be differentially expressed in Arabidopsis organs (Espunya et al. 2006, 2007). Higher GSNOR level was detected in the leaves and roots from the first stages of development; and transgenic plants, both GSNOR overexpressing and knock-down mutants, showed a short-root phenotype. In tomato, GSNOR gene expression and activities in cotyledons were found to be higher compared to roots during the early stage of development, whereas at later stages the expression was higher in roots and stem compared to leaves and the shoot apex. GSNOR expression in tomato (*Solanum lycopersicum*) also increased in the inflorescence, especially in the pistil and stamens, and during fruit development and ripening (Kubienová et al. 2013).

Experimental evidence obtained with Arabidopsis *hot5* mutants, identified as hot temperature-sensitive mutants, brought deeper understanding of GSNOR role in plant development (Lee et al. 2008). Both missense and null mutations in GSNOR gene displayed disturbed intracellular NO levels and increased nitrosative status. Beside the hot-sensitive phenotype, GSNOR mutant plants also failed to grow on nutrient plates and showed increased reproductive shoots and reduced fertility. Further study on *hot5* mutants reported also defects in stem and trichome branching (Xu et al. 2013). Using GFP-tagged GSNOR, the ubiquitous expression of GSNOR was confirmed, however, especially high signals were observed in the root tip, apical meristem and flowers.

Using the same Arabidopsis mutant, Loake's group observed the loss of GSNOR function in Arabidopsis affected apical dominance, leaf shape, germination, 2,4-D sensitivity, and reduced hypocotyl elongation in both light and dark grown seedlings (Kwon et al. 2012). It was concluded that Arabidopsis GSNOR may negatively regulate the autonomous and photoperiod flowering time pathways as GSNOR mutants showed early flowering compared to wild type, whereas GSNOR overexpression slightly delayed flowering time in both long and short days.

4.2 GSNOR in Plant Responses to Abiotic Stress

Published reports of the last decade point to an important role of GSNOR in plant responses to multiple abiotic stress conditions (reviewed in Salgado et al. 2013; Yu et al. 2014). In general, an opposite correlation between SNO or GSNO levels and GSNOR expression and activity has been demonstrated in different plants and conditions (Airaki et al. 2011; Espunya et al. 2012; de Pinto et al. 2013).

The expression of GSNOR gene in Arabidopsis (previously named as *ADH2*) was suppressed by wounding, and the GSNOR level and activity were decreased in tobacco plants following treatment with jasmonic acid, the plant hormone involved in the wounding signal transduction in plants (Díaz et al. 2003). Similarly, GSNOR activity and expression is strongly reduced by mechanical damage of sunflower seedlings (Chaki et al. 2011). In tobacco plants (*N. attenuata*) herbivore attack but not wounding transiently reduced GSNOR activity; moreover, virus-induced silencing of tobacco GSNOR decreased the accumulation of jasmonate and ethylene induced by feeding of the insect herbivore *Manducasexta* (Wunsche et al. 2011). GSNOR-silenced plants had also decreased activities of trypsin protease inhibitors and were more susceptible to herbivore attack in feeding bioassays. Interestingly, GSNOR was found to modulate only certain jasmonate-dependent responses, including the accumulation of defence secondary metabolites. Compromised herbivory-induced ethylene emissions in GSNOR-silenced plants might be related to the known effect of nitrosylation to methionine adenosyl transferases, the key ethylene biosynthesis enzymes inhibited by S-nitrosylation in vitro (Lindermayr et al. 2006). In conclusion, the transient decrease of GSNOR activity induced by herbivory insects might be required for optimal activation of specific defence responses, involving the modulation of the cellular nitrosylation status.

GSNOR is involved in plant thermotolerance mechanisms, and GSNOR activity was shown indispensable for the acclimatization of Arabidopsis plants to high temperatures (Lee et al. 2008). Arabidopsis plants with defect GSNOR gene, identified as *hot5* mutant, have decreased tolerance to high temperature stress resulting from disturbed homeostasis of cellular *S*-nitrosothiols and signalling pathways of reactive nitrogen species derived from NO. Constitutive expression of GSNOR during development and no evidence for a heat-induced protein increase was observed in this study. The lack of significant modulation of GSNOR of Arabidopsis at transcriptional or protein levels suggests redox regulation of GSNOR at the posttranscriptional level, possibly through the regulation of enzyme activity by Cys modification(s) (Lee et al. 2008). However, in later study GSNOR was found involved in the heat shock- or H_2O_2-induced programmed cell death (PCD) of tobacco Bright Yellow-2 cells. An increase in GSNO level due to increased GSNOR expression and activity occurred in cells exposed to H_2O_2, while heat-shocked cells had higher GSNOR activity and decreased GSNO (de Pinto et al. 2013).

On the other hand, cold stress by low temperatures was observed to induce GSNOR activity in pea plants, together with induced NOS-like activity and increased levels of nitrosothiols and nitrated proteins as marker of nitrosative stress

(Corpas et al. 2008a). However, the acclimatization of pepper plant to cold stress is associated with slight increase of GSNOR activity (Airaki et al. 2012). GSNOR was also proposed to play an important role in the control of seed germination under chilling stress by the removal of over-accumulated *S*-nitrosothiols. Chilling stress increased the levels of *S*-nitrosothiols and impaired the activity of GSNOR in embryos of recalcitrant *Baccaurea ramiflora* seeds, whereas the addition of exogenous CO or NO donor markedly reduced the chilling-induced increase in *S*-nitrosothiols and induced GSNOR activity (Bai et al. 2012). The studies of nitrosative responses in citrus plants exposed to six types of abiotic stresses revealed that GSNOR was rapidly and strongly down-regulated at transcript and especially at protein level by cold, and thus GSNOR was suggested as a marker of cold stress in citrus plants (Ziogas et al. 2013).

The modulation of GSNOR activity in response to light conditions was investigated in pea plants, when GSNOR was increased both under continuous light and darkness compared to control plants (Corpas et al. 2008b). Initial studies with Arabidopsis *hot5* mutant grown in the dark also observed differential GSNOR regulation by light conditions (Lee et al. 2008). Strong down-regulation of GSNOR was observed in hypocotyls of etiolated pea plants, which did not recover to values of green plants even 168 h after the transfer of etiolated plants to normal light regime (Kubienová et al. 2014).

Water stress in *Lotus japonicus* results in an increased NO content and a reduced GSNOR activity namely in roots, consequent increase of cellular nitrosothiols and stimulation of protein tyrosine nitration (Signorelli et al. 2013). GSNOR function under salinity stress was reported to be regulated by polyamines, well-known plant regulators involved in plant responses to abiotic stress. NaCl-induced GSNOR activity was strongly suppressed by polyamines in salinized citrus plants, therefore GSNOR was suggested to mediate molecular mechanisms by which plant polyamines orchestrate nitrosative signalling under salinity stress (Tanou et al. 2014).

Further evidence supported important GSNOR role in regulating cell death in plant cells through modulating intracellular NO level, when Arabidopsis GSNOR1/HOT5 mutant was shown to be identical to *paraquat resistant2-1(par2-1)* mutant with an anti-cell death phenotype. It was observed wild-type plants treated with an NO donor were also resistant to paraquat. GSNOR protein level was induced by paraquat but reduced by NO donors, while a steady-state mRNA level was maintained (Chen et al. 2009). The role of S-nitrosylation was also studied in light-dependent leaf cell death in *nitric oxide excess1 (noe1)*, an NO accumulation mutant, in rice (*Oryzasativa*). *NOE1* encodes rice catalase, OsCATC, and its mutation results in the increased leaf H_2O_2 levels, which consequently induced NO production by the activation of nitrate reductase. The overexpression of rice GSNOR gene resulted in the reduction of intracellular *S*-nitrosothiol levels and alleviation of leaf cell death in *noe1* plants (Lin 2012).

GSNOR might be involved in the control of a specific molecular switch including protein carbonylation and S-nitrosylation patterns to control antioxidant enzyme activities during the oxidative stress occurring in desiccated seeds. Treatment with GSNOR inhibitors of recalcitrant *Antiaris toxicaria* seeds, which show increased protein carbonylation and reduced S-nitrosylation of the ascorbate-glutathione cycle enzymes associated with deficient ROS scavenging, increased the seed germination rate (Bai et al. 2011). Exposure to GSNOR inhibitors prior to NO gas, a known inducer of seed germination, increased the S-nitrosylation status and the activities of antioxidant enzymes ascorbate peroxidase, glutathione reductase, and dehydroascorbate reductase (Bai et al. 2011).

Mechanisms of the toxic effect of heavy metals like cadmium involve disturbances in the antioxidant defence and induction of oxidative stress, which in plants results in a strong growth inhibition, decreased transpiration, photosynthesis, and chlorophyll content. GSNOR expression and activity were decreased about 30 % in pea leaves treated with 50 μM cadmium (Barroso 2006), and similar trend was reported in Arabidopsis plants grown in the presence of 0.5 mM arsenate (Leterrier et al. 2012). GSNOR expression and activity were increased by NO treatment in aluminium-treated rice plants, which are under nitrosative stress due to elevated levels of ROS and RNS. Using specific GSNOR inhibitors it was shown that suppressing GSNOR enzymatic activity increased the Al-induced accumulation of RNS and aggravated observed damage to rice plants (Yang et al. 2013). Arabidopsis mutant with decreased GSNOR activity and increased NO levels shows higher tolerance to selenite (Yang et al. 2013) and GSNOR is involved in the fine tuning of ROS/RNS in response to copper exposure in Arabidopsis (Yang et al. 2013). GSNOR might be involved also in the systemic responses to toxic metals inducing cross-resistance to pathogens, as GSNOR was found up-regulated in leaves and stems of potato plants exposed to aluminium, while the enzyme activity in roots was comparable to the non-treated control (Arasimowicz-Jelonek et al. 2014).

4.3 GSNOR in Plant Responses to Biotic Stress

Nitric oxide and *S*-nitrosothiols are now considered as important signal compounds involved in immune responses of animals and plants to microbial pathogens. It has been suggested that similarly to animals, modulations of *S*-nitrosothiol levels in vivo induced by changes in GSNO catabolism controlled by GSNOR can modulate plant defence response to pathogen challenge (reviewed in Malik et al. 2011).

An early study reported the induced expression of GSNOR gene in Arabidopsis plants treated with salicylic acid, a mediator of biotic stress. Salicylate also induces GSNOR levels and activity in tobacco (Díaz et al. 2003). In following experiments, Rusterucci et al. (2007) reported that the transgenic Arabidopsis plants with decreased GSNOR expression achieved by antisense strategy had increased basal resistance to biotrophic pathogen *Peronospora parasitica*. Decreased GSNOR levels correlated with higher intracellular levels of *S*-nitrosothiols and constitutive activation of *PR-1* gene. On the other hand, GSNOR overexpressors showed compromised systemic acquired resistance (SAR). Reported GSNOR localisation in the phloem was suggested to be related with the regulation of mobile signal transmission within plant systemic responses. Based on their experimental data, the authors suggested a hypothetical model, where NO and GSNO would play roles of positive regulators of hypersensitive reaction, defence gene activation, and plant resistance in general, and where decreased GSNOR and consequently increased intracellular *S*-nitrosothiol levels result in increased plant immunity both on local and systemic levels.

This conclusion has been supported by another study which further investigated the role of GSNOR in modulating GSNO levels in Arabidopsis leaves, with implications for wound and SAR responses. It was shown that GSNO accumulation, determined by immunohistochemical staining, was required to activate the jasmonate-dependent wound responses both locally and at systemic level. GSNO also acted synergistically with salicylate in the development of systemic responses, in agreement with the hypothesis of GSNOR as a key regulator of plant systemic responses in wounding and pathogenesis (Espunya et al. 2012).

However, other early studies on GSNOR function in plant defence responses reported rather contradictory results, when decreased GSNOR activity and increased *S*-nitrosothiols were associated with compromised basal and non-host resistance, as it was observed in the case of *A. thaliana* infected with bacteria (*Pseudomonas syringae* pv. *tomato* DC3000), powdery mildew (*Blumeria graminis*), or downy mildew (*Hyaloperonospora parasitica*) (Feechan et al. 2005) or in the case of sunflower (*Helianthus annuus* L.) resistant to pathogenic oomycete *Plasmopara halstedii* (Chaki et al. 2009).

Increased GSNOR (described as ADH class III) levels were identified by theproteomic analysis in maize treated with *Trichoderma harzianum*, a biocontrol symbiotic fungus which is known to induce maize seedling resistance against *Pythium ultimum* (Chen et al. 2005). Transcriptomic and proteomic studies revealed increased levels of both GSNOR and thioredoxin in transgenic *Nicotiana tabacum* plants overexpressing γ-glutamylcysteine synthetase with enhanced level of GSH, which show better tolerance to biotrophic *Pseudomonas syringae* pv. *tabaci* (Ghanta et al. 2011). To get deeper understanding of GSNOR involvement in the control of GSH levels, tobacco BY-2 cells were fed with exogenous GSH which resulted in up-regulation of *NtGSNOR* together with other genes of salicylate pathway dependent on transcriptional coactivator NPR1.

GSNOR is involved in NO-dependent biochemical modifications leading to the mobilization of an effective resistance of potato plants to *Phytophthora infestans*. Potato leaves exposed to a SAR inducers, β-aminobutyric acid or laminarin, accumulated NO and ROS, possibly mediated by temporal GSNOR down-regulation. However, after the pathogen inoculation, the authors observed an up-regulation of GSNOR activity tuned with the depletion of the nitrosothiol pool which enhanced potato defence responses to *Phytophtora* (Janus et al. 2013). Significantly higher GSNOR activity was found under normal conditions in leaves of the *Cucumis* spp. genotype sensitive to biotrophic pathogen *Golovinomyces cichoracearum*, when compared to resistant genotypes (Kubienová et al. 2014).

Conclusions

Current knowledge confirms the key role for GSNOR in the regulation of protein S-nitrosylation as a part of signalling pathways of nitric oxide and reactive nitrogen species. *S*-Nitrosothiols are involved in a wide array of cellular processes during normal development and stress responses in all types or organisms. The detailed characterization of molecular properties of GSNOR has contributed to uncover its important function in normal and pathological physiology of animal cells. Similar advances have been recently achieved also in our understanding of the properties, functions and regulation of plant GSNORs, illustrating the picture of GSNOR within the panorama of S-nitrosylation mechanisms in plant cells in a more complete manner as compared to earlier studies (reviewed in Leterrier et al. 2011). However, a still limited progress could be achieved in key issues related to this field, namely in the analysis of endogenous levels of GSNO as GSNOR substrate by specific and robust methods, and the mechanisms of the regulation of GSNOR expression and activity in vivo. Also, conflicting reports on GSNOR intracellular localization and the modulation of GSNOR levels by abiotic and biotic stress stimuli require further research in various plant models. Extensive application of transcriptomics and proteomics methods in the analysis of GSNOR downregulating or overexpressing plants will surely contribute to this goal (Holzmeister et al. 2011).

Acknowledgements This project was supported by Ministry of Education, Youth and Sport of Czech Republic through the grant Kontakt II - LH11013. We thank Prof. Elizabeth Vierling, University of Amherst Massachusetts, for her collaboration and valuable discussions of multiple aspects of plant GSNOR field. We also appreciate kind collaboration of Juan Barroso, Javier Corpas, and Christian Lindermayr in various projects concerning GSNOR role in plant development and stress responses.

References

Achkor H, Díaz M, Fernández MR, Biosca JA, Parés X, Martínez MC (2003) Enhanced formaldehyde detoxification by overexpression of glutathione-dependent formaldehyde dehydrogenase from Arabidopsis. Plant Physiol 132:2248–2255

Airaki MS-ML, Leterrier M, Barroso JB, Palma JM, Corpas FJ (2011) Detection and quantification of S-nitrosoglutathione (GSNO) in pepper (*Capsicum annuum* L.) plant organs by LC-ES/MS. Plant Cell Physiol 52:2006–2015

Airaki M, Leterrier M, Mateos RM, Valderrama R, Chaki M, Barroso JB, Del RíO LA, Palma JM, Corpas FJ (2012) Metabolism of reactive oxygen species and reactive nitrogen species in pepper (*Capsicum annuum* L.) plants under low temperature stress. Plant Cell Environ 35:281–295

Arasimowicz-Jelonek M, Floryszak-Wieczorek J, Drzewiecka K, Chmielowska-Bąk J, Abramowski D, Izbiańska K (2014) Aluminum induces cross-resistance of potato to Phytophthora infestans. Planta 239:679–694

Bai XG, Yang L, Tian M, Chen J, Shi J, Yang Y, Hu X (2011) Nitric oxide enhances desiccation tolerance of recalcitrant *Antiaris toxicaria* seeds via protein S-nitrosylation and carbonylation. PLoS One 6(6):e20714

Bai XG, Chen JH, Kong XX, Todd CD, Yang YP, Hu XY, Li DZ (2012) Carbon monoxide enhances the chilling tolerance of recalcitrant Baccaurea ramiflora seeds via nitric oxide-mediated glutathione homeostasis. Free Radic Biol Med 53:710–720

Barroso JB, Corpas FJ, Carreras A, Rodriguez-Serrano M, Esteban FJ, Fernandez-Ocana A, Chaki M, Romero-Puertas MC, Valderrama R, Sandalio LM, Del Río LA (2006) Localization of S-nitrosoglutathione and expression of S-nitrosoglutathione reductase in pea plants under cadmium stress. J Exp Bot 57:1785–1793

Barroso J, Valderrama R, Corpas F (2013) Immunolocalization of S-nitrosoglutathione, S-nitrosoglutathione reductase and tyrosine nitration in pea leaf organelles. Acta Physiol Plant 35:2635–2640

Benhar M, Forrester MT, Stamler JS (2009) Protein denitrosylation: enzymatic mechanisms and cellular functions. Nat Rev Mol Cell Biol 10:721–732

Bosworth CA, Toledo JC, Zmijewski JW, Li Q, Lancaster JR (2009) Dinitrosyliron complexes and the mechanism(s) of cellular protein nitrosothiol formation from nitric oxide. Proc Natl Acad Sci USA 106:4671–4676

Chaki M, Fernández-Ocaña AM, Valderrama R, Carreras A, Esteban FJ, Luque F, Gómez-Rodríguez MV, Begara-Morales JC, Corpas FJ, Barroso JB (2009) Involvement of reactive nitrogen and oxygen species (RNS and ROS) in sunflower-mildew interaction. Plant Cell Physiol 50(2):265–279

Chaki M, Valderrama R, Fernández-Ocaña AM, Carreras A, Gómez-Rodríguez MV, Pedrajas JR, Begara-Morales JC, Sánchez-Calvo B, Luque F, Leterrier M, Corpas FJ, Barroso JB (2011) Mechanical wounding induces a nitrosative stress by down-regulation of GSNO reductase and an increase in S-nitrosothiols in sunflower (Helianthus annuus) seedlings. J Exp Bot 62:1803–1813

Chen J, Harman GE, Comis A, Cheng GW (2005) Proteins related to the biocontrol of *Pythium* damping-off in maize with *Trichoderma harzianum* Rifai. J Integr Plant Biol 47:988–997

Chen R, Sun S, Wang C, Li Y, Liang Y, An F, Li C, Dong H, Yang X, Zhang J, Zuo J (2009) The Arabidopsis PARAQUAT RESISTANT2 gene encodes an S-nitrosoglutathione reductase that is a key regulator of cell death. Cell Res 19:1377–1387

Corpas FJ, Carreras A, Esteban FJ, Chaki M, Valderrama R, Del Río LA, Barroso JB (2008a) Localization of S-nitrosothiols and assay of nitric oxide synthase and S-nitrosoglutathione reductase activity in plants. Methods Enzymol 437:561–574

Corpas FJ, Chaki M, Fernandez-Ocana A, Valderrama R, Palma JM, Carreras A, Begara-Morales JC, Airaki M, del Río LA, Barroso JB (2008b) Metabolism of reactive nitrogen species in pea plants under abiotic stress conditions. Plant Cell Physiol 49(11):1711–1722. doi:10.1093/pcp/pcn144, Date of Input: 25 May 2009

Corpas FJ, Alché JD, Barroso JB (2013) Current overview of S-nitrosoglutathione (GSNO) in higher plants. Front Plant Sci 4

Crotty J (2009) Crystal structures and kinetics of S-nitrosoglutathione reductase from Arabidopsis thaliana and human. PhD Dissertation, The University of Arizona, Tuscon, AZ

de Pinto MC, Locato V, Sgobba A, Romero-Puertas MC, Gadaleta C, Delledonne M, De Gara L (2013) S-nitrosylation of ascorbate peroxidase is part of programmed cell death signaling in tobacco bright yellow-2 cells. Plant Physiol 163:1766–1775

Derakhshan B, Hao G, Gross SS (2007) Balancing reactivity against selectivity: the evolution of protein S-nitrosylation as an effector of cell signaling by nitric oxide. Cardiovasc Res 75:210–219

Díaz AH, Titarenko E, Martínez MC (2003) The gene encoding glutathione-dependent formaldehyde dehydrogenase/GSNO reductase is responsive to wounding, jasmonic acid and salicylic acid. FEBS Lett 543:136–139

Dolferus R, Osterman JC, Peacock WJ, Dennis ES (1997) Cloning of the Arabidopsis and dehydrogenase rice formaldehyde genes: implications for the origin of plant ADH enzymes. Genetics 146:1131–1141

Engeland K, Höög JO, Holmquist B, Estonius M, Jörnvall H, Vallee BL (1993) Mutation of Arg-115 of human class III alcohol dehydrogenase: a binding site required for formaldehyde dehydrogenase activity and fatty acid activation. Proc Natl Acad Sci USA 90:2491–2494

Espunya MC, Díaz M, Moreno-Romero J, Martínez MC (2006) Modification of intracellular levels of glutathione-dependent formaldehyde dehydrogenase alters glutathione homeostasis and root development. Plant Cell Environ 29:1002–1011

Espunya MC, Díaz CM, Martínez MC (2007) S-nitrosoglutathione reductase affords protection against pathogens in Arabidopsis, both locally and systemically. Plant Physiol 143:1282–1292

Espunya MC, De Michele R, Gómez-Cadenas A, Martínez MC (2012) S-Nitrosoglutathione is a component of wound- and salicylic acid-induced systemic responses in Arabidopsis thaliana. J Exp Bot 63:3219–3227

Feechan A, Kwon E, Yun BW, Wang Y, Pallas JA, Loake GJ (2005) A central role for S-nitrosothiols in plant disease resistance. Proc Natl Acad Sci USA 102:8054–8059

Foster MW, Hess DT, Stamler JS (2009) Protein S-nitrosylation in health and disease: a current perspective. Trends Mol Med 15:391–404

Frungillo L, de Oliveira JFP, Saviani EE, Oliveira HC, Martínez MC, Salgado I (2013) Modulation of mitochondrial activity by S-nitrosoglutathione reductase in Arabidopsis thaliana transgenic cell lines. Biochim Biophys Acta 1827:239–247

Gaston B (1999) Nitric oxide and thiol groups. Biochim Biophys Acta 1411:323–333

Ghanta S, Bhattacharyya D, Sinha R, Banerjee A, Chattopadhyay S (2011) Nicotiana tabacum overexpressing Î³-ECS exhibits biotic stress tolerance likely through NPR1-dependent salicylic acid-mediated pathway. Planta 233:895–910

Green LS, Chun LE, Patton AK, Sun X, Rosenthal GJ, Richards JP (2012) Mechanism of inhibition for N6022, a first-in-class drug targeting S-nitrosoglutathione reductase. Biochemistry 51(10):2157–2168

Groß F, Durner J, Gaupels F (2013) Nitric oxide, antioxidants and prooxidants in plant defence responses. Front Plant Sci 4:419

Handy DE, Loscalzo J (2006) Nitric oxide and posttranslational modification of the vascular proteome. Arterioscler Thromb Vasc Biol 26:1207–1214

Hanson AD, Gage DA, Shachar-Hill Y (2000) Plant one-carbon metabolism and its engineering. Trends Plant Sci 5:206–213

Hedberg JJ, Griffiths WJ, Nilsson SJF, Höög J-O (2003) Reduction of S-nitrosoglutathione by human alcohol dehydrogenase 3 is an irreversible reaction as analysed by electrospray mass spectrometry. Eur J Biochem 270:1249–1256

Hess DT, Stamler JS (2011) Regulation by S-nitrosylation of protein posttranslational modification. J Biol Chem 287:4411–4418

Holzmeister C, Frohlich A, Sarioglu H, Bauer N, Jr D, Lindermayr C (2011) Proteomic analysis of defense response of wildtype Arabidopsis thaliana and plants with impaired NO- homeostasis. Proteomics 11:1664–1683

Janus Ł, Milczarek G, Arasimowicz-Jelonek M, Abramowski D, Billert H, Floryszak-Wieczorek J (2013) Normoergic NO-dependent changes, triggered by a SAR inducer in potato, create more potent defense responses to Phytophthora infestans. Plant Sci 211:23–34

Jensen D, Belka G, Du Bois G (1998) S-Nitrosoglutathione is a substrate for rat alcohol dehydrogenase class III isoenzyme. Biochem J 331:659–668

Koivusalo M, Uotila L (1991) Glutathione-dependent formaldehyde dehydrogenase (EC 1.2.1.1): evidence for the identity with Class III alcohol dehydrogenase. In: Weiner H, Wermuth B, Crabb D (eds) Enzymology and molecular biology of carbonyl metabolism 3. Advances in experimental medicine and biology, vol 284. Springer, New York, pp 305–313

Koivusalo M, Baumann M, Uotila L (1989) Evidence for the identity of glutathione-dependent formaldehyde dehydrogenase and class III alcohol dehydrogenase. FEBS Lett 257:105–109

Kubienová L, Kopečný D, Tylichová M, Briozzo P, Skopalová J, Šebela M, Navrátil M, Tâche R, Luhová L, Barroso JB, Petřivalský M (2013) Structural and functional characterization of a plant S-nitrosoglutathione reductase from Solanum lycopersicum. Biochimie 95:889–902

Kubienová L, Tichá T, Jahnová J, Luhová L, Mieslerová B, Petřivalský M (2014) Effect of abiotic stress stimuli on S-nitrosoglutathione reductase in plants. Planta 239:139–146

Kwon E, Feechan A, Yun B-W, Hwang B-H, Pallas J, Kang J-G, Loake G (2012) AtGSNOR1 function is required for multiple developmental programs in Arabidopsis. Planta 236:887–900

Lee U, Wie C, Fernandez BO, Feelisch M, Vierling E (2008) Modulation of nitrosative stress by S-nitrosoglutathione reductase is critical for thermotolerance and plant growth in Arabidopsis. Plant Cell 20:786–802

Leterrier M, Chaki M, Airaki M, Valderrama R, Palma JM, Barroso JB, Corpas FJ (2011) Function of S-nitrosoglutathione reductase (GSNOR) in plant development and under biotic/abiotic stress. Plant Signal Behav 6:789–793

Leterrier M, Airaki M, Palma JM, Chaki M, Barroso JB, Corpas FJ (2012) Arsenic triggers the nitric oxide (NO) and S-nitrosoglutathione (GSNO) metabolism in Arabidopsis. Environ Pollut 166:136–143

Lindermayr C, Saalbach G, Bahnweg G, Durner J (2006) Differential inhibition of Arabidopsis methionine adenosyltransferases by protein S-nitrosylation. J Biol Chem 281:4285–4291

Lin A, Wan Y, Tang J, Xue P, Li C, Liu L, Hu B, Yang F, Loake GJ, Chu C (2012) Nitric oxide and protein S-nitrosylation are integral to hydrogen peroxide-induced leaf cell death in rice. Plant Physiol 158:451–464

Liu L, Hausladen A, Zeng M, Que L, Heitman J, Stamler JS (2001) A metabolic enzyme for S-nitrosothiol conserved from bacteria to humans. Nature 410:490–494

Luiking YC, Engelen MP, Deutz NE (2010) Regulation of nitric oxide production in health and disease. Curr Opin Clin Nutr Metab Care 13:97–104

Malik SI, Hussain A, Yun B-W, Spoel SH, Loake GJ (2011) GSNOR-mediated de-nitrosylation in the plant defence response. Plant Sci 181:540–544

Marino SM, Gladyshev VN (2010) Structural analysis of cysteine S-nitrosylation: a modified acid-based motif and the emerging role of trans-nitrosylation. Planta Med 395:844–859

Martínez MC, Achkor H, Persson B, Fernández MR, Shafqat J, Farrés J, Jornvall H, Parés X (1996) Arabidopsis formaldehyde dehydrogenase. Eur J Biochem 241:849–857

Martínez-Ruiz A, Lamas S (2004) S-nitrosylation: a potential new paradigm in signal transduction. Cardiovasc Res 62:43–52

Mur LAJ, Mandon J, Persijn S, Cristescu SM, Moshkov IE, Novikova GV, Hall MA, Harren FJM, Hebelstrup KH, Gupta KJ (2013) Nitric oxide in plants: an assessment of the current state of knowledge. AoB Plants 5:pls052

Reumann S, Babujee L, Ma C, Wienkoop S, Siemsen T, Antonicelli GE, Rasche N, Lüder F, Weckwerth W, Jahn O (2007) Proteome analysis of Arabidopsis leaf peroxisomes reveals novel targeting peptides, metabolic pathways, and defense mechanisms. Plant Cell 19:3170–3193

Sakamoto A, Ueda M, Morikawa H (2002) Arabidopsis glutathione-dependent formaldehyde dehydrogenase is an S-nitrosoglutathione reductase. FEBS Lett 515:20–24

Salgado I, Carmen Martínez M, Oliveira H, Frungillo L (2013) Nitric oxide signaling and homeostasis in plants: a focus on nitrate reductase and S-nitrosoglutathione reductase in stress-related responses. Braz J Bot 36:89–98

Sanghani PC, Stone CL, Ray BD, Pindel EV, Hurley TD, Bosron WF (2000) Kinetic mechanism of human glutathione-dependent formaldehyde dehydrogenase. Biochemistry 39:10720–10729

Sanghani PC, Bosron WF, Hurley TD (2002a) Human glutathione-dependent formaldehyde dehydrogenase. Structural changes associated with ternary complex formation. Biochemistry 41:15189–15194

Sanghani PC, Robinson H, Bosron WF, Hurley TD (2002b) Human glutathione-dependent formaldehyde dehydrogenase. Structures of apo, binary, and inhibitory ternary complexes. Biochemistry 41:10778–10786

Sanghani PC, Robinson H, Bennett-Lovsey R, Hurley TD, Bosron WF (2003) Structure–function relationships in human Class III alcohol dehydrogenase (formaldehyde dehydrogenase). Chem Biol Interat 143–144:195–200

Sengupta R, Holmgren A (2013) Thioredoxin and thioredoxin reductase in relation to reversible S-nitrosylation. Antioxid Redox Signal 18:259–269

Seth D, Stamler JS (2011) The SNO-proteome: causation and classifications. Curr Opin Chem Biol 15:129–136

Signorelli S, Corpas FJ, Borsani O, Barroso JB, Monza J (2013) Water stress induces a differential and spatially distributed nitro-oxidative stress response in roots and leaves of Lotus japonicus. Plant Sci 201–202:137–146

Staab CA, Ålander J, Brandt M, Lengqvist J, Morgenstern R, Grafström RC, Höög J-O (2008a) Reduction of S-nitrosoglutathione by alcohol dehydrogenase 3 is facilitated by substrate alcohols via direct cofactor recycling and leads to GSH-controlled formation of glutathione transferase inhibitors. Biochem J 413:493–504

Staab CA, Hellgren M, Höög JO (2008b) Medium- and short-chain dehydrogenase/reductase gene and protein families. Cell Mol Life Sci 65:3950–3960

Staab CA, Ålander J, Morgenstern R, Grafström RC, Höög J-O (2009) The Janus face of alcohol dehydrogenase 3. Chem Biol Interact 178:29–35

Sun X, Qiu J, Strong SA, Green LS, Wasley JWF, Colagiovanni DB, Mutka SC, Blonder JP, Stout AM, Richards JP, Chun L, Rosenthal GJ (2011) Structure-activity relationships of pyrrole based S-nitrosoglutathione reductase inhibitors: pyrrole regioisomers and propionic acid replacement. Bioorg Med Chem Lett 21:3671–3675

Tanou G, Ziogas V, Belghazi M, Christou A, Filippou P, Job D, Fotopoulos V, Molassiotis A (2014) Polyamines reprogram oxidative and nitrosative status and the proteome of citrus plants exposed to salinity stress. Plant Cell Environ 37:864–885

Uotila L, Koivusalo M (1979) Purification of formaldehyde and formate dehydrogenases from pea seeds by affinity chromatography and S-formylglutathione as the intermediate of formaldehyde metabolism. Arch Biochem Biophys 196:33–45

Veech RL, Guynn R, Veloso D (1972) The time-course of the effects of ethanol on the redox and phosphorylation states of rat liver. Biochem J 127:387–397

Wunsche H, Baldwin IT, Wu J (2011) S-Nitrosoglutathione reductase (GSNOR) mediates the biosynthesis of jasmonic acid and ethylene induced by feeding of the insect herbivore Manduca sexta and is important for jasmonate-elicited responses in Nicotiana attenuata. J Exp Bot 62:4605–4616

Xu S, Guerra D, Lee U, Vierling E (2013) S-nitrosoglutathione reductases are low-copy number, cysteine-rich proteins in plants that control multiple developmental and defense responses in Arabidopsis. Front Plant Sci 4:430

Yang L, Tian D, Todd CD, Luo Y, Hu X (2013) Comparative proteome analyses reveal that nitric oxide is an important signal molecule in the response of rice to aluminum toxicity. J Proteome Res 12:1316–1330

Yu M, Lamattina L, Spoel SH, Loake GJ (2014) Nitric oxide function in plant biology: a redox cue in deconvolution. New Phytol 202:1142–1156

Ziogas V, Tanou G, Filippou P, Diamantidis G, Vasilakakis M, Fotopoulos V, Molassiotis A (2013) Nitrosative responses in citrus plants exposed to six abiotic stress conditions. Plant Physiol Biochem 68:118–126

Interaction of Calcium Signalling with Reactive Oxygen and Reactive Nitrogen Species

Samir Sharma

1 Reactive Oxygen Species, Reactive Nitrogen Species and Ca^{2+}: Intrinsic Signals and Their Functional Position vis-à-vis Extrinsic Signals

The energy conserving pathway of respiration is an essential component of life and occurs in every living cell. It involves the harnessing of energy released during the spontaneous oxidation of reduced carbohydrates by a process that couples downhill electron transport to ATP production. Given the continuously changing redox status and the presence of significant amount of oxygen in the cell, it is inevitable that some of the transported electrons leak to generate reactive oxygen species (ROS). The situation is further complicated in green tissue, where photosynthetic electron transport operates vigorously. Photosynthesis, being a light-driven process, is less subject to demand for energy and/or fixed carbon and tries to optimize electron transport rates within a safe range using a collection of dissipatory processes. However, the main driver of photosynthesis, light, is an extrinsic factor beyond the control of the chloroplast and therefore despite best efforts, energization of the light-absorbing system cannot be avoided even if the reason to photosynthesize is not expressed. Chlorophyll is either involved in energy absorption and transfer or, less frequently, in photochemistry or charge transfer, making photosynthesis a major contributor of ROS. Although they certainly are major contributors of ROS, apart from respiration and photosynthesis, there are several other biochemical reactions that contribute significantly. ROS are therefore an inevitable spinoff of a number of redox reactions and have been implicated signalling in plants. Reactive nitrogen species (RNS), principally nitric oxide (NO), known for a long time to

S. Sharma (✉)
Department of Biochemistry, University of Lucknow, Lucknow 226007, India
e-mail: samiersharma@gmail.com

K.J. Gupta, A.U. Igamberdiev (eds.), *Reactive Oxygen and Nitrogen Species Signaling and Communication in Plants*, Signaling and Communication in Plants 23,
DOI 10.1007/978-3-319-10079-1_15

modify animal metabolism, has been implicated as a plant signalling entity controlling several aspects of plant metabolism. Both ROS and RNS are known to covalently modify biomolecules. This modification may be completely random and may lead to inactivation or structural impairment of the molecule, but at the same time, oxidative or nitrosative changes in molecules are perceived as markers of a changed metabolic state to which the plant must respond. Since a large number of these changes occur in proteins, they result in conformational changes, some of which are used as signals. Changes induced in lipid molecules trigger a wave of signalling through soluble or volatile lipid degradation products and affect several aspects of plant metabolism. The significant commonness in ROS- and RNS-based signalling is that these are signals that reflect the metabolic status of the cell as opposed to hormone, light or even pathogenesis (extrinsic signals)-induced signalling. Reactive species-based signalling, therefore, represents a different tier of signalling than extrinsic signals that reflect the status of metabolism and coordinate other activities of the cell within its own self.

Signalling elements having the simplest of structures seem to possess the most complicated and diverse signalling pathways and $Ca^{2+}$is a prime example of this class. This ion whose presence in the cytosol would wreak havoc due to formation of the almost insoluble calcium phosphate is kept out of the cytoplasm by a variety of Ca^{2+} pumps that remove the ion to internal stores or to the apoplast. Scrubbing the cytoplasm clean of Ca^{2+} provides a very low background for the ion, making it a remarkable signal, if moved to the cytoplasm in short, rapidly reversible bursts, the frequency and location of which conveys different messages within the cell. Ca^{2+} also belongs to the second tier of messengers, with the primary messages usually being of extracellular origin. A huge number of Ca^{2+}-binding proteins with different binding affinities as well different locations within the cell decipher Ca^{2+} signals to drive metabolic changes.

Lately, our understanding of signalling has moved significantly from perceiving them as linear pathways to seeing them as extensively branched pathways with a very high degree of cross-talk. The present chapter concerns itself with the three signals of the 'second tier' and their interactions that modify cellular metabolism far beyond what was understood earlier. One such aspect that becomes apparent with the current status of knowledge is that ROS and RNS production and removal maintains these species at almost continuously fluctuating levels that indicate the cell's redox state. This level appears to change with every alteration of conditions external to plants, including stress, and also changes with genetically imprinted internal cues, such as those governing development. Ca^{2+} dynamics being related to most, if not all signalling pathways, are also altered by ROS and RNS through covalent modification of channels, pumps, and Ca^{2+}-interacting proteins. Lastly, the fact that these signalling pathways are referred to as the 'second tier' does not imply lesser importance in the governance of cellular metabolism. Instead, they should be seen as the real effectors of intercellular signalling driven by growth hormones and extracellular growth-modulating factors.

2 Reactive Species: Controlling Cellular Metabolism from Within

Biological systems are known to form reactive oxygen species due to incomplete reduction of oxygen, as they perform redox reactions (Fig. 1). Oxygen per se is considered toxic and, perhaps taking this line of thought farther, even mutagenic (Halliwell and Gutteridge 2007). Oxygen owes its essentiality as well as the inherent risk associated with handling it, to its high reactivity. Driving biological oxidation is essential for energy production within any living system and is always associated with the production of reactive oxygen species (ROS) like superoxide anion (O_2^{-}), hydrogen peroxide (H_2O_2), singlet oxygen ($^1O_2$), and the hydroxyl radical ($HO^{\bullet}$). Being more reactive than their parent molecular oxygen, ROS react with large molecules (Mittler 2002; Buchanan and Balmer 2005; Møller et al. 2007) and produce a large number of covalent changes in them, prominent among these being oxidative modification of proteins. Most of these modifications are irreversible and lead to changes in the structure of the protein (Møller et al. 2007). Protein carbonylation is the most commonly occurring oxidative modification of proteins apart from reactions involving sulphur containing amino acids and appears to be essentially irreversible (Shacter 2000). Plants have a large arsenal of antioxidant defences comprising a set of enzymes such as superoxide dismutases, catalases, and peroxidases as well as a collection of small, diffusible molecules like glutathione, ascorbate, and a large number of secondary plant products capable of reacting sacrificially with ROS (Smirnoff 2005). ROS levels therefore are kept low by cells that produce them, minimizing oxidative damage in the process. However, what has

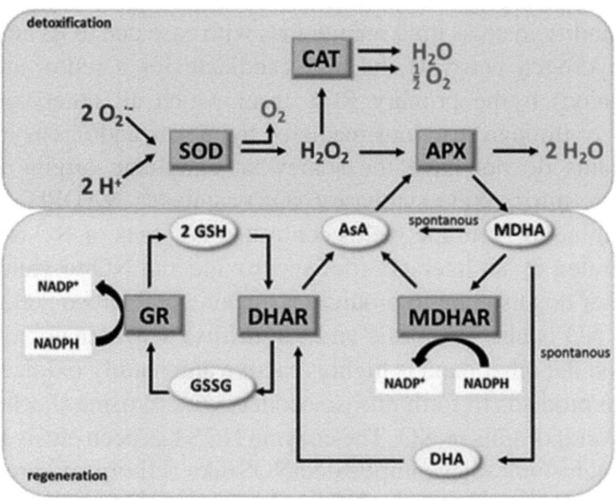

Fig. 1 Reactive oxygen species and their detoxification. SOD superoxide dismutase, CAT catalase, APX ascorbate peroxidase, GR glutathione reductase, DHAR dehydroascorbate reductase, MDHAR monodehydroascorbate reductase, AsA ascorbic acid [Reproduced with permission from Groß et al. (2013)]

come to the fore recently is that every organism tries to maintain ROS at a low albeit significant level instead of eliminating them completely (Chew et al. 2003; Navrot et al. 2006).

Understandably, plant cells involved in photosynthesis have significantly more robust defences against ROS as compared to other cells but even here, the status of redox as well as ROS concentration controls photosynthetic gene expression (Pfannschmidt et al. 2009). Oxidative modification of proteins has been considered one of the principle ways in which ROS control cellular metabolism, elevating ROS to the level of signalling entities from the mere destructive elements they have been known as since a long time. ROS are not alone in this action. They are known to start a cascade resulting in several other types of reactive species broadly classified as reactive electrophile species (RES) that includes a class of highly mobile molecules known as reactive lipid species (RLS) generated from lipid oxidation by ROS (Farmer and Davoine 2007; Wall et al. 2012; Farmer and Mueller 2013). ROS have been known to be produced in a highly controlled and localized pattern in biological stress responses also, integrating stresses and enabling the cell to respond (Fedoroff 2006). Apart from ROS and RES, there is a third, extremely important component of reactive species, very active in the control of cellular responses, the reactive nitrogen species (RNS). ROS, RES, and RNS are ubiquitous in their presence in living systems and are now established messengers that connect different aspects of signalling and metabolism and have been implicated in stress-induced changes in metabolism as well.

Considering all reactive species, their lifetimes understandably vary in inverse proportion with their reactivity. Among ROS, for example, $OH^{\bullet}$ has the greatest reactivity and consequently, the smallest lifetime, allowing it to react with molecules present in the immediate vicinity of its creation (Møller et al. 2007). H_2O_2 on the contrary, is least reactive amongst all ROS, has the longest lifetime and possesses the ability to cross lipid membranes with ease due to its non-polar nature. This makes hydrogen peroxide the best candidate for a redox messenger. O_2^- (superoxide anion) is the primary ROS from which all others originate either spontaneously or through reactions mediated by metals and/or enzymes.

RNS in plants do not have the somewhat simplistic origins as they do in mammals where nitric oxide synthase (NOS)-catalysed NADPH-dependent conversion of arginine to citrulline yields controlled amounts of NO (Fig. 2). Further RNS are generated by an interaction of superoxide and NO to yield peroxynitrite radical and other downstream to products (Durzan and Pedroso 2002). Like hydrogen peroxide, NO is also lipophilic and can diffuse across membranes with ease. Peroxynitrite on the other hand is highly reactive and readily oxidizes cysteine and its downstream products to form nitroso adducts with tyrosine (Pacher et al. 2007). Plants have several origins of NO. The enzyme NOS has been elusive so far in plant systems, although studies have implicated NOS-like activity and production of NO from Arg as one of the pathways for NO production. Apart from the oxidative origin just mentioned, plants have a distinct source of NO in reactions concerned with reduction of nitrate, a major plant nutrient (Gupta et al. 2010; Kocsy et al. 2013). NO_2^- yields NO through the action of nitrate reductase, nitrite-NO Reductase (Ni-NOR), or xanthine oxidase (XO) (Fig. 3). RNS have a very diverse effect on

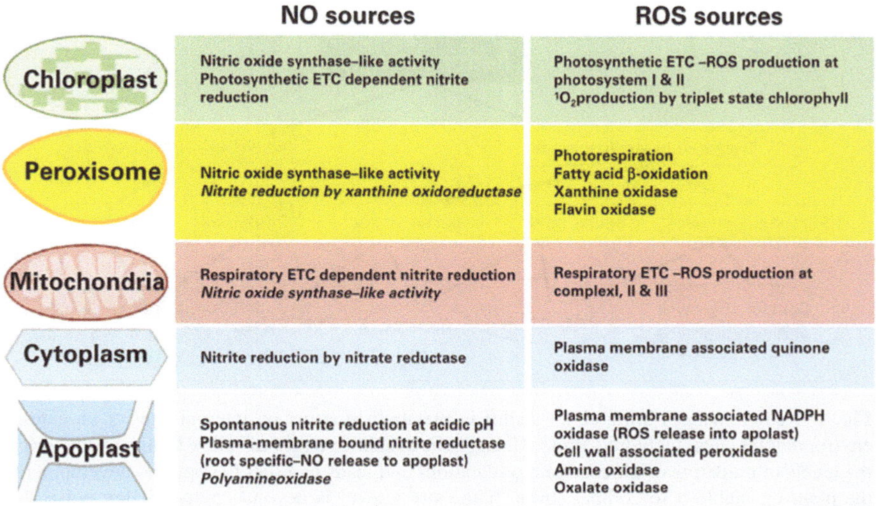

Fig. 2 Sources of ROS and RNS in plants [Reproduced with permission from Groß et al. (2013)]

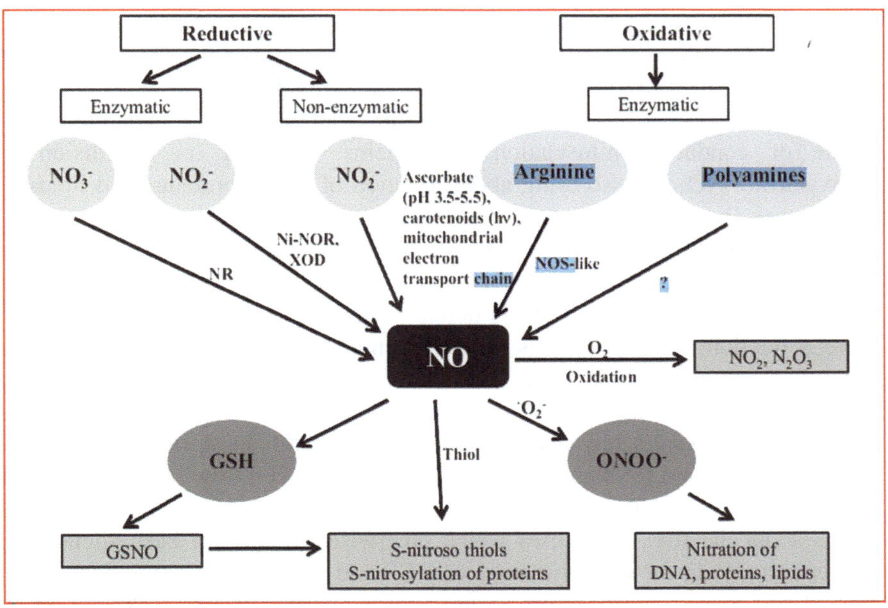

Fig. 3 Routes for production of NO and its central role in nitrosative stress in plants. Reactions of NO with oxygen and superoxide also emphasize the integration of oxidative and nitrosative stresses [Reproduced with permission from Kocsy et al. (2013)]

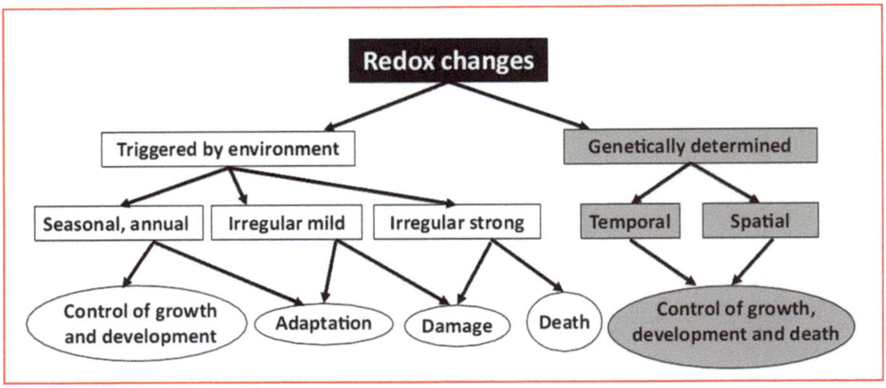

Fig. 4 Redox changes bring about a shift in metabolism either on internal cues or on external, environmental cues, including stress. Changes in dynamic ranges of redox brings about shifts in the levels of oxidative or nitrosative modifications and leads to either the optimal development of the plant or enable it to combat stress. If the stress level is beyond control, redox poise shifts drastically to oxidizing and leads to the demise of the cell, tissue, or perhaps the organism itself [Reproduced with permission from Kocsy et al. (2013)]

cellular metabolism. On one hand, NO modulates the concentration of ROS by combining with them and by conducting direct protein modification, while on the other, it yields RNS of different reactivities that modify target proteins different than the ROS do and give rise to an entirely different signalling scenario. RNS, therefore, also bring about a stress situation called nitrosative stress, which is somewhat similar to oxidative stress associated with ROS. The main player in this stress situation is NO that can lead to nitrosylation of thiols including glutathione or conduct S-nitosylation of proteins. Through its conversion to peroxynitrite, NO can also bring about nitration of DNA, proteins, and lipids to add another dimension to nitrosative stress.

At this point it is imperative to mention that the term ‘stress’ when used for oxidative and nitrosative changes, may be a bit of a misnomer. At any given point of time, plant metabolism, with all its defences against ROS and RNS, chooses to leave a fraction of these entities unaltered and reactive. There appears to be a dynamic range in which ROS and RNS convey the cell’s own metabolic status to concerned metabolic pathways and direct fine-tuning of metabolism to optimize productivity or plant’s response to stress (Foyer 2005). In stress situations, the balance or ROS and RNS shifts radically to move the cell to greater oxidative or nitrosative stress and the dynamic range of these reactive species changes. This brings about the stress responses which otherwise would have been silent. Any further increase in the dynamic range of ROS and RNS may take the stress level beyond tolerable limits, leading to the demise of the cell either directly or by triggering programmed cell death (Yu et al. 2014) (Fig. 4).

Recently, there has been accumulating evidence indicating large suites of kinases and E3 ligases from other organisms that either precisely regulate the extent of posttranslational modifications or degradation of transcriptional regulatory molecules. Such regulation would bring about a very precise qualitative and quantitative control of gene expression during development or response to stress (Kornberg et al. 2010).

3 Ca^{2+} Signalling in Higher Plants: Complex Interpretation of a Simple Message

Ca^{2+} (ionic calcium) is unequivocally the simplest and longest-lived messenger that is virtually ubiquitous in its presence. All other messages undergo reversal by chemical alteration of the messenger that may proceed to level of its complete destruction. Being an ion with a very low presence in the cytosol, Ca^{2+}generates signals that are reversible only by a change in the location of the ion. The primary domain of the messenger action of Ca^{2+} is the cytosol where it undergoes quick removal to either the apoplast or one of the intracellular Ca^{2+} stores, resulting in pulses or waves of Ca^{2+} (Trewavas 1999). A wide variety of biotic and abiotic stimuli cause rapid changes in Ca^{2+}concentration in the cytosol (Sanders et al. 2002). This influx of Ca^{2+} is driven by the high electrochemical difference at which the ion is held outside the cell or inside intracellular stores like the vacuole, endoplasmic reticulum, mitochondria, etc. (Stael et al. 2012). Without a doubt, the flux of Ca^{2+} into and out of the cytoplasm as well as the components involved in its decoding have been an area of extensive study and review (Clapham 2007; McAinsh and Pittman 2009; Dodd et al. 2010; Whalley and Knight 2013). A large number of Ca^{2+} permeable channels involved in generating Ca^{2+} signals have been identified in plants, with the word 'permeable' being used to reflect the thought that a significant number of these channels are likely to be non-selective cation channels (Sanders et al. 2002; Hetherington and Brownlee 2004). Unlike other ion channels that have been implicated in nutritional fluxes, Ca^{2+} channels have been implicated in signalling (Pineros and Tester 1997; White 2000). The most significant flux of Ca^{2+} in most situations is controlled by voltage-dependent and elicitor-activated Ca^{2+} channels of the plasma membrane and the depolarisation-activated, hyperpolarisation-activated, IP_3, and cADPR-dependent Ca^{2+} channels of the tonoplast. A classic example is that of ABA-induced closure of stomata guard cells in the presence of abscisic acid (ABA) where the coordination of Ca^{2+} activities drives a physiological response (White 2000).

Once the Ca^{2+} signal has been generated, it needs to be precisely and quickly interpreted in the cytoplasm. This decoding of the Ca^{2+} signal is done by Ca^{2+} sensor or adapter proteins, of which there are numerous and calmodulin (CaM) happens to be the prime and most worked-on example (Clapham 2007). Ca^{2+} biding induces conformational changes in CaM domains, inducing the so-called active form of the protein. In its Ca^{2+}-bound form, CaM triggers reversal of autoinhibition, remodelling of active sites, and even dimerization of proteins (Hoeflich and Ikura 2002). Hundreds of proteins possess CaM docking/binding sites and lend themselves to receive the Ca^{2+} message interpreted by CaM. The importance of CaM and the positive selection of Ca^{2+}-signalling pathways are underlined by the fact that the CaM has changed very little during the 1.5 billion years or more of evolution of Ca^{2+} signalling (Abzhanov et al. 2006). It is remarkable that extremely different eukaryotes have a remarkable similarity in their cytosolic Ca^{2+} fluxes as well as sensory toolkit (Fig. 5). However, some of these components are unique to photosynthetic organisms and offer explanation for the differences that higher plants have in Ca^{2+} signalling from their animal counterparts (Bothwell and Ng 2005).

Ca^{2+} flux into the cell is immediately reversed by membrane bound pumps. This results in a 'spike' of Ca^{2+} concentration in the cytosol. A collection of these comprises a Ca^{2+} wave, mentioned earlier. Different stimuli result in spikes arriving with different frequency and amplitude, so much so that hardly any two Ca^{2+} waves look identical. Apart from this, the influx of Ca^{2+} occurs neither uniformly nor randomly. Instead, there is stringent spatiotemporal regulation of the site of Ca^{2+} influx. A combination of all these features, wave frequency, spatial location as well as timing of Ca^{2+} flux constitutes a 'calcium signature' (Whalley and Knight 2013). These Ca^{2+} signatures, although difficult to obtain, are believed to be unique to every developmental cue such as hormones, to stresses or to combinations thereof (Fig. 6).

Plants thus use the Ca^{2+} signature-Calmodulin combination during development as well as during adverse environmental conditions to trigger cellular responses that are apt as well as timely. These responses are products of altered gene expression brought about by Ca^{2+} CaM signalling. Modification of the toolkit of Ca^{2+} signalling could be expected to lead to alteration of gene expression and can be construed as an attempt on the part of the plant to adjust to its developmental or environmental cues. Ca^{2+} signalling involves the simplest signalling entity of all and therefore possibly the largest toolkit among all signalling pathways is necessary to decode or synchronize Ca^{2+}-governed responses in plants. As nearly all components of the Ca^{2+} toolkit are proteins, this massive collection of proteins represents a prominent target for reactive agents capable of modifying protein structure and function.

4 Modification of Ca^{2+} Signalling by ROS and RNS

Being sessile, plants encounter a vast range of environmental conditions that sometimes simply guide plant form and function, but at other times may prove to be stressful. It has been established earlier in this chapter that ROS and RNS levels as well as their spatiotemporal characteristics changes with development and stress. Since these reactive species have a very dynamic presence in the cell and are highly reactive, modifications in macromolecules due to oxidative or nitrosative changes are bound to result in the vicinity of ROS or RNS. It is apparent that reactive species do not have any specific targets. Macromolecules in the immediate vicinity of ROS/RNS would stand the chance of maximum harm *prima facie*. However, the

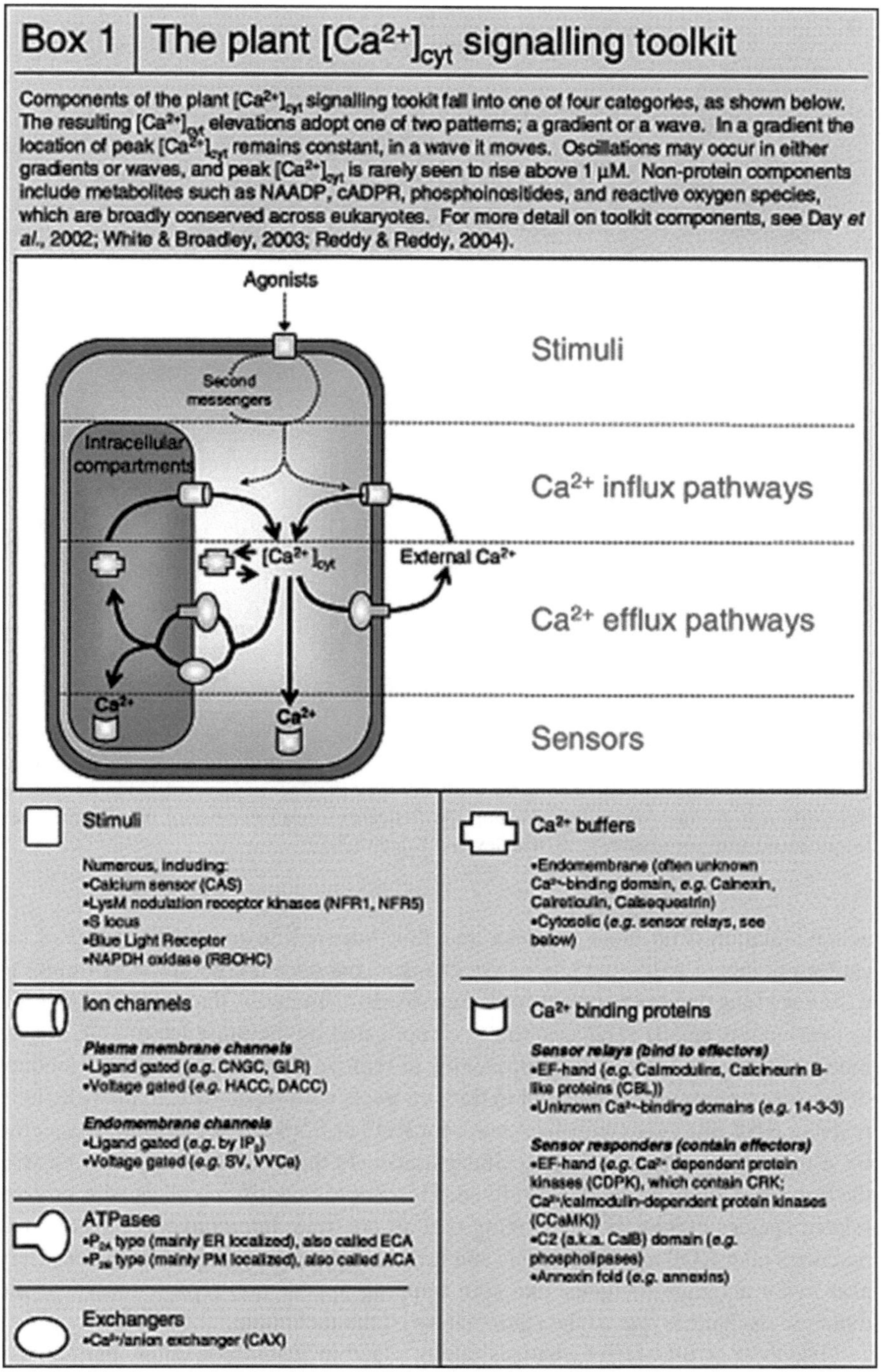

Fig. 5 The 'Ca^{2+} signaling toolkit', a collection of components comprising stimuli, channels, pumps, exchangers, Ca^{2+} buffers, and Ca^{2+}-binding proteins [Modified from Bothwell and Ng (2005)]

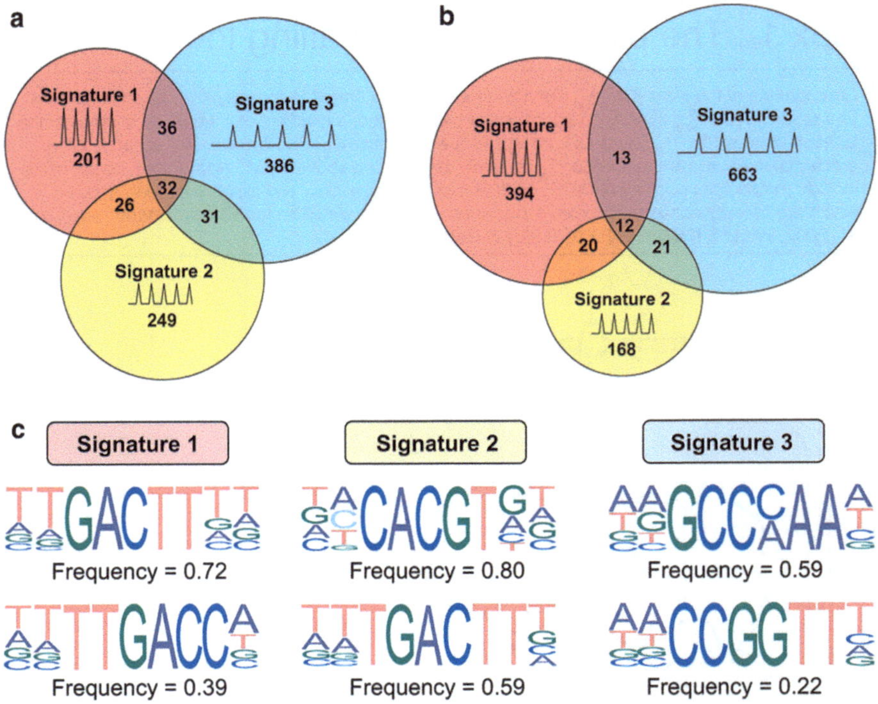

Fig. 6 Venn diagrams depicting the groups of genes whose expression increases (**a**) or decreases (**b**) by 1.5-fold or greater in response to three different calcium signatures. (**c**) Table showing consensus DNA sequences enriched in the promoter regions of genes up-regulated in response to three different calcium signatures, showing the frequency (mean number of motifs per gene). [Reproduced with permission from Whalley and Knight (2013)]

actual situation is far more complex than this. Superoxide appears to be one of the starting points of oxidative/nitrosative changes, but does not get the opportunity to exist very long due to the presence of superoxide dismutases, that quickly convert it to hydrogen peroxide. The scenario is complicated by the presence of NO, arising independently. NO reacts with superoxide to remove it and thus reduce the amount of hydrogen peroxide present, but in turn, interacts with superoxide to provide more reactive RNS like peroxynitrite. Assays for ROS or RNS are not absolutely specific for a particular species and intracellular transients that must be captured to assess the true situation in a cell all short lived. This makes quantifying all species or even related species extremely challenging if at all possible. Interconversions and cross reactions alter ROS and RNS levels and further complicate the quantification. They also make accurate methods like spin trapping appear non-representative of the dynamic exchanges due to the static nature of the technique.

Oxidative or nitrosative changes taking place in macromolecules, particularly proteins, are easier to quantify as well as assess qualitatively. Immunochemical detection, mass spectrometry and ultimately, genetically modified target proteins are generating vast volumes of data that is strongly indicative of at least some of these covalent changes being the game changers of metabolism. Collateral damage notwithstanding, specific alterations to proteins of discrete, established function is fast unravelling the non-random side of the web of changes brought about by ROS and/or RNS. It is becoming increasingly apparent that in one dynamic range of ROS/RNS, these changes seem deliberate and that there is a preference for certain molecules for oxidation/nitrosation. As the dynamic range shifts, as during transition to flowering, during germination or during perception of stress, the levels of ROS/RNS move up to a higher dynamic range and a new set of proteins undergoes modifications, sending out new messages.

4.1 ROS Interaction with Ca^{2+}

Versatility of Ca^{2+} as a messenger is as unparalleled as its simplicity. Given the vast protein machinery, referred to as the Ca^{2+} toolkit required for Ca^{2+} signalling and its presence in the apoplast as well almost all cellular compartments, the toolkit components present a prominent target for oxidative and nitrosative changes. Such changes would result in modification of the Ca^{2+} signature and consequently a change in metabolic status of the cell. Proteomics approach to identify components involved in Ca^{2+} signalling identified several components actively participating in ROS production or perception (Reddy and Reddy 2004). Early reports of Ca^{2+} interactions with ROS established a connection between the cation and ROS production and ROS in mitochondria, as Ca^{2+} is considered a key regulator of mitochondrial function. In animal systems, mitochondrial overload has been linked to massive overproduction of ROS (Brookes et al. 2004) and went on to implicate NO and MAP Kinases into the fold. In the marine algae *Ulva compressa* exposed to copper stress, co-occurrence of organellar calcium was observed (Gonzalez et al. 2010). Antioxidant enzyme activities were increased due to copper excess. Activation of ascorbate peroxidase was shown to be strictly ROS dependent, activation of phenylalanine ammonia lyase and lipoxygenase was strictly Ca^{2+} dependent, while glutathione reductase and glutathione-*S*-transferase required both Ca^{2+} and ROS as triggering signals.

Recent reports have increasingly reported both Ca^{2+} and ROS to be participatory in several cellular processes like stomata opening, programmed cell death, development, gravitropism, drought and salt stresses, or hormonal sensing (Gadjev et al. 2008; Hepler 2005; Hetherington and Brownlee 2004; Miller et al. 2009; Mittler and Berkowitz 2001; Quan et al. 2008). Although ROS signalling appears to be almost as complex as Ca^{2+} signalling, the concept of ROS signatures has not been proven so far (Fedoroff 2006). Plant plasma membrane resident NADPH oxidases possess a cytosolic terminal with two EF-hand domains to bind Ca^{2+} and its activation is thought to be controlled by Ca^{2+} binding (Takeda et al. 2008). Further, the activation of these oxidases is known to be activated by a Ca^{2+}-dependent protein kinase (Kobayashi et al. 2007). In another example of this synergy between ROS and calcium, Ca^{2+}/CaM in plants is known to activate a NAD kinase, producing NADPH to fuel NADPH oxidases to result in greater ROS production (Harding et al. 1997). The data supporting Ca^{2+}–ROS interactions as so far diverse but is accumulating at a fast pace, considering that ROS and NO signalling networks are also deeply entwined (Mazars et al. 2010). Plant annexins have recently been reported to act as multifunctional proteins that are capable of peroxidase activity (Laohavisit et al. 2009, 2010) thereby modulating ROS signatures also. Annexin expression responds to stress conditions like salinity, drought, nutrient deprivation, cold, etc. that are known to increase cytosolic Ca^{2+} and ROS as mentioned earlier. ROS have been observed to affect intracellular Ca^{2+}dynamics by activating a range of non-selective Ca^{2+}-permeable channels in plasma membrane ionic currents. Hydroxyl radicals were seen to induce K^+ and Ca^{2+} fluxes in pea (*Pisum sativum*) roots. Hydroxyl also activated a rapid Ca^{2+} efflux and a more slowly developing net Ca^{2+} influx concurrent with a net K^+ efflux (Zepeda-Jazo et al. 2011).

Membranes are key sources of signalling lipids (Farmer and Davoine 2007). As annexins formed an inwardly directed Ca^{2+}-permeable conductance in MDA-containing bilayers, it is feasible that annexins contribute to the inwardly directed and ROS-activated PM Ca^{2+} conductance identified in plants (Mori and Schroeder 2004; Demidchik et al. 2007) and the lipid peroxidation that is typified by increases in MDA content (Collin et al. 2008). This suggests that annexins could function in integrating ROS and Ca^{2+} in stress signalling. Additionally, annexins are found at growth points such as root hair and pollen tube apices (Mortimer et al. 2009; Laohavisit and Davies 2009) at which ROS can regulate $[Ca^{2+}]_{cyt}$, pointing to annexins as a putative ROS-regulated Ca^{2+} influx pathway for polar growth.

4.2 NO Interaction with Ca^{2+}

While the precise mechanism of the mechanism by which NO influences metabolism at the molecular level remains elusive, the discovery of interaction between NO, Ca^{2+}, cADPR, and protein kinases has been illuminating (Besson-Bard et al. 2008a, b). Nitric oxide has been shown to influence cellular events bringing about an increase in cytosolic Ca^{2+} in pathways for ABA, hyper-osmotic and elicitor transduction (Garcia-Mata et al. 2003; Gould et al. 2003; Lamotte et al. 2006). As a corollary, a role for Ca^{2+} in initiating NO production gas also been shown. The identification of NO as well as Ca^{2+} protein targets like S-nitrosylated proteins or new Ca^{2+}-dependent protein kinases through novel proteomic approaches (Lindermayr et al. 2006; Belenghi et al. 2007) is likely to yield answers. A cue could be forthcoming from the situation in animals where NO is one of the key regulators of cytosolic Ca^{2+} flux. In plants there has been considerable circumstantial evidence pointing to a similar situation. In the final analysis, NO is clearly thought of as being a Ca^{2+}-mobilizing messenger with considerable degree of cross-talk between NO, ROS, and Ca^{2+}-signalling pathways (Besson-Bard et al. 2008a, b) (Fig. 7).

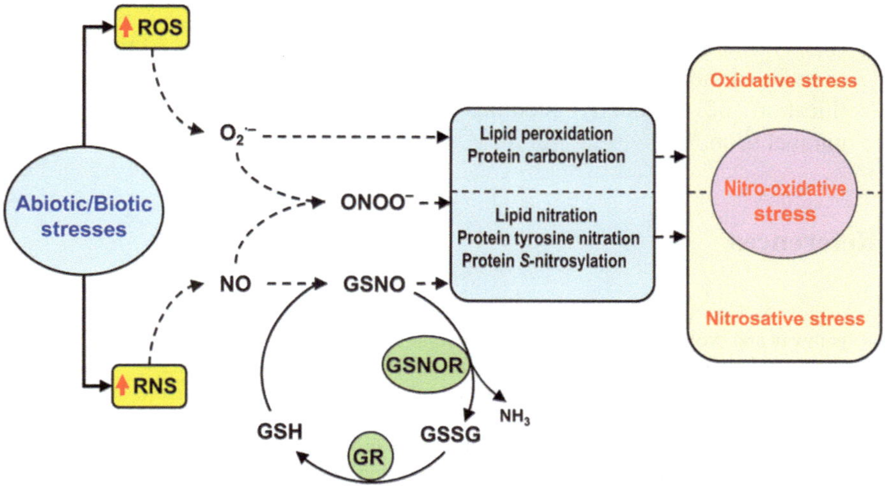

Fig. 7 Nitro-oxidative stress is the interplay between reactive oxygen species (ROS) and reactive nitrogen species (RNS) in response to biotic and abiotic stresses. The increase in the uncontrolled production of ROS and RNS provokes modifications in macromolecules that can act as markers for both oxidative stress (lipid peroxidation and protein carbonylation) and nitrosative stress (lipid nitration, protein tyrosine nitration, and S-nitrosylation). GR, glutathione reductase; GSH, reduced glutathione; GSNO, *S*-nitrosoglutathione; GSNOR, *S*-nitrosoglutathione reductase; GSSG, oxidized glutathione [Reproduced with permission from Corpas and Barroso (2013)]

Concluding Remarks

Paradoxically, the concluding section of this review presents more unanswered questions than validated statements. Unlike signalling pathways that involve proteolysis, reversible phosphorylation, nuclear translocation, oligomerization, etc. and depend strictly on ligand binding with the receptor, signalling involving ROS, RNS, and Ca^{2+} actually comes and goes in a flash. It is the balance between sustained production and continued removal of ROS or RNS and the quick change in location of Ca^{2+} as well as its binding to Ca^{2+} sensors that becomes a signal for the cell. Moreover, ROS and RNS levels are strongly interdependent, with ROS perhaps being more primary than RNS. Or perhaps not! It is the unsure status of knowledge regarding these three elements of signalling that makes this area challenging as well as exciting. The weakest link in our knowledge is the status of knowledge regarding NO. Multiple, yet not unequivocally established pathways of production, interaction with ROS and macromolecules make NO an enigma.

To be fair, NO or RNS related research as a while in plants has lagged behind ROS-related research. Besides, plant NO in particular, has one of its origins in nitrate metabolism, which is almost as mandatory as respiration itself, albeit occurring in fits and starts as per the requirement of the cells.

(continued)

However, with the advent of newer techniques of in situ measurement of short-lived molecules/reactive species and precise mass spectrometric identification of covalently modified proteins, the puzzle will begin to unravel soon.

References

Abzhanov A, Kuo WP, Hartmann C, Grant BR, Grant PR, Tabin CJ (2006) The Calmodulin pathway and evolution of elongated beak morphology in darwin's finches. Nature 442:563–567

Belenghi B, Romero-Puertas MC, Vercammen D, Brackenier A, Inzé D et al (2007) Metacaspase activity of Arabidopsis thaliana is regulated by S-nitrosylation of a critical cysteine residue. J Biol Chem 282:1352–1358

Besson-Bard A, Courtois C, Gauthier A, Dahan J, Dobrowolska G et al (2008a) Nitric oxide in plants: production and cross-talk with Ca2+ signaling. Mol Plant 1:218–228

Besson-Bard A, Pugin A, Wendehene D (2008b) New insight into nitric oxide signaling in plants. Annu Rev Plant Biol 59:21–39

Bothwell JHF, Ng CK-Y (2005) The evolution of Ca^{2+} signaling in photosynthetic eukaryotes. New Phytol 166:21–38

Brookes PS, Yoon Y, Robotham JL, Anders MW, Sheu S-S (2004) Calcium, ATP, and ROS: a mitochondrial love-hate triangle. Am J Physiol Cell Physiol 287:C817–C833

Buchanan BB, Balmer Y (2005) Redox regulation: a broadening horizon. Annu Rev Plant Biol 56:187–220

Chew O, Whelan J, Millar AH (2003) Molecular definition of the ascorbate-glutathione cycle in *Arabidopsis* mitochondria reveals dual targeting of antioxidant defences in plants. J Biol Chem 278:46869–46877

Clapham DE (2007) Calcium signaling. Cell 131:1047–1058

Collin VC, Eymery F, Genty B, Rey P, Havaux M (2008) Vitamin E is essential for tolerance of Arabidopsis thaliana to metal-induced oxidative stress. Plant Cell Environ 31:244–257

Corpas FJ, Barroso JB (2013) Nitro-oxidative stress vs oxidative or nitrosative stress in higher plants. New Phytol 199:633–635

Demidchik V, Shabala S, Davies JM (2007) Spatial variation in H_2O_2 response of Arabidopsis thaliana root epidermal Ca^{2+} flux and plasma membrane Ca^{2+} channels. Plant J 49:377–386

Dodd AN, Kudla J, Sanders D (2010) The language of calcium signaling. Annu Rev Plant Biol 61:593–620

Durzan DJ, Pedroso MC (2002) Nitric oxide and reactive nitrogen oxide species in plants. Biotechnol Genet Eng Rev 19:293–337

Farmer EE, Davoine C (2007) Reactive electrophile species. Curr Opin Plant Biol 10:380–386

Farmer EE, Mueller MJ (2013) ROS-mediated lipid peroxidation and RES-activated signaling. Annu Rev Plant Biol 64:429–450

Fedoroff N (2006) Redox regulatory mechanisms in cellular stress responses. Ann Bot 98:289–300

Foyer CH (2005) Redox homeostasis and antioxidant signaling: a metabolic interface between stress perception and physiological responses. Plant Cell 17:1866–1875

Gadjev I, Stone JM, Gechev TS (2008) Programmed cell death in plants: new insights into redox regulation and the role of hydrogen peroxide. Int Rev Cell Mol Biol 270:87–144

Garcia-Mata C, Gay R, Sokolovski S, Hills A, Lamattina L, Blatt MR (2003) Nitric oxide regulates K^+ and Cl^- channels in guard cells through a subset of abscisic acid-evoked signaling pathways. Proc Natl Acad Sci USA 100:11116–11121

Gonzalez A, Vera J, Castro J, Dennett G, Mellado M et al (2010) Co-occurring increases of calcium and organellar reactive oxygen species determine differential activation of antioxidant and defense enzymes in *Ulva compressa* (Chlorophyta) exposed to copper stress. Plant Cell Environ 33:1627–1640

Gould KS, Lamotte O, Klinguer A, Pugin A, Wendehenne D (2003) Nitric oxide production in tobacco leaf cells: a generalized stress response? Plant Cell Environ 26:1851–1862

Groß F, Durner J, Gaupels F (2013) Nitric oxide, antioxidants and prooxidants in plant defence. Front Plant Sci 4:419. doi:10.3389/fpls.2013.00419

Gupta KJ, Fernie AR, Kaiser WM, van Dongen JT (2010) On the origins of nitric oxide. Trends Plant Sci 16:160–168

Halliwell B, Gutteridge JMC (2007) Free radicals in biology and medicine, 3rd edn. Oxford University Press, New York

Harding SA, Oh SH, Roberts DM (1997) Transgenic tobacco expressing a foreign Calmodulin gene shows an enhanced production of reactive oxygen species. EMBO J 16:1137–1144

Hepler PK (2005) Calcium: a central regulator of plant growth and development. Plant Cell 17:2142–2155

Hetherington AM, Brownlee C (2004) The generation of Ca^{2+} signals in plants. Annu Rev Plant Biol 55:401–427

Hoeflich KP, Ikura M (2002) Calmodulin in action: diversity in target recognition and activation mechanisms. Cell 108:739–742

Kobayashi M, Ohura I, Kawakita K, Yokota N, Fujiwara M et al (2007) Calcium-dependent proein kinases regulate the production of reactive oxygen species by potato NADPH oxidase. Plant Cell 19:1065–1080

Kocsy G, Tari T, Vankova R, Zechmann B, Gulyás Z et al (2013) Redox control of plant growth and development. Plant Sci 211:77–91

Kornberg MD, Sen N, Hara MR, Juluri KR, Nguyen JV et al (2010) GAPDH modulates nitrosylation of nuclear proteins. Nat Cell Biol 12:1094–1100

Lamotte O, Courtois C, Dobrowolska G, Besson A, Pugin A, Wendehenne D (2006) Mechanisms of nitric-oxide induced increase of free cytosolic Ca^{2+} concentration in *Nicotianaplumbaginifolia*cells. Free Radic Biol Med 40:1369–1376

Laohavisit A, Davies JM (2009) Multifunctional annexins. Plant Sci 177:532–539

Laohavisit A, Mortimer JC, Demidchik V, Coxon KM, Stancombe MA, Macpherson M, Brownlee C, Hofmann A, Weber AAR, Miedema H, Battey NH, Davies JM (2009) *Zea mays* annexins modulate cytosolic free Ca^{2+} and generate a Ca^{2+}-permeable conductance. Plant Cell 21:479–493

Laohavisit A, Brown TA, Cicuta P, Davies JM (2010) Annexins: components of the calcium and reactive oxygen signaling network. Plant Physiol 152:1824–1829

Lindermayr C, Saalbach G, Bahnweg G, Durner J (2006) Differential inhibition of Arabidopsis methionine adenosyltransferases by protein S-nitrosylation. J Biol Chem 281:4285–4291

Mazars C, Thuleau P, Lamotte O, Bourque S (2010) Cross-talk between ROS and calcium in regulation of nuclear activities. Mol Plant 3:706–718

McAinsh MR, Pittman JK (2009) Shaping the calcium signature. New Phytol 181:275–294

Miller G, Suzuki N, Cifti-Yilmaz S, Mittler R (2009) Reactive oxygen species homeostasis and signaling during drought and salinity stresses. Plant Cell Environ 33:453–467

Mittler R (2002) Oxidative stress, antioxidants and stress tolerance. Trends Plant Sci 7:405–410

Mittler R, Berkowitz G (2001) Hydrogen peroxide, a messenger with too many roles? Redox Rep 6:69–72

Møller IM, Jensen PE, Hansson A (2007) Oxidative modifications to cellular components in plants. Annu Rev Plant Biol 58:459–481

Mori IC, Schroeder JI (2004) Reactive oxygen species activation of plant Ca^{2+} channels: a signaling mechanism in polar growth, signal transduction, stress signaling and hypothetically mechanotransduction. Plant Physiol 135:702–708

Mortimer JC, Coxon KM, Laohavisit A, Davies JM (2009) Heme dependent soluble and membrane-associated peroxidase activity of a *Zea mays* annexin preparation. Plant Signal Behav 4:428–430

Navrot N, Collin V, Gualberto J, Gelhaye E, Hirasawa MI et al (2006) Plant glutathione peroxidases are functional peroxiredoxins distributed in several subcellular compartments and regulated during biotic and abiotic stresses. Plant Physiol 142:1364–1379

Pacher P, Beckman JS, Liaudet L (2007) Nitric oxide and peroxynitrite in health and disease. Physiol Rev 87:315–424

Pfannschmidt T, Brautigam K, Wagner R, Dietzel L, Schroter Y, Steiner S, Nykytenko A (2009) Potential regulation of gene expression in photosynthetic cells by redox andenergy state: approaches towards better understanding. Ann Bot 103:599–607

Pineros M, Tester M (1997) Calcium channels in higher plants: selectivity, regulation and pharmacology. J Exp Bot 48:551–577

Quan L-J, Zhang B, Shi W-W, Li H-Y (2008) Hydrogen peroxide in plants: a versatile molecule of the reactive oxygen species network. J Integr Plant Biol 50:2–18

Reddy VS, Reddy ASN (2004) Proteomics of calcium-signaling components in plants. Phytochemistry 65:1745–1776

Sanders D, Pelloux J, Brownlee C, Harper JF (2002) Calcium at the crossroads of signaling. Plant Cell Suppl 2002:S401–S417

Shacter E (2000) Quantification and significance of protein oxidation in biological samples. Drug Metab Rev 32:307–326

Smirnoff N (2005) Antioxidants and reactive oxygen species in plants. Blackwell, Oxford. ISBN 978-1-4051-2529-1

Stael S, Wurzinger B, Mair A, Mehlmer N, Vothknecht UC, Teige M (2012) Plant organellar calcium signalling: an emerging field. J Exp Bot 63:1525–1542

Takeda S, Gapper C, Kaya H, Bell E, Kuchitsu K, Dolan L (2008) Local positive feedback regulation determines cell shape in root hair cells. Science 319:1241–1244

Trewavas A (1999) Le calcium. C'est la vie: calcium makes waves. Plant Physiol 120:1–6

Wall SB, Oh J-Y, Diers AR, Landar A (2012) Oxidative modification of proteins: an emerging mechanism of cell signaling. Front Plant Sci 3:369. doi:10.3389/fphys.2012.00369

Whalley HJ, Knight MR (2013) Calcium signatures are decoded by plants to give specific gene responses. New Phytol 197:690–693

White PJ (2000) Calcium channels in higher plants. Biochim Biophys Acta 1465:171–189

Yu M, Lamattina L, Spoel SH, Loake GJ (2014) Nitric oxide function in plant biology: a redox cue in deconvolution. New Phytol 202:1142–1156

Zepeda-Jazo I, Velarde-Buendía AM, Enríquez-Figueroa R, Bose J, Shabala S et al (2011) Polyamines interact with hydroxyl radicals in activating Ca^{2+} and K^{+} transport across the root epidermal plasma membranes. Plant Physiol 157:42167–42180

MIX
Papier aus verantwortungsvollen Quellen
Paper from responsible sources
FSC® C105338

If you have any concerns about our products,
you can contact us on
ProductSafety@springernature.com

In case Publisher is established outside the EU,
the EU authorized representative is:
Springer Nature Customer Service Center GmbH
Europaplatz 3, 69115 Heidelberg, Germany

Printed by Libri Plureos GmbH
in Hamburg, Germany